ESSENTIALS
of Industrial Hygiene

THOMAS P. FULLER

SCD, CIH, MSPH, MBA

PRESS

National Safety Council
Itasca, IL

NSC Press Editor: Deborah Meyer
Senior Director, NSC Press: Jeff Kraft
Product Line Manager: Kelly Buterbaugh
Cover Design, Interior Design, and Composition: Jennifer Villarreal
Cover Photos: (Top, from left) leszekglasner/iStock, sdlgzps/iStock, dra_schwartz/iStock; (bottom, from left) used with permission from MSA, tunart/iStock, Leslie Achtymichuk/iStock

COPYRIGHT, WAIVER OF FIRST SALE DOCTRINE

The National Safety Council's materials are fully protected by the United States copyright laws and are solely for the noncommercial, internal use of the purchaser. Without the prior written consent of the National Safety Council, purchaser agrees that such materials shall not be rented, leased, loaned, sold, transferred, assigned, broadcast in any media form, publicly exhibited or used outside the organization of the purchaser, reproduced, stored in a retrieval system or transmitted in any form or by any means, electronic, mechanical, photocopying, recording, or otherwise. Use of these materials for training for which compensation is received is prohibited, unless authorized by the National Safety Council in writing.

DISCLAIMER

Although the information and recommendations contained in this publication have been compiled from sources believed to be reliable, the National Safety Council makes no guarantee as to, and assumes no responsibility for, the correctness, sufficiency, or completeness of such information or recommendations. Other or additional safety measures may be required under particular circumstances.

Copyright © 2015 by the National Safety Council
All Rights Reserved
Printed in the United States of America
24   10 9

Library of Congress Cataloging-in-Publication Data
Fuller, Thomas P.
  Essentials of industrial hygiene / Thomas P. Fuller—1 Edition.
     pages cm
  Includes bibliographical references and index.
  ISBN 978-0-87912-336-9
  1. Industrial hygiene. 2. Industrial safety. I. Title.
  HD7261.F855 2015
  658.4'08—dc23
            2014045955

Product Number: 151500000

Contents

Preface v

1. Introduction to Industrial Hygiene 1
2. The Occupational Safety and Health Act and Industrial Hygiene . . . 23
3. Management Systems 45
4. Basic Concepts in Industrial Toxicology 65
5. Occupational Exposure Limits and Assessment of Workplace Chemical Risks 95
6. Gases and Vapors 113
7. Aerosols 137
8. Ventilation 159
9. Respiratory Protection 191
10. Dermal Hazards 213
11. Noise 237
12. Radiation 259
13. Thermal Stressors 293
14. Ergonomics 319
15. Biological Hazards 349

Index 373

PREFACE

The study of industrial hygiene is based upon several core sciences including chemistry, physics, physiology, anatomy, and biology. In the past several decades, undergraduates with science and math backgrounds have pursued advanced degrees in industrial hygiene that build on their education and experiences. Most industrial hygiene textbooks are thus written at the advanced reading and technical levels of postgraduates and working professionals.

With the growing recognition of the need for additional safety professionals in the United States and around the world, an increasing number of undergraduate programs in safety and occupational health are being developed. A course in industrial hygiene is fundamental to an undergraduate degree in these programs. The clear and concise delivery of the material makes *Essentials of Industrial Hygiene* a useful resource for new safety professionals as well as for those students taking an introductory course on the subject. Professionals with years of experience may also find the book a quick and easy reference to have at the ready.

Graduates of safety or occupational health degree programs may begin their careers in a broad array of industries. In manufacturing, energy, health care, and agriculture, young professionals need a sound industrial hygiene foundation in order to effectively and accurately evaluate working conditions and protect employees from chemical and physical hazards. The textbook used in such a program must address all of the key subject areas in sufficient detail to ensure that readers can understand and apply fundamental principles, but not provide so much information that it distracts and overwhelms them.

Decades of working experience in industrial hygiene have provided the authors of this textbook with a clear understanding of what levels of proficiency early industrial hygiene professionals should have when they enter the field. The authors also have years of undergraduate teaching experience, which allows them a precise understanding of the reading skills, capabilities, and interests of today's occupational health students. In addition, this book presents the core body of knowledge that those graduating with advanced degrees in related fields such as public health, ergonomics, epidemiology, and environmental health should understand.

Each chapter contains learning objectives, examples, and hypothetical case studies, providing instructors with a straightforward basis for lecture material. Sample homework problems appear at the end of each chapter to help streamline courses using the text. Faculty can access supplemental materials that accompany the text on the National Safety Council's Faculty Portal (www.nsc.org/facultyportal). The goals are laid out, the path is clear, and the reward is a solid foundation in industrial hygiene.

Because this is the first edition of this book, comments, criticisms, and suggestions for improvement are encouraged and will be actively considered for subsequent revisions. Please send your comments and suggestions to the National Safety Council, 1121 Spring Lake Drive, Itasca, IL 60143, attn. Deborah Meyer or deborah.meyer@nsc.org.

The ever-expanding need for occupational safety and health professionals in the United States and abroad ensures a need for well-educated and competent graduates with a solid understanding of the industrial hygiene field. In a realistic, efficient, and timely manner, this book is a way to increase the number of such graduates and provide them with a valuable resource they can continue to use once in the field.

CONTRIBUTING AUTHORS

The National Safety Council very much appreciates the dedication and expertise of the following authors who contributed to this textbook.

Dr. Thomas P. Fuller is Associate Professor of Safety at Illinois State University. He is a Certified Industrial Hygienist with more than 34 years of experience in occupational safety and industrial hygiene. He also has experience in health care, nuclear power plants, labor organizations, biopharmaceutical labs, manufacturing, and universities. He earned his doctorate from the University of Massachusetts Lowell, a Master of Science in Public Health from the University of North Carolina, and a Master of Business Administration from Suffolk University. Dr. Fuller is a Contributing Editor for the *American Journal of Nursing*, and is on the Editorial Advisory Board of the National Safety Council. Dr. Fuller is a past President of the Board of Directors of the *Journal of Occupational and Environmental Health*. E-mail: tfulle2@ilstu.edu.

Dr. Farhang Akbar (Chapter 13) is a professor in the Department of Public Health and Preventive Medicine at the University of Toledo. He is an environmental engineer and occupational health and safety scientist. His areas of expertise include exposure assessment regarding physical agents, hazardous chemicals, and mechanical agents; health and safety programs, auditing, and regulations; safety management; and hazard control methods. His current research activities are focused on exposure to nonionizing radiation; human chemical exposure assessment and biological changes; and exposure modeling. E-mail: farhang.akbar@utoledo.edu.

Dr. Sergio Caporali Filho (Chapter 7) has more than 21 years of manufacturing, industrial hygiene, safety, and ergonomics experience in the field. He has been a Certified Safety Professional since 2005 and a Certified Industrial Hygienist since 2011. Dr. Caporali Filho earned his BSc in Industrial Engineering from the University of Lima in 2004, his ME in Manufacturing Systems Engineering from the University of Puerto Rico at Mayaguez in 1998, his MSc in Occupational Hygiene and Occupational Safety in 2001, and his Ph.D. in Industrial Engineering–Ergonomics in 2002. He currently holds a professor and industrial hygiene program coordinator position at the Graduate School of Public Health in the University of Puerto Rico's Medical Sciences Campus, and coordinates the occupational hazards control courses as well as the industrial hygiene and sampling courses at UPR's IH graduate program. Dr. Caporali Filho's most recent research interests are assessing hearing protection performance in real-world applications and welding fume control effectiveness through the correct use of Local Exhaust Ventilation. E-mail: sergio.caporali@upr.edu.

Dr. Andrew Maier (Chapters 4 and 5) is an Associate Professor of Environmental and Industrial Hygiene at the University of Cincinnati. He serves as a Science Fellow at the National Institute for Occupational Safety and Health. He is board certified in industrial hygiene and toxicology and maintains an active research program and field practice in occupational risk assessment methods. E-mail: maierma@ucmail.uc.edu.

Dr. Margaret Levin Phillips, CIH, (Chapter 12) is an Associate Professor of Occupational and Environmental Health at the University of Oklahoma Health Sciences Center and a member of the American Industrial Hygiene Association's Nonionizing Radiation Committee. Prior to joining the faculty of the University of Oklahoma, Dr. Phillips worked as a field industrial hygienist in the private sector. She received her doctorate in Physical Chemistry from the University of Illinois and her Master of Health Science degree in Industrial Hygiene from the Johns Hopkins University School of Hygiene and Public Health. E-mail: margaret-phillips@ouhsc.edu.

Paul Ronczkowski (Chapter 9) teaches several courses in the Safety program at Illinois State University, including Agricultural Safety and Health, Accident/Incident Investigation, Construction Safety, Safety Fire Protection and Prevention, and General Industry Standards. He also coordinates the professional practice (internship) program. Mr. Ronczkowski has close to 20 years of college teaching experience in safety and health and 35 years of experience in the safety and health profession in various industries, including production agriculture, construction, safety consulting, insurance (loss control), and the Veterans Affairs Medical Center. He has also published articles on student professional practice (internship) experiences. In addition to his teaching responsibilities, he is the student advisor and liaison for the student section of the American Society of Safety Engineers and the Central Illinois Chapter of the American Society of Safety Engineers. E-mail: pjroncz@ilstu.edu.

ACKNOWLEDGMENTS

The National Safety Council and author also want to thank the following for their contributions: Leo J. DeBobes, Dr. Katy Ellis, Dr. Janvier Gasana, Dr. David Christian Grieshaber, Philip E. Hagan, Dr. Michelle Homan, Michael Husarek, Dr. Farman A. Moayed, Jil Niland, and Dr. Allen Sullivan.

1

Introduction to Industrial Hygiene

LEARNING OBJECTIVES

After completing this chapter, readers should be able to do the following:

- Briefly explain the relevance and importance of the practice of industrial hygiene in society today and throughout history
- Explain the types of sciences and studies that comprise the field of industrial hygiene
- Describe the differences between pathways of exposure and routes of exposure to hazardous agents
- List general types of workplace hazards
- Identify the basic health effects of hazardous exposures in the workplace
- Describe the concept of risk and the fundamental principles that it is based upon
- Discuss the basic tenets of industrial hygiene hazard assessment and control
- Recognize the types of careers and industries available for professional industrial hygienists
- Briefly describe the various government agencies and professional organizations with roles in industrial hygiene

Photo credit: Reprinted Courtesy of Caterpillar Inc.

INTRODUCTION

Industrial hygiene, which has been practiced in one form or another for centuries, is the science of anticipating, recognizing, evaluating, and controlling workplace hazards. Building upon fundamental principles of chemistry, mathematics, physics, physiology, anatomy, and biology, industrial hygienists develop analytical methods to identify and measure workplace hazards, with the goal of reducing workplace injuries, illnesses, and fatalities.

HISTORY

Workplace hazards have been recognized for thousands of years. In 400 BC, Hippocrates was the first to make note of work-related disease in lead miners. Around AD 50, Pliney the Elder observed negative health outcomes in zinc and sulfur miners. He created the first known form of respiratory protection by using animal bladders to filter lead fumes from the air. In the second century AD, Greek physician Galen was one of the earliest to describe the pathology of lead poisoning in miners.

In 1556, the German scholar Agricola wrote *De Re Metallica*, in which he described the diseases associated with mining and smelting, such as silicosis. He also presented recommendations for reducing exposures, including the use of ventilation.

In 1700, Bernardo Ramazzini wrote the first comprehensive book on industrial medicine, *De Morbis Artificum Diatriba (The Diseases of Workmen)*. The book linked diseases to many of the period's occupations. Ramazzini is often referred to as the father of occupational medicine.

In the 1700s, Ulrich Ellenborg published information on the occupational diseases and injuries in gold and asbestos miners in which he described the toxic effects of lead, carbon monoxide, mercury, and nitric acid on the workers.

In the 18th century in England, Percival Pott noted that a high number of chimney sweepers had scrotal cancer. He also discovered that the workers' poor personal hygiene habits were linked to their excessive exposure to coal soot. Pott's findings led to improved occupational health measures for chimney sweeps and other workers in that era.

Dr. Alice Hamilton, in the early 20th century, studied workplace conditions in mines and factories in the United States and identified relationships between exposures to toxic materials and worker illness. In so doing, she advanced the scientific study of toxicology and occupational health in the United States and led efforts to improve industrial hygiene in workplaces through research and advances in worker protection legislation.

In the late 1800s, U.S. federal and state agencies became interested and involved in workplace health and safety. In 1875, for example, Massachusetts began drafting legislation on machine guarding, fire protection, and improved ventilation (DOL 2015). Some states passed laws in 1911 covering workers' compensation. The New York Department of Labor and the Ohio Department of Health established the first state industrial hygiene programs in 1913. By the end of the first half of the century, most states had enacted workplace safety regulations.

Despite this progress at the state level, many years passed before more comprehensive federal regulations for occupational health and safety were developed. These regulations were a response to concerns about environmental hazards such as air and water pollution and to public interest in occupational safety and health. In 1966, Congress passed the Metal and Nonmetallic Mines Safety Act, which required improved working conditions for miners. The Federal Coal Mine Safety and Health Act provided more protections in 1969. Finally, the Occupational Safety and Health Act (OSH Act) became law in 1970, with significant impacts on a broad range of industries (OSHA 1998). (See Chapter 2, The Occupational Safety and Health Act and Industrial Hygiene, for more information.)

In the past several decades, industrial hygiene applications and methods have been increasingly used in new areas, including asbestos abatement and control, environmental radon measurement and control in buildings and homes, and control of mold to improve indoor air quality. Industrial hygiene techniques are also being applied to worker protection from exposure to nanoparticles and silica dust in the construction industry.

INDUSTRIAL HYGIENE TODAY

Industrial hygiene has grown to include a variety of discrete areas of practice and scientific study, and is the culmination or combination of a broad range of sciences including the following:

chemistry	epidemiology
physics	engineering
physiology	psychology/organizational behavior
anatomy	social science
toxicology	ergonomics
mathematics	risk assessment
biology	management
statistics	ethics

Practicing industrial hygienists have training in many, if not all, of the fundamental sciences listed. They may or may not practice each science equally in the field. Essentially, industrial hygiene uses these scientific methods to analyze workplace hazards and identify or create solutions to reduce those hazards.

University programs in industrial hygiene are available at the bachelor's, master's, and doctoral levels. Academic programs for bachelor's and master's programs may be accredited by the Accreditation Board for Engineering and Technology (ABET).

The Occupational Safety and Health Administration (OSHA) is mandated to provide training programs on various aspects of safety and industrial hygiene. There are currently 27 education centers in the United States that teach a wide variety of safety courses.

The National Institute for Occupational Safety and Health (NIOSH) is the branch of the Centers for Disease Control and Prevention (CDC) responsible for health and safety research. NIOSH provides information and guidance to OSHA on regulatory development and for dissemination to the general public. NIOSH also conducts research to find the most effective ways to reduce workplace injuries and illnesses through training and supports academic degree programs and research training opportunities in areas such as industrial hygiene, occupational health, occupational medicine, and safety. In addition, NIOSH provides a number of short-term training programs for professionals (NIOSH 2014).

OCCUPATIONAL EXPOSURE

Workers in different industries and jobs are exposed to a variety of hazards. A **hazard** is a condition that poses some level of potential damage or harm to life, health, property, or the environment. A hazardous exposure of long duration is described as **chronic**. An example of a chronic exposure is a painter who dips his brush in paint and applies it to walls all day long for months, thus breathing in the paint vapors for an extended period of time. In contrast, an **acute** exposure involves a worker being subjected to a hazard over a short duration. For example, when the painter cleans his brushes at the end of each day with a strong solvent, he is exposed to the hazard for only a short period of time.

PATHWAYS OF EXPOSURE

Exposure to hazards takes place through different environmental **pathways**. One of the most common such pathways is the air. Gases, vapors, and dusts from chemical and manufacturing processes move through the air, while liquids evaporate and can become concentrated in the air. Another common pathway is liquid itself. Hazardous liquids can splash onto a worker, or a worker may submerge his or her hands into the chemical or other hazardous liquid. Powders or dusts in the work environment can also be a form of exposure when they come in contact with the skin.

Vectors are living organisms that transport a hazardous agent through the environment and into contact with a worker. Mosquitoes, ticks, and rats are well known as vectors that move infectious agents from a source to a person and spread disease. Other vectors include the leaves of poisonous plants such as poison ivy and poison oak. Inanimate objects can also act as vehicles that transmit hazards. For instance, if an asbestos abatement technician wears contaminated coveralls home, she would expose family members or anyone else in the household to asbestos.

ROUTES OF EXPOSURE

Routes of exposure are the ways a hazard enters the body. **Inhalation** into the respiratory tract is a common route of bodily exposure to hazardous agents. The worker breathes in the chemical or dust, which is then absorbed in the lungs or another part of the respiratory system. Workers can expose their internal organs to chemicals and other hazardous materials by **ingestion** when a hazardous material enters the mouth and when they eat contaminated food or swallow mucous from their respiratory tract that contains hazardous substances. Hazardous materials can also enter the body via the **dermal** route of exposure by coming in contact with the skin. Damage can occur directly to the skin, and if sufficient quantities are absorbed into the skin, internal organs can also be damaged. Exposure by **injection** takes place when the contaminant or hazardous agent is injected directly into the body by a sharp object or enters through open wounds in the skin.

> ### Case: Routes and Pathways of Exposure
>
> During the Severe Acute Respiratory Syndrome (SARS) outbreak in 2002, there was a lot of confusion about how the infectious agent was spread. The respiratory virus was first observed in Southeast Asia and then later emerged in 2003 in Toronto, Canada. As with most flulike respiratory diseases, members of the health care community assumed the virus was spread by patients' sneezes, which produced large droplets that were presumably breathed in by nearby individuals. That is, they assumed that the route of exposure to the infectious agent was by contact and that the pathway through the environment was droplets from nearby people.
>
> After the outbreak was under control and industrial hygienists and infection control specialists could devote more time to studying the disease, it was learned that the infectious agent could actually live in and be transported through ventilation ducts to people in remote parts of buildings, which meant that the pathway was airborne. In addition, it was discovered that the SARS virus could live for long periods outside the body on surfaces—a second pathway.
>
> The primary route of exposure was inhalation. But the second route of exposure was direct contact with mucous membranes. If people touched a contaminated surface or item and then rubbed their eyes or ingested contaminated food, they also could become infected.
>
> It is important to be able to differentiate between an exposure *pathway*, such as a path through the woods where the wind can carry hazardous agents, and an exposure *route* into the body, such the lungs, gastrointestinal tract, or skin.

TYPES OF HAZARDS

CHEMICAL

Chemical agents in the workplace are the primary focus of most practicing industrial hygienists. Many chemicals produced or used in manufacturing are toxic to humans. **Toxic** effects can be immediate, or *acute*, symptoms such as shortness of breath, dizziness, or headaches. The exposure can also cause *chronic* toxic effects such as cancer or asbestosis, which may not be apparent or diagnosed until years after the exposure. Different chemicals can also act on the environment to create a hazard via **oxygen displacement** (simple asphyxiation), making the air uninhabitable. Some gases, such as carbon dioxide or argon, can displace oxygen in the environment, which can result in asphyxiation when a worker is not equipped to work in an oxygen-deficient environment. For example, many confined-space incidents result from oxygen displacement in an underground vault or large vessel. Depending on the chemical and its concentration, volatility, and other characteristics, it can possibly create a fire or explosion hazard in the workplace. Many solids (e.g., dusts) are explosive and can cause catastrophic equipment failures, explosions, and fires.

PHYSICAL

Physical hazards can cause damage through direct contact with the worker. Excessive levels of **noise energy** can travel through the air, come in direct contact with the worker's ears or hearing apparatus, and cause irreversible damage. (See Chapter 11, Noise.) Ionizing and nonionizing radiation sources can emit hazardous energy that causes damage when it comes in contact with living tissue. (See Chapter 12, Radiation.) Extreme hot or cold temperatures affect exposed workers in a variety of ways, including death. (See Chapter 13, Thermal Stressors.) Although less common than other hazards, vibration also has been shown to cause adverse health outcomes in exposed workers. In addition, extreme variations in environmental pressure can be hazardous to workers.

MUSCULOSKELETAL

Musculoskeletal hazards have been shown to exist when workers are required to perform physical movements beyond the capabilities of their bodies. Four of the primary factors that have been shown to play a role in the likelihood of musculoskeletal injuries are the **force** needed to do a task, the **rate** or **repetitiveness** at which the task is performed, the **duration** the task is performed, and the **posture** of the worker as he or she performs the task. Studies have shown that for a given anthropometric size and type of worker, lifting a certain *weight* of box a certain *number* of times per hour for a certain *number hours* of the day is likely to cause injury.

$$\text{Force} \times \text{Rate} \times \text{Duration} \times \text{Posture} = \text{Injury}$$

The study of the work process and the attempt to design the workplace to fit the worker is called **ergonomics**. Other factors included in the study of ergonomics and the relationship between the workplace and injuries are **environmental factors** (e.g., noise, lighting, chemicals) and **psychosocial conditions** (e.g., work hours and duration, control over work, staffing levels) of the workplace. (See Chapter 14, Ergonomics.)

BIOLOGICAL

Biological hazards include infectious agents such as those that workers in health care, food processing, and sanitation fields may be exposed to regularly. (See Chapter 15, Biological Hazards.) Animal handling and veterinary workers are also regularly exposed to biological agents, as are many laboratory workers in research and health care. A synopsis of different types of infectious workplace agents and their associated hazards is provided in Table 1–1.

Table 1–1. Workplace Agents and Associated Hazards

Chemical	*Hazards*
Flammable	Fires, burns, explosions
Corrosive	Corrosion of skin or containers
Combustible	Fires, burns, explosions
Reactive	Fires, burns, explosions
Carcinogenic	Cancer
Sensitizing	Hypersensitivity, allergies
Asphyxiant	Suffocation, loss of consciousness
Physical	
Ionizing radiation	Burns, cancer, genetic mutations, malformations
Nonionizing radiation	Burns, cancer, cataracts, loss of vision
Thermal conditions	Syncope, exhaustion, stroke, hypothermia, frostbite
Noise	Hearing loss, stress
Vibration	Cardio-circulatory damage, spinal damage, gastrointestinal damage, vibration syndrome, loss of muscle function and strength
Musculoskeletal	Muscle and bone injuries, sprains, strains, tears
Biological	Infections, illnesses

HEALTH EFFECTS

INJURY VERSUS DISEASE

The term *injury* typically refers to direct physical or emotional harm or pain to someone that leads to impaired function or loss of abilities. **Disease** is a pathological process that arises from an exposure or environmental stress and leads to an adverse health outcome in structure or function.

ACUTE HEALTH EFFECTS

An **acute health effect** becomes evident shortly after exposure. For example, skin reddening from exposure to excessive levels of ionizing radiation and increased sweating from being in a hot environment are considered acute health effects. A worker who breathes in toxic vapors and begins to feel nauseated or dizzy is experiencing an acute health effect. Such health effects can be serious and permanent. For example, short-term exposure to excessive noise can cause irreversible hearing degradation, and an acute exposure to a toxic chemical such as a poison can be immediately fatal.

CHRONIC HEALTH EFFECTS

Chronic health effects manifest over a long period of time. For example, asbestosis may take up to 20 years to become evident in an exposed worker.

Cancer also tends to have a long latency period before it is diagnosed. Cancer is considered a special category of worker illnesses because it is commonly considered irreversible and fatal. However, this categorization is losing significance because many types of cancer, such as certain forms of skin cancer, are often treatable, and some types are even considered curable. However, many other occupational illnesses, such as silicosis or asbestosis, are fatal at certain exposure levels.

Chronic health effects can be temporary or permanent. A skin rash from a dermal exposure may be treatable and completely disappear. However, the same chemical exposure that caused the rash may cause permanent disfiguration or discoloration at a higher dose or in a different worker.

RISK

Much of the day-to-day practice of industrial hygiene involves assessing **risk**. Risk is the product of the probability of a harmful event occurring in a given condition or scenario and the severity of the outcome of that event.

PROBABILITY

Probability is a branch of mathematics that measures and describes the relative likelihood or frequency of an event and looks at the distributions of the event's occurrences within a given population. In industrial hygiene, the likelihood that an exposure or hazardous event will occur may be estimated, and probabilities are then assigned to the likely outcomes.

SEVERITY

In many areas of industrial hygiene, the acceptable levels of risk have already been determined. Government organizations and other standard-setting bodies have identified the safest levels of worker exposure to a certain chemical or physical agent and set required or recommended limits for that exposure. These limits are generally set at the exposure level that would be assumed to be safe if a healthy worker, using appropriate safety measures, is exposed to a particular amount or concentration of the agent for 40 hours a week over 40 years of work.

In other cases, there may be no regulations or recommended limits, and industrial hygienists must do their own investigations to determine the hazard and likelihood of exposure and make their own decisions about the risks to their employees and the acceptable level of exposure to the agent. These decisions are often situational in that the industrial hygienist reviews the work practices and levels of additional controls that may be required. For many of these cases, recommended or industry-generated guidelines are available.

THE INDUSTRIAL HYGIENE APPROACH

Industrial hygiene involves applying the sciences listed previously to the improvement of working conditions in any workplace or industry. At its most basic, industrial hygiene involves four discrete activities that take place in a set chronological order. These activities are the anticipation, recognition, evaluation, and control of workplace hazards.

ANTICIPATION

Anticipation involves identifying potential or actual hazards through knowledge of materials, operations, processes, and conditions in the workplace.

For any given work setting, it is typically possible to anticipate what types of hazards exist for the people doing the work. Even without visiting a factory, for example, one might consider that its environment would likely be noisy. In addition, there are often machines with dangerous moving parts, or there might be hazardous chemicals used to make materials.

When considering the hazards that might exist at a large construction site, falling off a scaffold or girder might come to mind. In fact, falls are the largest source of fatalities in the construction industry. In addition to falls, moving machines, musculoskeletal injuries, extreme noise, and heat stress are common hazards associated with construction.

As an aid in anticipating the hazards associated with an industry, an extensive information base is available from a broad variety of sources. OSHA maintains and distributes an excellent database of hazards for numerous industries. In many cases, OSHA even divides the hazards into discrete tasks or activities within the industries. NIOSH is another helpful government source of information about specific industry workplace hazards.

On the other hand, for some industries or jobs, there may be little information about the hazards. Nanoparticles, for example, have been growing in use and popularity. But because of the newness of this technology, information about the health effects is not yet fully determined or readily available. In this case, the industrial hygienist should take a precautionary approach to protecting the workers and build upon his or her previous experiences in the field, understanding of the core sciences from their training, and research into the information about the health effects as it becomes available.

Nanoparticles are smaller than other materials typically encountered in the workplace, and they may react differently from how larger particles do when they are ingested or inhaled. They might be more toxic than other materials. In addition, they might be difficult to measure in the workplace, or methods used to measure them accurately may not have been developed yet.

Without full information on the hazards associated with nanoparticles and other materials, the industrial hygienist must anticipate the hazards and take a precautionary approach to how materials are handled and used in the manufacturing process.

RECOGNITION

Recognition is the observation and discovery of the hazardous materials and conditions in the workplace. Industrial hygienists should have a preliminary understanding of possible workplace hazards so that they can approach a site and review the conditions appropriately. As they enter the worksite, industrial hygienists will use their training, experience, and any additional available data to recognize the existing hazards.

Upon entering a facility, an industrial hygienist should immediately take note of any high noise levels, huge machines with dangerous moving parts, and workers lifting heavy equipment, all of which are hazards that should be anticipated (Figure 1–1). He or she might also smell chemicals upon entering the facility and should review safety data sheets (SDSs) and other facility information to find out what chemicals are being used in addition to the quantity and manner of their use.

During the recognition phase of the investigation, the industrial hygienist learns as much as possible about the work processes, input materials, wastes, and byproduct materials. In many cases, process mapping is used to diagram the production process. In these maps, all worker activities are broken down into steps and tasks. For each of these steps and tasks, the hazards are identified and prioritized to the greatest extent possible. A sample process map is shown in Figure 1–2.

Figure 1–1. A typical heavy-duty tractor manufacturing facility.

(Reprinted Courtesy of Caterpillar Inc.)

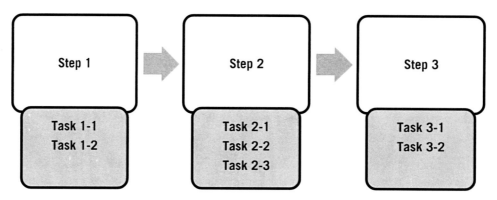

Figure 1–2. A sample process map.

EVALUATION

Evaluation involves measuring exposures and doses, and comparing the measurements to regulatory, professional, and other exposure standards and limits. Once the hazards have been identified, the industrial hygienist can more quantitatively evaluate the levels of worker exposure and risk in each of the job areas, steps, or tasks. Because it is typically not possible to immediately evaluate the level of hazard in all jobs, in between recognition and evaluation, the industrial hygienist needs to prioritize the jobs that seem the most hazardous.

Jobs evaluated first are typically those with the most significant risk of serious injury or death. If conditions that could lead to death are found, they need to be corrected and remedied immediately. Other conditions that could lead to injury or illness over a period of time can be evaluated and addressed in order of importance.

Evaluation may take the form of a review of workplace injury and illness rates. Looking at incident rates, an industrial hygienist might identify slips and trips as the number-one injury at a given work area, or at a meat-processing plant, an industrial hygienist might use medical records of on-site accidents to determine that cuts and punctures are the most common injuries.

Evaluation can also be performed by administering surveys or taking measurements, such as monitoring noise levels or sampling air concentrations. These levels would then be compared with recognized exposure limits for the given physical or chemical agents.

However, because hazards are sometimes difficult to quantify, ergonomic assessments of hazardous exposures might instead be used to measure the number of times a day that a worker is required to lift a heavy load. These measurements would then be compared with guidelines for what is considered a safe workload. Sometimes the work conditions need to be quantified in another form to make an assessment of the hazards (e.g., "How slippery are the floors?").

Evaluation is a key part of the industrial hygienist's responsibility in health and safety. It is a responsibility for which he or she has specialized training. Skills in all the technical rubrics should be combined to accurately identify and quantify the most significant hazards to workers. No workplace has only one hazard, and the industrial hygienist should not only evaluate the hazards but also prioritize hazards that he or she finds.

CONTROL OF HAZARDS

Control involves the elimination or reduction of identified and measured hazards. Perhaps the most challenging aspect of industrial hygiene work is creating controls to reduce the identified hazards and risks to make the workplace safer and, more importantly, to bring the levels of risk down to acceptable levels. Control is accomplished through the methodical use of a hierarchy of controls. In order of decreasing importance, the hierarchy of controls includes hazard elimination, hazard substitution, and engineering, administrative, and personal protective equipment controls. The most effective tools for accomplishing hazard controls are strong leadership and communication skills and the ability to effect change within an organization.

ELIMINATION AND SUBSTITUTION

The obvious way to make a workplace safer is to completely eliminate the hazard. Often, simple and creative measures can be taken to change a process or reorient the workspace to completely eliminate a hazard. Imagine a scenario where 50 bags of cement are delivered to a loading dock. Workers then have to lift each 50-pound bag onto a dolly to move the bags into the factory. By having the bags delivered to the loading dock and directly onto a waiting dolly, the workers would no longer need to lift the bags, saving time and eliminating the risk of back injuries to the workers. In a different scenario, if workers are exposed to toxic vapors in a specific part of a room, reorienting the workspace to maximize the ventilation system can reduce the workers' inhalation of the vapors.

Another example of elimination is discontinuing the use a toxic chemical in a particular process or product. For many years, lead was added to paint to improve its durability and resistance to moisture. But after the lead was found to cause toxic effects in children exposed to it in their home environments, its use was discontinued. In addition, the workers were no longer exposed to the lead in the manufacturing process, making the workplace safer.

Sometimes eliminating a hazard can be as simple as using less of a certain material in the product or process so that the concentration does not reach dangerous levels. Another way that significant volumes of toxic or hazardous chemicals can be eliminated from the workplace is to store them off-site, in a remote location, or to order just what is needed for the immediate processes and activities.

When complete elimination is not feasible, it is often possible to substitute a less toxic or hazardous chemical for the toxic one. Formaldehyde, used for many years as a preservative, was found to be a carcinogen. In recent years, alternative chemicals that do not have hazardous health effects have been shown to be comparable substitutes for formaldehyde.

ENGINEERING CONTROLS

Another means of protecting workers from known hazards is the use of engineering controls. The most fundamental definition of an engineering control is a control that physically separates, or protects, the worker from the hazard. For instance, a metal screen over a fan can prevent workers from sticking their

---

Case: Value of Controls

Josh recently graduated with a degree in industrial hygiene and took a job at a metalworking plant where machines and assembled parts are produced. He noticed one area of the factory where workers wear half-face respirators because the solvents they work with are toxic.

When Josh evaluated the working conditions and the air concentrations of the solvent, he confirmed that respirators are needed to bring the levels within the Occupational Exposure Limits (OELs). But when Josh investigated further, he found that with the installation of a new area exhaust system, the solvent concentrations could be reduced by 80% and brought well below the Action Level for the agent, at a purchase and installation cost of $10,000.

In order to build his case, Josh showed that the Respiratory Protection Program needed to support the workers' use of the respirators actually costs the company $5,600 per year. This cost includes annual medical clearance evaluations of all workers, purchasing and maintaining respirators, and annual worker fit-testing and training. It does not include worker time off to attend training or receive medical evaluations after exposure to a dangerous level of the solvent.

Josh was able to show that within two years, the installation of the new ventilation system could save the company thousands of dollars. And the life expectancy of the fan system is 10 years.

Despite decades of workers wearing heavy, uncomfortable respirators, Josh was able to convince management that better engineering controls are the most efficient (and safest) option in the long run.

fingers between the fan blades, or positioning a worker inside a protective booth can reduce the worker's exposure to noise or heat in a work area.

After elimination and substitution, engineering controls are the second choice for protecting workers because they tend to be the most protective. Unless the engineering control is disabled or removed, it is impossible for the worker to be injured.

Over the years, engineering controls have become more sophisticated. Ventilation systems and fume hoods now protect workers from hazardous aerosols. Motion-sensing curtains using lasers or infrared radiation shut down hazardous systems before any part of a worker's body can enter the point of operation. Sound barriers or machine enclosures are also frequently used to protect workers from hazards.

ADMINISTRATIVE CONTROLS

Administrative controls are the third level of protection for workers. These controls consist of policies, programs, procedures, and other managerial tools. To be most effective, administrative controls should be in print, easily understood, and available to all workers.

Policies—High-level and overreaching directives that set the corporate culture and goals for worker health and safety.
Programs—Broad areas of activity that address particular topics. Programs might identify the means used to ensure that particular regulations or other goals will be met. They typically assign responsibilities for procedure development and implementation.
Procedures—Methods through which the goals of the program will be achieved. Procedures provide specific operational instructions to workers. A program such as Hearing Conservation may have several supporting procedures including conducting surveys, selecting hearing protection, performing audiometric testing, and training workers.

Training is essential for administrative controls to be effective. Once programs and procedures have been developed, workers must be educated on their use to ensure their adherence to them and their safety.

Warning signs in dangerous areas and on equipment and labels on containers of hazardous materials are also considered administrative controls. These signs and labels should alert workers to hazards and inform them of the appropriate controls or actions to be taken to avoid coming in contact with those hazards.

Smaller companies may have only one health and safety representative, and this person should have an extensive background in the expertise necessary for the facility. If the representative does not have such a background, it may be necessary to engage people from outside the company who have the specific areas of expertise to develop and implement safety programs. It may also be possible to send the health and safety representative for specific training provided elsewhere.

Often, a large company has an occupational safety and health team, which is a group of professionals such as industrial hygienists, occupational medicine physicians, occupational health nurses, safety professionals, human resources professionals, and management. In many workplaces, the group members work together to identify where programs are needed and then to develop and implement the programs to improve workplace safety and minimize injuries and illnesses. The main mission of the safety team is to protect the health of the workers. In doing so, the members should maintain an objective view and look at situations from a variety of viewpoints. They may interview and counsel employees but should always respect confidences.

Health and safety committees can also be an effective tool to move the safety agenda forward in an organization. These committees can offer a forum for information exchange, policy making, program planning, implementation, and evaluation.

In any organization, team members and employees are responsible to their employer or client. However, these obligations are secondary to their responsibilities to protect the health and safety of employees. Sometimes industrial hygienists are put in difficult positions and have hard decisions to make. Effective communication skills and the ability to use diplomacy are often required to ensure worker safety while maintaining management respect and support.

PERSONAL PROTECTIVE EQUIPMENT

Personal protective equipment (PPE) is commonly considered the last line of defense between a hazard and the worker. When other control methods are not feasible or do not completely eliminate the hazard, PPE can be used to provide some protection for the worker. In some instances, PPE is used as a backup or redundant method of control to ensure worker safety. In other cases, such as in a spill of hazardous materials, PPE is the only available defense.

One way to think about the effectiveness of PPE is to consider that there often are only a few millimeters between the worker and the hazard. Gloves can be used to reduce possible skin exposure to toxic chemicals, but gloves are not very thick and can develop holes that would allow the toxic chemical through. Earmuffs can reduce exposure to hazardous levels of noise, but they are effective only if they are fitted properly and the worker wears them. Workers often complain that the appropriate PPE is uncomfortable, interferes with the work process, makes it difficult to communicate with coworkers, and even makes the task they perform more hazardous in other ways because, for instance, they are not able to manipulate materials as easily. An effective PPE program thus depends on a training program that impresses on the workers the importance of wearing PPE, and on supervisory oversight that ensures workers comply with the PPE program requirements.

CAREERS

With a bachelor's or master's degree in industrial hygiene or safety, graduates can work in a variety of industries and organizations. Many graduates work for the government specifically at OSHA or NIOSH. Many jobs are also available in state OSHA offices throughout the country. Other federal agencies that have job opportunities in safety or occupational health include the following:
- Environmental Protection Agency
- Federal Communications Commission
- Nuclear Regulatory Commission
- National Institutes of Health's Center for Disease Control and Prevention
- Public Health Service
- Chemical Safety Board
- National Institute for Environmental Health Studies
- Consumer Product Safety Commission
- Department of Energy
- Mine Safety and Health Administration
- National Transportation Safety Board

Employment in industrial hygiene can be found in almost every industry. Anywhere that chemicals are used or workers come in contact with a hazardous work environment, there is a need for industrial hygienists to evaluate and control workplace hazards. Common industries where industrial hygienists are employed include the following:
- manufacturing
- health care
- research/education
- transportation
- insurance
- agriculture
- public utilities
- construction

- energy
- waste management
- mining

The jobs that industrial hygienists perform are as varied as the industries themselves. Most industrial hygiene positions include a variety of responsibilities from management to technical evaluation or collection of data. However, some positions may be more management-focused, while many others may consist of much more scientific and technical responsibilities. Some of the jobs that industrial hygienists perform include the following:
- consulting
- laboratory analysis
- management
- exposure assessment and control
- research
- policy and regulation development and analysis
- quality assurance
- ergonomics
- noise assessment and control
- radiation safety

PROFESSIONAL ORGANIZATIONS

Several professional organizations represent, support, and disseminate information about the field of industrial hygiene and safety. In some cases, the roles of these organizations may overlap, and individuals may belong to more than one organization.

AMERICAN INDUSTRIAL HYGIENE ASSOCIATION

The American Industrial Hygiene Association is a nonprofit organization created in 1939 with the mission of compiling and disseminating knowledge to protect worker health. It is one of the largest international associations, with approximately 10,000 members, and serves as a resource for professionals and the public on a wide range of health and safety topics.

The AIHA has more than 40 volunteer groups that collect and prepare information on a wide variety of topics. AIHA also has approximately 62 chapters in the United States, called local sections, which provide educational and networking opportunities. The organization prepares two monthly publications on the field of health and safety. *The Synergist* is a news magazine on the industrial hygiene profession and related topics, and the *Journal of Occupational and Environmental Hygiene* is a peer-reviewed publication focused on the science and management of industrial hygiene and occupational health.

AMERICAN SOCIETY OF SAFETY ENGINEERS

The American Society of Safety Engineers (ASSE) was founded in 1911 to promote workplace health and safety by identifying hazards and promoting the implementation of safety advances. Today there are more than 35,000 members, with chapters and sections in over 80 countries.

ASSE publishes a journal called *Professional Safety* each month. It also hosts a safety conference each year on technical and management topics and with thousands of attendees from across the nation and around the world.

AMERICAN CONFERENCE OF GOVERNMENTAL INDUSTRIAL HYGIENISTS

The American Conference of Governmental Industrial Hygienists (ACGIH®) was created in 1946 with the mission of improving occupational and environmental health. Several ACGIH committees work together to share information and create guidance documents on a variety of topics.

The Threshold Limit Values for Chemical Substances Committee is the best known of the ACGIH committees and, as its name suggests, documents the Threshold Limit Values (TLVs®) of chemical substances. This document uses the latest health effects data on over 700 chemical substances and physical agents to create recommended worker exposure levels to ensure workers' health and safety. The group also prepares more than 50 Biological Exposure Indices (BEIs®) for selected chemicals.

The ACGIH publishes a monthly news magazine called *Action Level* and, with the AIHA, copublishes the *JOEH* and cosponsors the American Industrial Hygiene Conference and Exposition (AIHce) every year. This international conference, with approximately 5,000 annual attendees, disseminates information and educates participants on the latest issues and developments in the industrial hygiene field.

NATIONAL SAFETY COUNCIL

The National Safety Council was founded in 1913 as a nonprofit organization whose mission is to save lives by preventing injuries and deaths at work, in homes, in communities, and on the road through leadership, research, education, and advocacy. NSC advances this mission by partnering with businesses, government agencies, elected officials, and the public to have the greatest influence where the most preventable injuries and deaths occur. NSC uses research to determine optimal solutions to safety issues and then provides education and outreach to change behaviors by building awareness and sharing best practices (NSC 2014).

The council's annual Congress & Expo has approximately 14,000 attendees from around the world, who teach and attend educational seminars, view exhibitions, and develop professional and business relationships. NSC publishes a monthly magazine entitled *Safety+Health* that provides approximately 86,000 professionals with safety news and analysis of industry trends. It also publishes a peer-reviewed journal called *Journal of Safety Research* five times a year.

HEALTH PHYSICS SOCIETY

The Health Physics Society (HPS) was formed in 1956 as a scientific organization of professionals specializing in radiation safety. The mission of the group is to promote excellence in the practice of radiation safety. There are nearly 5,000 HPS members working in a variety of industries throughout the United States. The society also promotes education and training in the field through publication of the peer-reviewed *Health Physics Journal* and sponsorship of professional conferences.

HUMAN FACTORS AND ERGONOMICS SOCIETY

The Human Factors and Ergonomics Society (HFES) was created in 1957 with the mission of promoting the discovery and exchange of knowledge about the relationships among human beings and the design of systems and devices. Two of the organization's technical groups have particular interest in workplace safety and industrial hygiene: the Safety Technical Group and the Occupational Ergonomics Technical Group.

The HFES publishes a variety of peer-reviewed and news journals, in addition to an annual series in *Reviews of Human Factors and Ergonomics*. Other educational resources produced by the organization include webinars, books, and scientific conferences and training courses on ergonomics.

AMERICAN NATIONAL STANDARDS INSTITUTE

The American National Standards Institute (ANSI) was founded in 1918 and is a nonprofit organization that develops guidelines and standards used in business. The mission of the organization is to enhance the global competitiveness of U.S. businesses; however, many ANSI standards specifically address the safe practices and operations of numerous business activities. Such standards range from guidance on the safe use of lasers to safety management systems, quality assurance and environmental protection, and the safe design and use of ladders. ANSI is integral to occupational health and safety, and OSHA refers to it throughout its regulations as a way for companies to meet OSHA compliance. As regulatory agencies lose their significance because of government cutbacks, ANSI will become even more necessary.

SUMMARY

Industrial hygiene has been practiced in one form or another for hundreds of years. It is the science of anticipating, recognizing, evaluating, and controlling hazardous exposures to workers. It is built upon a variety of fields of study including chemistry, physics, math, anatomy, biology, psychology, and management. Professionals with expertise in this field work in a variety of industries and workplaces to make them safer. It is a challenging and rewarding career with many opportunities for making a difference in the world.

REVIEW QUESTIONS

1. Define the term *industrial hygiene*.
2. Explain what is meant by *anticipation* as it is used in industrial hygiene.
3. Explain what is meant by *recognition* as it is used in industrial hygiene.
4. Explain what is meant by *evaluation* as it is used in industrial hygiene.
5. Explain what is meant by *control* or *intervention* as it is used in industrial hygiene.
6. What is the hierarchy of controls?
7. Give examples of four types of environmental factors or stresses.
8. Name four major professional organizations closely related to the practice of industrial hygiene.
9. Name two major governmental organizations closely associated with the practice of industrial hygiene and explain the role of each.
10. Describe the requirements for effective occupational safety and health (OSH) programs.
11. Describe the role of the industrial hygienist on the OSH team.

REFERENCES

American Conference of Governmental Industrial Hygienists (ACGIH). *2012 Threshold Limit Values for Chemical Substances and Physical Agents and Biological Exposure Indices*. ACGIH. 2012. ISBN: 978-1-607260-48-6.

Department of Labor. "Government Regulation of Worker's Safety and Health." www.dol.gov/dol/aboutdol/history/mono-regsafepart02.htm.

National Institute of Occupational Safety and Health. *Training and Workforce Development.* www.cdc.gov/niosh/training/.

National Safety Council. *Improving Workplace Safety* Itasca, IL: NSC, 2014. www.nsc.org/learn/pages/safety-at-work.aspx?var=mnd.

Occupational Safety and Health Administration. *OSHA 3143: Informational Booklet on Industrial Hygiene.* Washington DC: U.S. Department of Labor, 1998. www.osha.gov/Publications/OSHA3143/OSHA3143.htm#Industrial.

2

The Occupational Safety and Health Act and Industrial Hygiene

LEARNING OBJECTIVES

After completing this chapter, readers should be able to do the following:

- Understand the various branches of the U.S. government and their respective roles in the development of industrial hygiene regulations.
- Identify federal agencies and commissions and their responsibilities for occupational health and safety.
- Describe how the Occupational Safety and Health Act (OSH Act) was created and what the various sections of the OSH Act require.
- Explain the different types of rules, regulations, and standards and how they are created.
- Understand OSHA inspection and enforcement programs and practices.
- Identify other sources of important information available from OSHA.

Photo credit: iStock/Alina Solovyova-Vincent

INTRODUCTION

Federal and state regulations and laws play key roles in the development and practice of industrial hygiene. Therefore, it is extremely important to understand the governmental processes and rationale by which such laws are made in order to fully understand the statutes and regulations that affect industrial hygiene.

In 1970, the **Occupational Safety and Health Act (OSH Act)** was passed in order to prevent workers from being killed or seriously harmed while at work. Prior to that time, very few safeguards existed to ensure workers' safety.

Although the OSH Act has become the core of health and safety regulation and enforcement, industrial hygienists need to understand the fundamental principles of the U.S. government system and how they relate to numerous areas of occupational health and safety. It is also necessary to understand how and why laws are created and what these professional roles are in creating and interpreting those laws. For example, industrial hygienists are often called upon to write regulations and need to be able to work with lawmakers to draft laws to protect workers, a process that often involves explaining technical concepts to people who do not have industrial hygiene training or experience. In addition, industrial hygienists frequently are called upon to describe worker protections and the applicability of laws to those protections. A brief review of the fundamental principles of government and law creation is provided in the following sections.

REVIEW OF U.S. GOVERNMENT

Through a series of steps and processes from 1787 to 1790, the U.S. Constitution was written, debated, revised, and ratified by the first 13 states in the colony. The Constitution is the foundation of the U.S. governmental system and protects the rights of its citizens. The Constitution was designed to give enough power and funding to the government so that it can be effective at a national level, with power divided among three branches: executive, judicial, and legislative.

To further distribute governmental powers, the legislative Congress is divided into two parts, the **Senate** and the **House of Representatives**. In the House of Representatives, each state's number of representatives and number of votes from each state are based on the state's population. In the Senate, each state has two representatives, giving states equal significance in this house of Congress.

Amendments to the Constitution are proposed by a two-thirds vote of both houses of Congress, or by two-thirds of the states by convention. Amendments must then be ratified by three-fourths of the state legislatures or three-fourths of the state conventions.

Protections that the Constitution provides were clarified and enumerated in the Bill of Rights in 1789, which later became the first 10 amendments to the Constitution. Some of these amendments are applicable to the practice of industrial hygiene. For example:

The Fourth Amendment protects citizens from unreasonable search and seizure. This means that the government may not conduct any searches without a warrant, and such warrants must be issued by a judge and based on probable cause. Because this right also extends to organizations and businesses, a federal organization such as OSHA cannot enter property or conduct a search of an operation without a warrant or permission from the owner.

The Fifth Amendment prohibits citizens from being subject to criminal prosecution and punishment without due process. The amendment also establishes the power of eminent domain, which ensures that private property cannot be seized for public use without just compensation. This amendment protects a company's operations or products from harm and sets specific limits on when the government can interfere with operations to protect the public health and safety.

The Sixth Amendment ensures the right to a speedy trial by a jury of one's peers, to be informed of the crimes with which one is charged, and to confront the witnesses brought by the government. The amendment also provides the accused the right to compel testimony from witnesses and to legal representation. In addition, expert witnesses, such as industrial hygienists, must be a part of any judicial processes brought against a company. Cases are decided by facts and information through due process.

EXECUTIVE BRANCH

The executive branch of the U.S. government gives power to the president, who is the head of state and commander in chief of the armed forces. The president is responsible for implementing and enforcing the laws written by Congress and appointing the heads of the federal agencies, including the Cabinet. The Cabinet and independent federal agencies enforce and administer laws. Many of these Cabinet departments have designated health and safety responsibilities for both government workers and private citizens. The Department of Labor (DOL), for example, is responsible for collecting data on all worker accidents to improve working conditions. OSHA is a part of the DOL. Other large departments of the executive branch such as the Department of Health and Human Services, the Department of Transportation, the Department of Energy, the Department of Veterans Affairs, and the Department of Homeland Security are responsible for implementing their own industrial hygiene programs to protect their workers. The Environmental Protection Agency (EPA), which has significant oversight on workers and public health and safety, is another department in which industrial hygienists are commonly involved. The EPA is responsible for its thousands of employees and also conducts significant research in chemical toxicity and other agents that can cause environmental degradation and affect worker and public health.

LEGISLATIVE BRANCH

The House of Representatives and the Senate together comprise the U.S. Congress. The Constitution gives Congress the authority to create laws and considerable power to conduct investigations.

Legislative Process

The first step in the legislative process is the introduction of a bill to Congress. After a bill is introduced, it is sent to the appropriate congressional committee for consideration. Each committee oversees a specific area of the government, such as finance or transportation. Subcommittees take on more specialized work within the major policy areas. Industrial hygienists may be part of the initiation of a law, or they may be called upon to testify at any point of the legislative process.

A bill is first considered in a subcommittee, which may accept, amend, or fully reject it. If the members of the subcommittee agree to move a bill forward, the bill is sent to the full committee, where the process is repeated. During this phase, the committees and subcommittees meet to discuss the pros and cons of new bills. During hearings they listen to the opinions of experts such as industrial hygienists and other interested or affected parties.

It is important for industrial hygienists to stay current with the regulations affecting worker health and safety because these regulations are constantly changing and new ones are written regularly. When new regulations or laws are passed, they are published in the *Federal Register*, which compiles the most current federal regulations and is available online. The major types of federal publications include the following:
- General notices (meetings, documents, etc.)
- Proposed rulemaking/advance notices of proposed rulemaking
- Regulatory text that is used in specific laws

Eventually the laws are printed in the *Code of Federal Regulations* (CFR), which categorizes general and permanent rules published in the *Federal Register* by the departments and agencies of the federal government. Each government agency is assigned a title within the CFR. For example, OSHA is assigned CFR 29, and the Environmental Protection Agency has Title 40 of the CFR.

JUDICIAL BRANCH

The judicial branch is the third part of the U.S. government. These members are not elected but are appointed by the president and confirmed by the Senate.

The Constitution gives authority to Congress to determine the structure of the federal judicial system. The first level of the court system is made up of district courts, which try cases where one party brings suit against another party. These courts interpret the law, make sure laws abide with the requirements of the Constitution, and apply laws to the individual cases to issue a verdict. Because industrial hygienists are often called upon to provide expert witness testimony about the exposures or hazards at a workplace that led to a perceived negative outcome, it is important for them to understand the judicial process.

When a district court verdict is appealed, the case goes to one of the 13 U.S. Appeals Courts. Sometimes appeals from cases in the appellate court system proceed all the way to the Supreme Court for review and decision. Typically, these are cases where controversial laws need additional interpretation and consideration or where the Court needs to determine a law's relevance to a particular case. Supreme Court decisions are final and often become precedent for similar future cases.

FEDERAL AGENCIES AND COMMISSIONS

Of the hundreds of federal agencies and commissions, many are charged with writing and enforcing regulations that affect industrial hygiene activities. OSHA, part of the Department of Labor, is charged with protecting the health and safety of workers. The National Institute for Occupational Safety and Health (NIOSH) is tasked with conducting research in workplace safety and issuing reports and recommendations to OSHA. NIOSH is funded and supported by the Centers for Disease Control and Prevention, which is part of the Department of Health and Human Services.

Other federal agencies and commissions are associated with workplace safety and industrial hygiene include the following:

- **Federal Communications Commission**—The Federal Communications Commission (FCC) is an independent agency, overseen by Congress, that regulates the telecommunications industry. Part of its role is to ensure public safety and protection from harmful effects of radiation from communication systems.
- **Nuclear Regulatory Commission**—The Nuclear Regulatory Commission (NRC) has five members who are appointed by the president. Their primary role is to regulate and ensure the safe operation of the nation's commercial use of nuclear materials. The members' main activities are to create regulations, inspect nuclear facilities, and ensure emergency preparedness.
- **Mine Safety and Health Administration**—The Mine Safety and Health Administration (MSHA) is part of the Department of Labor and is charged with ensuring the health and safety of mine workers.
- **Environmental Protection Agency**—The Environmental Protection Agency (EPA) is charged with writing and enforcing laws to protect the health of the environment, the public, and workers. The EPA administrator is appointed by the president.

STATE GOVERNMENT

The Tenth Amendment of the U.S. Constitution grants powers of governance to the states when there are no applicable federal powers. Most state governments are modeled after the federal system and have three branches: executive, legislative, and judicial. Because OSHA allows states to promulgate and enforce their own regulations for occupational health and safety, states must enact the same, or more conservative, regulations than what the federal government enacts. Some states choose to assume OSHA responsibilities and are called "Agreement States." In those states, any regulations written by OSHA must be enforced at the state level and follow the same protocols and practices that are at the federal level.

OCCUPATIONAL HEALTH AND SAFETY REGULATIONS

Until 1970, few federal regulations provided protections to workers by ensuring safe workplaces. The first state law providing some safe workplace conditions was enacted in Massachusetts in 1877. By 1890, an additional 21 states had provisions for worker protection from health hazards.

In 1911, the Triangle Shirtwaist building in New York City caught fire, and 146 workers were killed. Despite prior demands for better working conditions, the building owners made few changes; on the day of the fire, in fact, the building exits were locked, and workers were caught in the flames and smoke.

In the aftermath of the fire, public outrage and activism persuaded the New York State legislature to create the Factory Investigating Commission. Findings from the commission led to state laws regarding building egress, fireproofing, fire alarms, and fire extinguishers.

In 1913, the Bureau of Labor began collecting data and publishing statistics on occupational illnesses and fatalities. Dr. Alice Hamilton, an employee of the Bureau of Labor, studied workplaces and prepared reports that clearly linked hazardous workplaces to worker illnesses. By 1940, all states had some laws regarding workplace safety (DOL 2014).

The Federal Compensation Act created the Office of Workers' Compensation in 1918 to provide benefits to workers who are injured or contract illnesses in the workplace. Because of this office, employees who are injured at work receive compensation from funds paid by employers as insurance premiums. With the creation of this program, employers could, for the first time, see a relationship between a safer workplace and better profits.

The Bureau of Labor Standards was created in 1934. This permanent federal agency was the first to focus primarily on worker health and safety. Also in 1934, the United States joined the International Labor Organization and began to take part in efforts to improve working conditions worldwide.

In 1935, the National Labor Relations Act provided workers with the right to bargain collectively. This right led to improved working conditions in numerous industries. In 1936, the Walsh-Healy Act required basic health and safety programs for any federal contracts exceeding $10,000. The 40-hour workweek, paid overtime, minimum wage, and limited child labor became law in 1938 with the creation of the Fair Labor Act (DOL 2014).

During the late 1950s and 1960s, labor organizations grew in strength and number. They worked to overturn several long-standing antilabor laws, including the following:
1. **The Fellow Servant Rule,** which stated that employers were not liable for workplace injuries that resulted from the negligence of other workers;
2. **Contributory Negligence,** which relieved employers of responsibility if the actions of their employees contributed to their own injuries; and
3. **Assumption of Risk,** the notion that workers who accept payment for work should assume that there will be risks involved in doing that work.

During this period, labor organizations also let the public and workers know about the risks and hazards in the workplace. They fought for safer working conditions and compensation for injured workers. In addition to,

> **Case: Antilabor Laws**
>
> Bill worked in a bicycle factory on the assembly line putting spokes into the wheels. One day while walking to the adjacent work area to retrieve additional spokes, Bill was struck by a moving handcart. He suffered a broken ankle and couldn't work for six weeks.
>
> The employer refused to pay Bill's medical bills, and subsequently fired Bill for not being at work. The employer claimed that it was not responsible because it was an employee at the firm who had driven the cart into Bill and caused the injury. In addition, Bill was partly responsible for the injury because he wasn't watching where he was going. At any rate, Bill should have expected hazards to exist at a fast-paced work environment like a bicycle factory. This employer used all three antilabor practices to argue its case: fellow servant rule, contributory negligence, and assumption of risk.

and partly because of, the efforts of labor organizations, businesses began to understand the relationship between safer workplaces and increases in productivity and profitability. At the same time, however, worker injury and illness rates continued to increase.

Legislators began lobbying for laws to protect workers from hazards. In New Jersey, Senator Harrison A. Williams Jr. pushed for workplace safety and health legislation. He called attention to the need to protect workers from a variety of hazards including noise, cotton dust, and asbestos. In Congress, Representative William A. Steiger worked for passage of a bill to improve working conditions.

The **Williams-Steiger Occupational Safety and Health Act** (OSH Act) was signed into law on December 29, 1970, by President Richard M. Nixon. The act requires every employer to provide its employees a place of employment that is free of recognized hazards that are causing or are likely to cause death or serious physical harm. The act authorizes the Department of Labor to conduct inspections and to issue citations and penalties for alleged violations (OSHA 2015).

The act is divided into 34 sections. Some of the most significant aspects of some of the sections are described below.

SECTION 2—PURPOSE

In Section 2 of the act, Congress declares its purpose:

> To assure so far as possible every working man and woman in the Nation safe and healthful working conditions and to preserve our human resources

The act encourages employers to reduce workplace hazards by implementing safety and health programs. It encourages OSHA to conduct research on how to solve the most hazardous occupational safety and health problems, and to promulgate regulations for mandatory workplace standards. The act gives OSHA the right to enforce the regulations that it creates and also requires OSHA to conduct training and public outreach programs. This section of the act also makes provisions for the development, evaluation, and approval of state occupational safety and health programs.

SECTION 4—THE OSH ACT'S COVERAGE

The OSH Act and applicable regulations apply to employment performed in a workplace in a state, the District of Columbia, and territories of the United States. The regulations do not apply to working conditions in industries and areas in which other federal agencies exercise statutory authority to prescribe or enforce standards or regulations affecting occupational safety and health. OSHA regulations also do not apply to sole proprietors when the owner is the only worker, to farms with fewer than 10 workers, or to businesses in which workers are family members.

The act states that other existing federal agencies are responsible for developing and enforcing regulations for worker safety in their respective industries. For example, the Federal Railroad Administration (FRA) is responsible for worker safety on train tracks and on the trains themselves. Other related facilities, like train stations, are regulated by OSHA. The Federal Aviation Administration covers worker safety anywhere on the tarmac, while OSHA covers the safety of workers up to the tarmac.

Two additional examples are the Nuclear Regulatory Commission, which covers worker safety during the use of radiation and radioactive byproduct materials and nuclear power generation. The Federal Communications Commission covers worker exposure to hazardous radiofrequency and microwave radiation, whereas OSHA regulations apply to fall hazards for tower climbers.

SECTION 5—DUTIES

Section 5 specifies the duties of employers to provide a safe workplace. When a workplace condition exists that is not covered specifically under a regulation or standard, OSHA can always cite the employer under the **General Duty Clause**. Simply put, Section 5(a)(1) of the act states that each employer shall

> furnish to each of his employees employment and a place of employment which are free from recognized hazards that are likely to cause death or serious physical harm to his employees.

It is not always easy for OSHA to use this clause to issue violations or citations. However, if there is a death of a worker or if a significant number of workers are being injured, it becomes more difficult for an employer to claim that the workplace meets the standard and provides a workplace free from recognized hazards.

SECTION 7—ADVISORY COMMITTEES

Section 7 of the act provides for the creation of committees to advise OSHA and make recommendations to the Secretary of Labor and the Secretary of Health and Human Services about implementing the act. The currently 12-member National Advisory Committee for OSHA has members representing management, labor, occupational health professions, and the public.

SECTION 8—INSPECTIONS, INVESTIGATIONS, AND RECORD KEEPING

Section 8 of the act describes inspections, investigations, and record keeping. It also requires a system for distributing information regarding the inspection process and results and allows employees to be involved in, and gives them the right to request, inspections.

SECTION 9—CITATIONS

This section describes how citations are issued. Citations must be in writing and refer to specific provisions of the act and applicable regulations. Violations and citations from OSHA must be prominently posted at or near the place of the violation.

SECTION 10—ENFORCEMENT

When an OSHA citation is issued, the employer has 15 days to contest the citation, penalty, or time allowed for abatement and must do so in writing. Appeals are sent to the Occupational Safety and Health Review Commission (OSHRC), which reviews the case and makes a decision.

SECTION 11—JUDICIAL REVIEW

If employers are not satisfied with the decision of the OSHRC, they can further appeal to the U.S. Courts of Appeals and up to the Supreme Court. Paragraph 11(c) of this section provides special protections to *whistleblowers* who report unsafe conditions or practices at a workplace. Employers cannot take action against employees for notifying OSHA of workplace hazards, nor can employers retaliate against or reprimand workers who bring unsafe conditions or activities to the attention of management.

SECTION 12—OCCUPATIONAL SAFETY AND HEALTH REVIEW COMMISSION
The OSHRC acts independently of OSHA. The mission of the OSHRC is to provide fair and timely review of workplace safety and health disputes between the Department of Labor and employers.

SECTION 13—PROCEEDINGS TO COUNTERACT IMMINENT DANGERS
Section 13 of the act allows OSHA to petition to U.S. District Courts for restraining orders to stop business activities or close facilities in cases of **imminent danger**.

SECTION 17—PENALTIES
This section allows for criminal and civil penalties for OSHA violations. Penalties can be in the form of monetary fines and fees and can include imprisonment.

Criticisms of the existing penalty system have been raised recently. Some of the criticism stems from statistics that demonstrate OSHA seldom assigns the total possible fines and penalties for observed violations. In addition, many of the fines are negotiated to much lower sums during the appeals process. Another criticism is that the fines themselves were set in 1970 dollars and four decades later, in terms of inflation and the value of a dollar, the current penalties—and thus the deterrent benefit—are greatly reduced from the initial intent of the fee system.

The maximum civil penalty in the OSH Act for a serious violation of OSHA regulations was originally—and remains—$7,000. That figure has not changed in many years and is still the same today. If the fine was adjusted to follow the Consumer Price Index, which accounts for inflation and the change in monetary value, the actual penalty in today's dollars would be six times higher than the original amount, or $42,000.

SECTION 18—STATE PLANS
This section describes the right of the states to have their own workplace health and safety regulations as long as they are at least as stringent as the corresponding federal regulations. The review, evaluation, and approval process for state plans is also described.

SECTION 19—FEDERAL AGENCY SAFETY PROGRAMS AND RESPONSIBILITIES
This section discusses the requirement of all federal agencies to have occupational health and safety programs consistent with OSHA standards.

SECTION 20—RESEARCH AND RELATED ACTIVITIES
This section describes OSHA's role in conducting research on workplace health and safety. It requires OSHA to explore new workplace problems and

devise improved methods of control. It also calls on OSHA to review toxicological information and evaluate worker exposures to determine the need for additional protections. In this section, the role for conducting research is delegated to the Director of NIOSH.

SECTION 21—TRAINING AND EMPLOYEE EDUCATION
One responsibility OSHA has is educating qualified personnel to carry out duties specified by the act. The other responsibility is supplying information to educate workers and employers in the specifics of the OSHA regulations and in general health and safety practices and controls. This section also describes training grants OSHA may award to groups conducting outreach training.

SECTION 22—NATIONAL INSTITUTE FOR OCCUPATIONAL SAFETY AND HEALTH
Most OSHA research is carried out by NIOSH, under the Department of Health and Human Services' Centers for Disease Control and Prevention. By gathering information, conducting scientific research, and translating the knowledge into products and services, NIOSH provides leadership to prevent work-related illness, injury, disability, and death.

NIOSH is responsible for setting national strategic goals for workplace health and safety and for conducting a focused program of research to reduce injuries and illnesses among workers in high-priority areas and high-risk sectors. NIOSH is also responsible for implementing and maintaining a system of surveillance of major workplace illnesses, injuries, exposures, and health and safety hazards.

SECTION 27—NATIONAL COMMISSION ON STATE WORKMEN'S COMPENSATION LAWS
This section is designed to ensure adequate worker compensation in the event of disabling work-related injuries and illnesses.

OSHA STANDARDS

During the first few years of its existence, OSHA conducted research to determine which workplace hazards were most hazardous and which ones could benefit from controls or regulations. The creation of a rule typically follows the general activities described in this section.

RULE DEVELOPMENT
A variety of sources can indicate where new health and safety rules may be warranted. Initially, OSHA conducts an assessment to determine the extent of the hazard and whether a rule could help control the hazard.

After it is decided that a new rule is the best way to control the hazard, OSHA expands its research into the hazard and potentially effective controls. Economic analyses are performed to evaluate possible financial effects of the proposed regulations. The technical processes and materials used in affected industries are studied to determine whether controls to reduce the hazards are feasible.

Eventually, all the available information is compiled and reviewed. A proposed rule is written based on the information obtained in the analysis. The regulatory text and a short explanation of its purpose, called a preamble, are prepared for the publication of the Notice of Proposed Rulemaking in the *Federal Register* for a public comment period of typically 60 to 90 days.

After the comment period, OSHA uses all previously collected information and the results from the public comment period to draft a final rule. Once the rule is approved by the White House Office of Management and Budget, it is published in the *Federal Register*. In order for the regulation to become fully effective, it must be filed with Congress and the Government Accountability Office.

TYPES OF RULES

A variety of rules may be created depending on the particular industry or the hazards that need to be controlled.

Specification standards are rules that say exactly what engineering or design criteria must be met to protect workers from a particular hazard. For example, a specification standard might describe the exact type of ventilation system that must be installed to protect workers from toxic vapors during a specific industrial process.

Performance standards, on the other hand, specify only the levels of toxic chemical in the air that are acceptable. Although employers have to be sure the levels are not exceeded, they can use any ventilation system design to accomplish this.

Vertical standards typically apply only to the industry where a particular hazard exists. Grain-handling facilities, for example, have a discrete set of particular hazards that are specific to that industry and thus specific design and implementation rules have to be met (1910.272). Vertical control requirements do not always involve engineering controls, but can be administrative controls or require certain personal protective equipment.

Horizontal standards can apply to nearly any workplace or industry. For example, 1910 Subpart D lists requirements for walking and working surfaces that any workplace with floors or stairs must comply with.

OSHA's current standards comprise several thousand pages of regulations. The first set of rules provides the structure of OSHA and describes the operating procedures that support the intent of the original OSH Act. Some of the OSH Act topics, such as record keeping of injuries and illnesses and procedures for conducting inspections and enforcement, are described in their own sections in great detail. In addition, the original rules are separated into General Industry, Construction, and Agriculture. Other industries such as Longshoring and Marine Terminals also have their own sets of standards.

The main body of OSHA regulations for general industry is contained in Chapter 29 of the *Code of Federal Regulations* Part 1910. These regulations take up more than 1,000 pages and describe numerous specific activities and workplace conditions that fall under the purview of industrial hygiene professionals. The most accessed OSHA standards are shown in Table 2–1.

Table 2–1. The Top 10 Most Accessed General Industry Standards

Bloodborne Pathogens—1910.1030
Hazard Communication—1910.1200
Respiratory Protection—1910.134
Occupational Noise Exposure—1910.95
Powered Industrial Trucks—1910.178
Permit-required Confined Spaces—1910.146
Lockout/Tagout—1910.147
Hazardous Waste Operations and Emergency Response—1910.120
Guarding Floor and Wall Openings and Holes—1910.23
Personal Protective Equipment—1910.132

Source: Occupational Safety and Health Administration

The first three sections of the standards indicate how the standards meet the OSH Act. Specific requirements, definitions, and scope of the standards are provided. In addition to the main body sections of the regulations, numerous subparts have been added that pertain to special conditions or provide more detail on certain topics. Some of the more common and relevant subparts are briefly described in the following paragraphs.

In Subpart D, Walking and Working Surfaces, detailed requirements for workplace floors, stairs, ladders and scaffolding are provided. In Subpart E, the requirements for emergency exit routes, emergency action plans, and fire prevention plans for organizations are defined.

Some regulations directly involving industrial hygiene fall in Subpart G, on occupational and environmental control. In this subpart, the requirements for the design and operating parameters of ventilation systems to protect workers are described. This subpart also provides rules for occupational noise exposure, including audiometric testing of employees and surveying workplace noise levels.

Subpart H covers the storage and handling of hazardous materials, including requirements for process safety management and for hazardous waste operations and emergency response.

All aspects of the use of personal protective equipment are discussed in Subpart I. This subpart also provides detailed requirements for the development and implementation of a respiratory protection program, including the medical evaluation of workers and fit-testing procedures.

Subpart J, on general environmental controls, provides requirements for workplace postings. Section 146 of this subpart provides detailed requirements for the identification and control of confined spaces. Another very important section in this subpart, 1910.147, covers the control of hazardous energy, otherwise known as lockout/tagout.

Requirements for the availability of medical services and first aid are described in Subpart K, and a long list of detailed rules on fire protection and associated systems and procedures is provided in Subpart L.

Many of OSHA 1910 subparts M through Y are associated with safety activities and controls. The industrial hygienist should still be familiar with these subparts, however.

The last subpart, Z, provides extensive detail on the hazards of and necessary controls for occupational exposure to toxic substances. Permissible exposure limits for approximately 1,200 chemicals or elements are provided in section 1910.1000. Additional limits and required controls for another 32 chemicals and elements, including asbestos, formaldehyde, cotton dust, lead, and cadmium, are included in this subpart.

Three other important topics in Subpart Z are the requirements for training and hazard communication (in 1910.1200), for bloodborne pathogens (in 1910.1030), and for occupational exposure to hazardous chemicals in laboratories (in 1910.1450).

ENFORCEMENT ACTIVITIES

A significant amount of OSHA activities involve inspection and enforcement of the regulations. Compliance officers typically have training and experience in health and safety or industrial hygiene. Except in certain cases, inspections are conducted without advanced notice. There are two types of inspections: programmed and unprogrammed. Programmed inspections are discussed below.

Programmed inspections occur as part of planned goals and initiatives. Site-specific targeting programs are designed to reduce the number of injuries or illnesses in a particular worksite where data show that many hazards or risks exist. **Special emphasis programs** are developed to provide additional inspections in industries with higher-than-average injury and illness rates. Special emphasis programs include **national emphasis programs**, which look at industries across the country, and **regional emphasis programs**, which increase inspections in a particular region of the country. The OSHA special emphasis programs are based partly on industry trends and accident statistics and are used to identify areas that warrant more OSHA oversight, investigation, training, and enforcement. Past and current special emphasis programs have included lead in construction, silica, hazardous machinery, and shipbreaking.

Local emphasis programs are especially designed to correct weaknesses and dangerous working conditions common to a particular geographic area of the country. These programs are intended to address hazards or industries that pose a particular risk to workers in the jurisdictions of one of the 10 OSHA areas. Some examples of local emphasis programs include fish processing in Region I, dairy farm operations in Region II, logging in Region III, and casino hotels in Region IX.

Severe violator enforcement programs are programmed inspections that focus on employers that have committed willful or repeated violations.

OSHA inspections are typically scheduled in decreasing order of importance. First-priority inspections take place when there is an **imminent danger of injury or illness**. Second-priority inspections occur when there has been a **fatality** or **catastrophe** at a site or facility. Third-priority inspections take place when there is a **complaint** or **request** for or a suggestion of an inspection. Programmed inspections have the lowest priority.

In addition to reviewing applicable regulations and laws during inspections, OSHA compliance officers refer to the **Field Operations Manual** for detailed instructions on how to conduct inspections and evaluate programs, activities, conditions, and documents. An additional **Technical Manual** provides guidance on evaluating and recognizing hazards during inspections.

Inspections can take a variety of forms. Comprehensive inspections may look at an entire facility or all potential hazards at a site. Partial inspections, on the other hand, may focus on only one area of an activity at a facility.

Inspections are conducted in accordance with established codes and protocols. Compliance officers will typically begin by reviewing the industry and the background of the particular site they plan to inspect. They may create an inspection checklist of specific items or areas and may gather and calibrate sampling instruments or other materials they will need to conduct the inspection.

When arriving at the inspection site, the compliance officers hold an opening conference to explain the reason for and scope of the inspection. They may ask for detailed information about operations or systems and may also request records of operational activities, employee training, or injuries and illnesses.

Inspectors may walk through various parts of the facility to observe systems and workers performing tasks. They may review medical records or data regarding engineering controls in place, or the use of personal protective equipment. They may also ask to speak with employees.

At the end of the inspection, the compliance team will conduct a closing conference during which they will indicate any areas of nonconformance and allow the site supervisor to provide any additional or supporting information to refute the findings. They will not indicate the specifics of any possible penalties but will describe the report process and timeline for responses and appeals.

Any citations for noncompliance with OSHA standards must be issued in writing by the area director no later than six months after the occurrence of the violation. If the employer wishes to contest the citation, it must submit a written objection to the area director, who then forwards the objection to the Occupational Safety and Health Review Commission (OSHRC) for consideration. The area director often will informally negotiate citations and penalties directly with employers. If employers do not contest the citation or penalties, they must typically complete corrective actions by an agreed-upon abatement date. For three days or until the violation conditions have been abated, whichever is longer, employers must post citations in a conspicuous place near where the violation occurred.

OTHER OSHA INFORMATION

Other information that OSHA publishes includes statistics on illnesses and injuries. A broad range of industries and business sectors are listed, and injury/illness rates are provided for comparison. See Figures 2–1, 2–2, and 2–3 for fatality and injury/illness rates provided by OSHA.

The statistics on where accidents and violations occur most frequently are useful because they often indicate common weak spots in organizations. Those

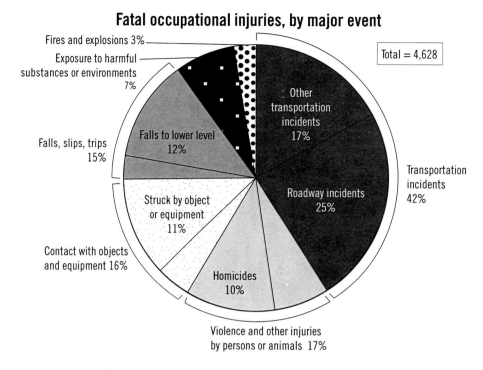

Figure 2–1. Fatal occupational injuries, by major event.

(Source: U.S. Bureau of Labor Statistics, U.S. Department of Labor, 2013. Percentages may not add to 100 due to rounding. Reference year 2011 constitutes a series break from earlier years for event.)

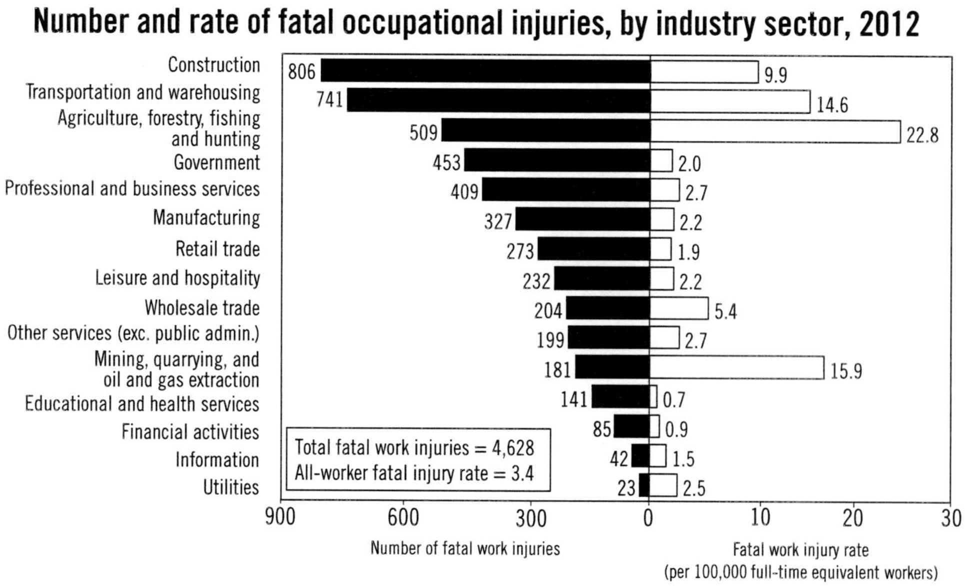

Figure 2–2. Number and rate of fatal occupational injuries, by industry sector, 2012.

(Source: U.S. Bureau of Labor Statistics, U.S. Department of Labor, 2013)

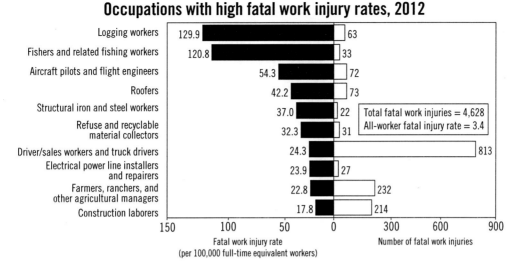

Figure 2–3. Occupations with high fatal work injury rates, 2012. *Data for 2012 are preliminary. NOTE: Fatal injury rates exclude workers under the age of 16 years, volunteers, and resident military. The number of fatal work injuries represents total published fatal injuries before the exclusions.

(Source: Occupational Safety and Health Administration. Special Emphasis Programs)

weak spots also tend to become OSHA special enforcement areas. The injury and illness rates for a given company can be compared with the OSHA statistics for the industry to see how an individual company is doing and where improvements may be needed. These statistics can be found on the OSHA website under Newsroom/News Releases/Enforcement. At the end of the news releases are links to the citations given to the organizations.

OSHA also publishes news releases and statistics on enforcement. All enforcement citations are available on the OSHA website for review. These citations are very informative and give the reader great insight into how and why OSHA enforces particular working conditions.

The top 10 most frequently cited OSHA standards for the period of October 1, 2013, through September 30, 2014, are shown on Table 2–2.

Table 2–2. Top 10 Most Frequently Cited OSHA Standards for the Period of October 1, 2013, through September 30, 2014 (as of October 28, 2014)

1. 1926.501—Fall Protection
2. 1910.1200—Hazard Communication
3. 1926.451—Scaffolding
4. 1910.134—Respiratory Protection
5. 1910.178—Powered Industrial Trucks
6. 1910.147—Lockout/Tagout
7. 1926.1053—Ladders
8. 1910.305—Electrical, Wiring Methods
9. 1910.212—Machine Guarding
10. 1910.303—Electrical, General Requirements

Source: Occupational Safety and Health Administration (OSHA 2015)

In addition to the OSH Act and the voluminous written regulations, OSHA has created many useful training and educational tools that can be used to educate employers and employees about the hazards and controls for various workplaces and activities. Many of the guides also provide useful information on how to comply with individual standards or laws. Some OSHA guides are provided in the following list:

- All About OSHA (OSHA 3302-01R–2013)
- OSHA Inspections (OSHA 2098–2002)
- Aerial Lift Quick Card (OSHA 3267–2011)
- Tree Care Work Hazards—Hazard Bulletin (OSHA HB 3731–2014)
- Laboratory Safety: Latex Allergy Quick Facts (OSHA 3411–2011)
- Scaffolding (OSHA 3150–2002)
- Safe Patient Handling–Preventing Musculoskeletal Disorders in Nursing Homes (OSHA 3708–2014)
- Personal Protective Equipment (OSHA 3151–2003)
- Materials Handling and Storage (OSHA 2236–2002)
- Ladder Safety: Falling Off Ladders Can Kill: Use Them Safely (OSHA 3625–2013)
- Hazard Communication (OSHA 3084–1998)
- Recommendations for Workplace Violence Prevention Programs in Late-Night Retail Establishments (OSHA 3153-12R–2009)

The OSHA website (www.osha.gov) provides quick access to the regulations and most of OSHA's educational and informational materials. The OSHA e-tool is a web-based training tool that can be used to make training on a variety of topics more interactive and interesting. Some of the OSHA e-tool devices and systems can be found at the following links.

- Evacuation Plans and Procedures
  www.osha.gov/SLTC/etools/evacuation/index.html
- Eye and Face Protection
  www.osha.gov/SLTC/etools/eyeandface/index.html
- Hospital
  www.osha.gov/SLTC/etools/hospital/index.html
- Lockout/Tagout
  www.osha.gov/dts/osta/lototraining/index.html
- Machine Guarding
  www.osha.gov/SLTC/etools/machineguarding/index.html
- Noise and Hearing Conservation
  www.osha.gov/dts/osta/otm/noise/index.html

- Nursing Home
  www.osha.gov/SLTC/etools/nursinghome/index.html
- Powered Industrial Trucks
  www.osha.gov/SLTC/etools/pit/index.html
- Respiratory Protection
  www.osha.gov/SLTC/etools/respiratory/index.html
- Safety & Health Management Systems
  www.osha.gov/SLTC/etools/safetyhealth/index.html
- Scaffolding
  www.osha.gov/SLTC/etools/scaffolding/index.html

FATALITY AND CATASTROPHE REPORTS

OSHA's records of all workplace fatalities and catastrophes are accessible through its website. These records are updated weekly, and related reports and analyses are published periodically. The records include the date, company, location, and description of the incident.

SUMMARY

Prior to the OSH Act, worker health and safety protections were based on a patchwork of state and federal regulations. Through the *Code of Federal Regulations*, OSHA now promulgates and enforces hundreds of regulations to protect workers from hazards in the workplace. Because industrial hygienists are sometimes called to draft laws or explain how the laws apply to a particular work situation in a court of law, industrial hygienists need to understand how and why laws are created and how laws are enforced. In addition, numerous other federal and state organizations share responsibilities with OSHA for worker, public health, and environmental protection, and practicing industrial hygienists may need to draw upon these other organizations to support their activities.

REVIEW QUESTIONS

1. What does the Fourth Amendment protect? Does this same protection apply to organizations and businesses?
2. If a neighborhood group sues a company for allowing lead paint chips to blow off its property and into the neighborhood, what constitutional amendment allows the company to have a jury in the civil suit?

3. Under what branch of the U.S. government does the Department of Labor fall?
4. At what phase of the preparation of a proposed bill is an industrial hygienist likely to be called to testify before Congress?
5. What is the *Federal Register*?
6. What type of information do appeals courts typically review?
7. What is the role of federal agencies and commissions? What are two agencies whoes responsibilities involve industrial hygiene?
8. When was OSHA created?
9. To whom or what did the fellow servant rule apply, and how was it used?
10. Name two instances where OSHA regulations do not apply.
11. What is the General Duty Clause?
12. What is NIOSH's role?
13. Explain the difference between a performance standard and a specification standard.
14. Explain the difference between a horizontal standard and a vertical standard.
15. Explain the difference between an OSHA special emphasis program and a regional emphasis program.

REFERENCES

Bureau of Labor Statistics. U.S. Department of Labor. "National Census of Fatal Occupational Injuries in 2012 (preliminary results)." News release, August 22, 2013.

Department of Labor. *U.S. Department of Labor Historical Timeline*. 2014. www.dol.gov/100/timeline/#57.

Occupational Safety and Health Administration. *Fatality and Catastrophe Reports*. 2014. www.osha.gov/dep/fatcat/fy14_federal-state_summaries.pdf.

———. *OSHA at 30: Three Decades of Progress in Occupational Safety and Health*. 2015. www.osha.gov/as/opa/osha-at-30.html.

———. *Top Ten Most Cited Standards*. 2015. www.osha.gov/Top_Ten_Standards.html.

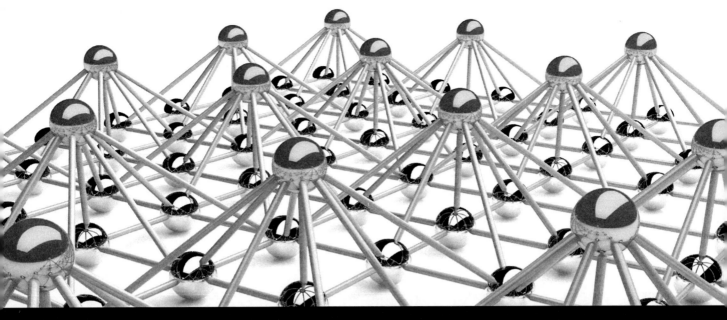

3
Management Systems

LEARNING OBJECTIVES

After completing this chapter, readers should be able to do the following:

- Describe the basic business activities and tools necessary to implement successful industrial hygiene programs.
- Discuss the relationship between quality assurance and industrial hygiene management.
- Understand the importance of industry standards and their relevance to industrial hygiene.
- Use various business models such as Total Quality Management, *Value Strategy Manual*, and Six Sigma to improve industrial hygiene program management.

Photo credit: iStock/leszekglasner

INTRODUCTION

Any organization, whether it is a for-profit company, a not-for-profit company, a government entity, an educational institution, a health care network, or a labor union, needs to operate according to basic management principles in order to be efficient and effective. Proper management also leads to the ethical operation of the organization, which includes overseeing financial assets for stakeholders and shareholders, protecting the environment and public, and ensuring the health and safety of organization employees.

Management defines the organizational structure of a business or company and delineates lines of authority and responsibility. It also plays a key role in designing and maintaining lines of communication among all groups and employees, the government, and the public at large.

Concerning industrial hygiene, management is responsible for identifying hazards in terms of their probabilities of occurrence and their severity. This responsibility includes providing medical surveillance of employees and assessing workplace exposures to hazards. Hazards might be defined in terms of environmental or public health or in terms of the cost of insurance premiums related to accidents or lawsuits.

Management should also provide and support a culture of safety. This

> **Case: Industrial Hygiene in a Supporting Role**
>
> Industrial hygiene and safety should be considered service functions in an organization. Functional groups within a business each have their own jobs to accomplish, and are rated by how well they do those jobs. Industrial hygiene needs to support those activities, rather than hold them back. The importance of this relationship is evident in the following example.
>
> A hospital's housekeeping department and safety office received numerous complaints about noxious odors when the floor wax was stripped twice a year in the hospital corridors. The odors were a problem not only for the workers, but also for the immunocompromised patients, many of whom had respiratory diseases and were on ventilators to support their breathing.
>
> After reviewing the processes and products used, the industrial hygienist researched alternative methods and products. She found a floor wax remover with much safer chemicals that would be less likely to cause the respiratory distress to that the original product had. Once the product was purchased and tried in a small area of the hospital, it was found to work as well as the original remover and cost the same, and the housekeeping director officially implemented the change to the new floor wax remover.

means setting the values and attitudes toward safety issues concerning all work groups and individuals within the organization. An effective safety culture transcends written programs and rules and becomes ingrained in the thought processes and actions of all employees in their everyday activities. In a successful safety culture, workers do not fear being reprimanded for reporting unsafe or potentially illegal conditions, accidents, injuries, illnesses, and near misses.

In order for a safety culture to be established in an organization, management must demonstrate its commitment to the safety principles that they are trying to encourage workers to comply with, and management must demonstrate those beliefs in its behaviors. Everyone in the organization, particularly those in management, must be adaptive to change and responsive to new information. New information must be collected and disseminated to all groups and individuals who need to learn and apply it. To promote the safety culture, management must operate in a just and fair manner, and employees must perceive management as being fair.

REGULATIONS

As described in Chapter 1, Introduction to Industrial Hygiene, much of the industrial hygiene that is practiced today incorporates regulations and industry standards for employee protection and safety. These rules have been written to protect workers in certain activities and industries that have been shown to be hazardous. If employers do not comply with these regulations, they will be issued citations and penalties.

In recent years, a safe and healthy workplace has been linked to organizational efficiency and industry profits. Effective industrial hygiene management leads to a safer workplace and few injuries, illnesses, and regulatory penalties. Plus, industrial hygiene practices have been shown to minimize environmental damage because they reduce the toxicity and volumes of hazardous materials used in operational processes and the associated effluents and waste streams.

In addition to issuing regulations and requiring compliance with them, industrial hygiene adheres to ethical principles of practice related to protecting worker health and safety and health of the environment. Ethical industrial hygiene management practices seek to minimize worker exposures to hazardous conditions and chemicals. Management systems should thus be designed to address the ethical responsibilities and goals of the industrial hygiene profession in addition to existing laws and regulations.

> **Case: Ethical Responsibility**
>
> Robert is the industrial hygienist at a large hospital where glutaraldehyde was originally used as a high-level disinfectant at 21 clinics around the facility. Glutaraldehyde is a respiratory sensitizer that can cause nausea and asthma. Although there are no OSHA permissible exposure lmits for the chemical, the National Institute for Occupational Safety and Health's Recommended Exposure Limit (REL) is 0.2 ppm.
>
> Whenever Robert performed annual surveys at the glutaraldehyde sites, it was not unusual to find glutaraldehyde levels exceeding the 2.0 ACGIH limit, and they almost always were above the NIOSH REL. In addition, about once or twice a year, workers in these areas would report to the Occupational Medicine Department with symptoms of respiratory distress, nausea, and headaches as a result of breathing in the vapors.
>
> When a high-level disinfectant called Ortho-pthalaldehyde was approved, Robert investigated its performance and found it to be just as effective as glutaraldehyde but without glutaraldehyde's negative health effects. Robert was eventually able to persuade 19 of the hospital's clinics to change to the new chemical. As a result, annual air measurements were no longer required, there were no airborne limits to meet, and workers no longer reported negative health effects or symptoms.
>
> In this case, a change to a safer chemical was not required but led to lower exposures to a toxic chemical and eliminated health problems for the workers.

POLICIES

The first step in designing any management system is to create straightforward and concise management policies that reflect the company's or organization's goals and objectives. The highest levels of management should create these policies to guide the direction and future of the organization.

Management policies for industrial hygiene should state how worker health and safety and environmental protection coordinate with and support the overall institutional objectives and goals. In many instances, the safety policies parallel and specifically address governmental regulations. Forward-thinking organizations that have a thorough understanding of the relationship between effective industrial hygiene practices and the goals of the organization move beyond regulations and incorporate accepted industry standards and strive to incorporate into company policies higher-level methods such as sustainable development and pollution prevention.

> **Example**
>
> **Sample Safety Policy**
>
> The organization is dedicated to providing a safe and healthy workplace for all employees. We are also committed to minimizing risks and impacts to the environment. We manage company activities to ensure sustainable development and continuous improvement. A culture of safety and health is promoted at all levels of the organization to set the highest standards to protect workers and the communities where we operate.
>
> We demonstrate our commitment to safety by adhering to the following policy objectives:
>
> - Fully evaluating all workplace and environmental risks, and developing systems, programs, and procedures to minimize them
> - Developing and improving programs and procedures to ensure compliance with all applicable laws and regulations
> - Ensuring that personnel are properly trained and provided with appropriate safety and emergency equipment
> - Following sustainable development principles for continuous improvement
> - Communicating our desire to improve our performance continuously

PROGRAMS

An organization can create a wide variety of programs to address different functional areas of industrial hygiene within the organization. These programs should describe the scope of safety in the organization as well as where and when the safety policy is to be applied or adhered to. Usually, programs use general terms to indicate what actions need to be taken to comply with regulations for health and safety or the environment.

Programs should describe the objectives of the company in specific areas such as functional areas of industrial hygiene, geographical areas of the company, or even departments within the company. Examples of commonly used industrial hygiene programs include:

- Respiratory Protection
- Hearing Conservation
- Bloodborne Pathogens
- Chemical Hygiene
- Radiation Protection
- Laser Safety
- Emergency Management
- Asbestos Management

Industrial hygiene programs should clearly delegate responsibility for each task. However, not all directives in a program are the responsibility of industrial hygienists or the safety program. Often, line supervisors and line workers also have responsibilities. For example, it may be the responsibility of the industrial hygienist to sample the air for toxic chemical concentrations and to select the appropriate respiratory protections, but it is the responsibility of the employees that need to wear the protection to schedule medical evaluations and acquire adequate training in wearing the respiratory protection effectively. Line supervisors are responsible for having management systems in place to ensure that workers adhere to the program requirements and wear the respirators according to the program.

PROCEDURES

Procedures provide details for day-to-day operations that support program requirements and describe in a step-by-step, easily understood manner how the job is to be done.

Any given program may have a couple or several procedures that support it. A hearing conservation program, for example, may be composed of procedures that describe the area surveys to be performed, when to perform them, and which worker exposures to measure with personal dosimetry. The program may incorporate another set of procedures that define what type of training workers need to attend, who is to conduct the training, and where the training records are to be maintained. Individual procedures may be developed on the basis of the operations manual provided by the manufacturer of each piece of noise-monitoring meters and equipment. Another procedure may describe when noise-monitoring equipment needs to be maintained or calibrated, who performs these procedures, and where the maintenance records are to be kept.

The following list is a sample of procedures that support a respiratory protection program:

- Selecting appropriate proper respirators for particular hazards
- Issuing respirators to workers
- Indicating when and how respirators should be used to perform routine work activities
- Indicating when and how respirators should be used to perform infrequent activities and in emergencies such as spill response, rescue, or escape situations

- Specifying medical evaluations and the physical requirements to wear a respirator
- Fit-testing (qualitative or quantitative)
- Cleaning, maintaining, storing, and disposing of respirators
- Training in airborne hazards and the respiratory protection program, including donning, doffing, and understanding the limitations of the selected respiratory protection
- Periodically evaluating the program

In procedures, the necessary activities are often numbered sequentially to ensure the proper order is followed.

QUALITY ASSURANCE

Quality assurance is a standard business practice in which an independent oversight body reviews company policies, programs, and procedures to evaluate whether they meet regulatory requirements and company objectives in its day-to-day operations.

Quality assurance activities may be delegated to an independent department within an organization, or they may be allocated to outside vendors or consultants specializing in those services. Similar to an accounting audit of company finances, an industrial hygiene audit will investigate whether all required practices are being conducted appropriately. Quality assurance uses the methods described in the following sections to fulfill specific objectives and tasks and to examine operating practices at various times, durations, and operational levels.

Each organization determines its own methods for implementing a quality assurance program. The program methods should incorporate various levels of severity for noncompliance with regulations or policies. Audit activities, responsibilities of participants, presentation of the findings, and procedures to apply necessary corrective actions are all part of a comprehensive quality assurance program.

PROGRAM REVIEWS

At the highest level of quality assurance, program reviews determine whether company policies are fully satisfied. Program reviews also evaluate whether all requirements of regulations and laws for worker health and safety and in industrial hygiene are included in company documents.

Program reviews may be conducted by several reviewers or by auditors who each investigate different functional and operational departments in the organization. A program review can take a day or up to several weeks, depending on the size and complexity of the organization.

AUDITS

Audits typically evaluate a specific aspect of a program over a specific time frame. For example, an audit may look at the activities performed in a respiratory protection program over one year. The audit investigators review the procedures that make up the program and then conduct interviews, observations, reviews of records, and measurements to determine whether all the procedural and program requirements are being met.

Audit teams are typically composed of a team leader and two to six members. The team should arrive at a facility with a predetermined audit plan that describes exactly what the team will be looking at and for how long. The audit plan should also describe the scope of the audit, the departments involved, and the specific audit objectives.

Prior to arriving on-site, the audit team leader should create an audit checklist. Different portions of the checklist are assigned to different team members, who seek to answer the checklist questions to determine whether all aspects of the program and procedures are being conducted effectively.

Audit team members should be experienced in audit management and operations; they are also often experts in the fields they have been assigned to audit. Although the auditors might ask a lot of questions, they tend to be extremely knowledgeable about the technical area to which they are assigned. An effective and efficient auditor will not only determine where procedures are not being followed, but will also find ways to improve the program and procedures.

SURVEILLANCE

Surveillance is a quality assurance function that tends to take place on a frequent, sometimes scheduled basis in order to ensure ongoing compliance with specific operating parameters. Surveillance activities are usually short, routine inspections that document whether required activities are being performed. Surveillance inspections are often conducted by professionals independent of the department or functional group performing the specific operational tasks.

An example of a surveillance activity is a quality assurance professional weekly reviewing a company's records to determine whether the safety release valves on an operating system have been tested daily, as required by company procedures. Surveillance is another assessment of the most important safety systems within a facility and another management tool to ensure safe and efficient ongoing operations.

ABOVE AND BEYOND—NEW DIRECTIONS IN MANAGEMENT

New safety management systems are continuously being developed to improve safety and efficiency in business and organizational operations. These systems provide orderly methods to identify and evaluate hazards and limit risks. The most effective and successful organizations attempt to integrate new methodologies into their management programs wherever possible to improve industrial hygiene.

Safety management systems are comprehensive processes for setting goals, making plans, and measuring performance. Safety is managed the same way other organizational departments are managed, except that safety is integrated throughout the entire company. Safety management systems define how an organization manages risks and implements controls to reduce those risks. Communication and methods to ensure continuous improvement are cornerstones of an effective safety management system. A few of these systems are described in the following sections (AIHA 2001).

AMERICAN NATIONAL STANDARDS INSTITUTE

The American National Standards Institute (ANSI) was founded in 1918 as a private nonprofit organization. Its goal is to promote and "facilitate voluntary consensus standards and conformity assessment systems" to improve organizational efficiency (ANSI 2015). It also promotes the safety and health of workers and the public and the protection of the environment.

Standards are created through a participatory process whereby representatives of various professional organizations meet to collaborate on a given topic to build a consensus. The process is meant to incorporate the interests of a broad range of stakeholders and participants. There are currently several hundred participating organizations and more than 10,000 standards in the ANSI system (ANSI 2015).

ANSI standards are often referred to in Occupational Safety and Health Administration (OSHA) regulations. An ANSI standard is a consensus standard and by itself does not have the force of law. However, when it is incorporated by reference into the OSHA regulations, it can become legally enforceable. Some ANSI standards that are closely related to environmental health and safety are the following:

- ANSI Z535.1, American National Standard for Safety Colors
- ANSI Z136.1–2014, Safe Use of Lasers in Manufacturing Environments
- ANSI Z535.2, American National Standard for Environmental and Facility Safety Signs

- ANSI Z535.3, American National Standard for Criteria for Safety Symbols
- ANSI Z535.4, American National Standard for Product Safety Signs and Labels
- ANSI Z535.5, American National Standard for Safety Tags and Barricade Tapes (for Temporary Hazards)
- ANSI Z535.6, American National Standard for Product Safety Information in Product Manuals, Instructions, and Other Collateral Materials
- ANSI N43.17, Radiation Safety for Personnel Security Screening Systems Using X-Ray or Gamma Radiation
- ANSI Z10, Occupational Health and Safety Management Systems

A very useful consensus standard guideline for organizing health and safety management programs was created by ANSI and the American Industrial Hygiene Association (AIHA) in ANSI/AIHA Z10. The Z10 document builds upon common business management practices but emphasizes environmental health and safety. This document provides a template for management programs focusing on employee participation, teamwork, communication between employees and management, and continuous improvement.

The ANSI Z10 document includes discussions of wide-ranging management and leadership ideals. But it also includes fundamental management tools such as models of how to assess occupational hazards and exposures by setting objectives and gathering data. For example, the standard describes the criteria that should be met by any health and safety program using the SMART Objective approach. In this approach, objectives should be Specific, Measureable, Actionable, Realistic, and Time-Bound, as shown in Figure 3–1. Other very detailed management tools provided in Z10 are management performance scorecards and an incident investigation form (ANSI 2005).

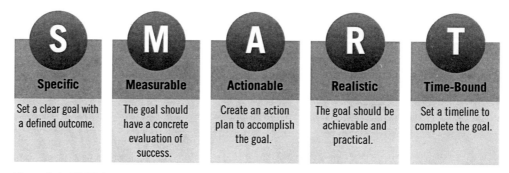

Figure 3–1. SMART Objective approach.

TOTAL QUALITY MANAGEMENT

Total quality management is a business system that began in Japan in the 1950s. It uses a strategic set of organizational and process management approaches to attain the highest possible levels of service and product quality. Consistent use of TQM leads to improved business profits (Deming 1982).

Throughout this business system, TQM uses data to provide indications of the levels of quality. Indicators of performance include measures of customer satisfaction and documentation of product defects. Results are used to make improvements continuously throughout the business process, from production to customers.

TQM can be implemented horizontally across the business organization, with input from both management and line workers. It can even include information from suppliers (Youngless 2000).

Continuous improvement is a key feature of TQM. As soon as problems with quality are identified, solutions can be determined and implemented quickly. Effective communication and a supportive management are critical. Workers and supervisors need to feel part of the team and comfortable raising issues and making suggestions (Inc.com 2015).

According to the American Society for Quality Control, "TQM consists of seven basic elements: (1) policy, planning, and administration; (2) product design and design change control; (3) control of purchased material; (4) production quality control; (5) user contact and field performance; (6) corrective action; and (7) employee selection, training, and motivation."

It is important to implement fundamental TQM principles throughout the company. Implementing an effective TQM program might be a painful process for some workers or departments, particularly the ones who have to change the most (Montgomery 2004). However, achieving the best results requires all company departments and workers to be involved in the TQM program.

Some of the essential business principles associated with TQM include the following:

Customer-oriented. The satisfaction of the customer or client, in products or services, is at the center of TQM. An organization must satisfy customers' expectations of quality in order to retain their business. Measures of quality must be quantifiable to be able to make comparisons.

Employee involvement. Another core principle of TQM is the full engagement of all employees and productive communication. Workers are self-managed, self-motivated, and self-empowered. Everyone works toward high performance and the success of the operation.

Communications. Effective and timely communications also play a major role in achieving objectives and goals by improving efficiencies and the continual engagement of employees.

Process-centered. TQM focuses on the processes that take inputs from suppliers (internal or external) and transform them into outputs that are delivered to customers. Detailed analysis of the steps and tasks in production processes leads to continuous improvement. Business activities must be performed in a controlled manner.

Integrated systems. The vertical and horizontal structures and systems in the organization are designed to ensure maximum efficiency in production processes, as shown in Figure 3–2. Functional specialties are interconnected in ways that reduce duplication and minimize delays and miscommunication. All business units operate with the same vision, mission, quality policies, and business objectives. If manufacturing can make 50 units per day but shipping can send only send 45 units out the door, then the systems need to be adjusted so the numbers match.

Quality control. As the name implies, TQM is dedicated to ensuring quality. This dedication requires collecting and information related to consistently high levels of product and service performance and continuously improving deficient areas. The quality chain must also incorporate the suppliers of materials used in the product or process.

Systematic approaches. Strategic, long-term planning and management are key to the TQM process. Decisions are made with a critical eye and an orientation toward the organization's vision, mission, and goals.

Continuous improvement. Quality management is a long-term investment, and continuous improvement is a way for a company to differentiate itself from competitors. The most effective companies are able to find and correct the operation inefficiencies that lead to increased costs and reduced quality.

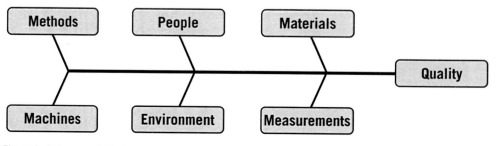

Figure 3–2. Cause and effect.

Part of the TQM approach is the plan-do-study-act (PDSA) cycle, which describes how a company incorporates continuous improvement into its operations (Figure 3–3). This is sometimes referred to as a Shewhart cycle or Deming wheel, in honor of its inventors.

Plan. The first step in the PDSA cycle is to document all current processes and activities, collect data, and identify deficiencies. This information is then studied and used to develop improvement plans.

Do. The next step in the cycle is to implement the plan while collecting new data regarding the plan's impacts on process and quality.

Study. The third step is to study the data collected to see whether the plan is achieving the strategic goals.

Act. The last step of the cycle is to act based on the results and to implement changes in procedures and processes to achieve continuous improvements. New program elements are communicated to employees. At the end of the act step, the cycle begins again.

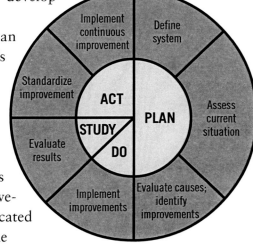

Figure 3–3. Plan-do-study-act cycle.

SIX SIGMA

Six Sigma is a quality improvement system that uses various management techniques and statistical analyses to evaluate and guide organizational performance. Although typically used to improve process or product quality, the methods can be similarly applied to health and safety risks and practices to make improvements.

The term *Six Sigma* comes from the statistical corollary that at six standard deviations from product specifications, there is a 99.9999 percent chance that the product will not have any defects. Put another way, there is a chance of 3.4 defects per 1 million opportunities. In safety, the corollary might be used to indicate that a hazard has a 3.4 out of 1 million chance of occurring (Basu and Wright 2003).

Six Sigma is based on the belief that accurate statistical information can be used to continuously improve business processes. It follows that Six Sigma practices can lead to quantifiable and substantial financial returns.

The Six Sigma program provides guidelines on how to evaluate and improve projects by using a series of progressive management phases. Here are some examples of those phases:

- Define process goals or requirements.
- Measure key aspects of the existing process and collect data.
- Analyze the data to examine and verify cause-and-effect relationships. Determine the root cause of the defect under investigation.
- Use the results of data analysis to make changes to improve existing processes.
- Pilot proposed changes and collect new data to ensure feasibility of new processes.
- Continuously reevaluate and monitor the process, and make changes accordingly.

Six Sigma uses a variety of sophisticated statistical analysis tools, and these tools combined with the need to have many data points to work from, make the methods most suitable for large manufacturing companies that have dedicated staff who are experts in the techniques. However, Six Sigma consultants can be engaged for special projects in smaller organizations.

Some large corporations such as Motorola and General Electric have internal Six Sigma employees who are specially trained and experienced in the methods. Other companies and some nonprofit organizations provide their own certifications and credentialing in Six Sigma for a fee.

VALUE STRATEGY MANUAL

The *Value Strategy Manual* is a management health and safety model developed by the AIHA. This system provides a framework for evaluating environmental health and safety programs, systems, and equipment in financial terms. Industrial hygienists can use the *Manual* to develop a business rationale, supported by financial and nonfinancial information, for health and safety projects (AIHA 2010).

The *Manual* provides step-by-step instructions for analyzing risk, identifying alternatives, and quantifying the financial and nonfinancial benefits of making changes or improvements. The *Manual* is also useful when presenting health and safety alternatives to management stakeholders and decision makers.

INTERNATIONAL STANDARDS ORGANIZATION

The International Standards Organization (ISO) is a nonprofit network of international standards–developing bodies. Since its inception in 1946, the organization has created more than 19,500 international standards on almost all aspects of technology and manufacturing. More than 162 countries and 3,368 different technical bodies are associated with developing standards through the ISO. The following ISO standards are pertinent to environmental health and safety.

ISO 9001—Quality Management
The 9000 series of standards provides a framework for developing and implementing management systems and programs to ensure process and product quality. These standards do not specifically address environmental health and safety, but because of the close relationship among effective management systems, quality, and safety, the standards are a useful foundation for quality health and safety management systems.

ISO 14001—Environmental Management
The ISO 14001 standard specifically addresses certain areas of environmental management. It provides practical tools and methods companies can use to measure and control their environmental impacts and meet applicable regulations. Companies can use the standard and possible certification to demonstrate to stakeholders and the general public their commitment to environmentally safe practices.

ISO 31000—Risk Management
ISO 31000 provides principles and a framework for identifying and managing a wide variety of risks within an organization. The standard also addresses health and safety issues and how they should be included in the overall risk management system. Standard ISO 31010 provides details on risk assessment methods that can be used in business operations.

The ISO/IEC Guide 51—Safety Risk Assessment document specifically addresses health and safety risk and provides recommendations for reducing risk for people, property, and the environment all at once.

Other ISO Standards and Guides
Numerous other guides have been developed regarding safe practices or operation of equipment. The following is a sample of currently available guides:

ISO 1819:1977, Continuous mechanical handling equipment—Safety code—General rules
ISO 5388:1981, Stationary air compressors—Safety rules and code of practice
ISO 8456:1985, Storage equipment for loose bulk materials—Safety code
ISO 11449:1994, Walk-behind powered rotary tillers—Definitions, safety requirements and test procedures
ISO 5031:1977, Continuous mechanical handling equipment for loose

bulk materials—Couplings and hose components used in pneumatic handling—Safety code

In 2013, the ISO announced plans to develop a standard specifically designed to address occupational health and safety. A standards preparation committee was selected to produce a draft document by 2016. The standard is intended to build upon British occupational health and safety guideline OHSAS 18001—Occupational Health and Safety Management, which is currently promoted by the British Standards Institute.

OSHA VOLUNTARY PROTECTION PROGRAMS

In addition to its many required programs, OSHA has created a series of Voluntary Protection Programs (VPPs) meant to promote effective worksite health and safety. In these programs, management, labor, and OSHA coordinate to create broad safety and health management systems that go beyond what was typically required at the worksite. Through a variety of incentives and recognitions, these programs encourage employers to reduce work hazards.

General requirements for participation in the incentive program include developing a safety and health management system based on employee involvement, self-evaluations, and continuous improvement. VPP sites are exempt from programmed OSHA inspections, but are expected to maintain a management system that ensures compliance and continuous improvement.

There are three possible VPP levels (Table 3–1). Participation in the **Star Program** is limited to exemplary worksites that have comprehensive and successful safety and health management systems. These companies have injury and illness rates at or below the national average of their respective industries. Star sites are reevaluated every three to five years, and their incident rates are reviewed annually.

Merit VPP sites have good safety management systems but need improvement in some areas. These sites demonstrate the potential and commitment to achieve Star quality within three years. They are evaluated every 18 to 24 months to determine whether problems have been addressed adequately to allow them to move up to the Star rating.

VPP Demonstration sites are worksites that have Star-quality safety and health protection but also have management systems that differ from currently approved VPP approaches. These sites are evaluated every 12 to 18 months.

In order to become a VPP site, a company undergoes an on-site OSHA review that lasts about four days. If OSHA approves the site, it becomes a Star or Merit site.

Table 3–1. Voluntary Protection Program Criteria

Star	These companies have exemplary systems for the prevention and control of health and safety hazards. They implement continuous improvement as routine parts of their management programs.
Merit	Organizations at the Merit level have good health and safety management systems but need to complete additional actions to achieve Star status.
Demonstration	These companies operate effective health and safety programs, but their management systems differ from currently approved VPP approaches.

THE ROLE OF INDUSTRIAL HYGIENISTS IN ALTERNATIVE SOLUTIONS

In recent years, new approaches to industrial hygiene have included toxic materials or chemicals in the workplace. Industrial hygienists can play a key role in this process by suggesting ways to improve company productivity and efficiency while making the workplace safer.

Often, by analyzing a hazardous work chemical or condition, the industrial hygienist can find ways to improve profitability and efficiency, improve company competitiveness, reduce legal liabilities, improve public relations and perceptions, protect the environment, and improve the health and safety of the employees.

When searching for an alternative to a hazardous process, chemical, or environment, the industrial hygienist should follow these basic steps:

1. Become familiar with all work processes, steps, and tasks. Compile relevant exposure information, data on given chemicals and processes, and lists of the most hazardous chemicals or conditions to be replaced or eliminated.
2. Inventory and prioritize the hazards to identify which projects to begin first.
3. Identify alternative processes or chemicals that can replace the existing process or chemical. Whenever possible, substantiate the alternatives with data or peer-reviewed results from independent agencies or researchers.
4. Assess the possible alternatives and prioritize them according to the effectiveness, performance, cost, safety, and toxicity criteria.
5. Select an alternative according to how well it satisfied the criteria.
6. Test the alternative in the field. Evaluate its performance and employees' responses to the alternative.
7. If the pilot is successful, implement the alternative on a larger scale and continue to evaluate its performance. If conditions change, make necessary improvements.

SUMMARY

Ultimately, the transition to safer processes or chemicals requires a team effort. But the industrial hygienist can be a catalyst for program changes. Often, using alternative methods and chemicals can lead to improved efficiencies, productivity, work safety, and environmental safety. These improvements can then reduce legal liabilities and insurance premiums. The advantage an industrial hygienist provides is his or her ability to tie hazardous conditions and properties to management systems and principles. That is, by analyzing the data, evaluating the costs, and researching alternatives, the industrial hygienist should be able to persuade management to make a change that ultimately makes the workplace safer.

REVIEW QUESTIONS

1. What is a "culture of safety," and how is it created and developed within an organization?
2. Why is the management of industrial hygiene and safety a good business practice even when no regulations apply to a certain area or work hazard?
3. Create a short policy describing a beauty salon's safe operations and commitment to safety.
4. Why are established procedures necessary, and how do they relate to worker safety?
5. How does quality assurance apply to safety?
6. How does a program review differ from an audit?
7. What are three of the basic business principles associated with TQM methods?
8. What are the four parts of the Deming wheel used to describe the TQM process?
9. What type of safety management system is designed around the collection and statistical analysis of data?
10. What does the International Standards Organization (ISO) do?
11. What is the highest level a company can attain in OSHA's Voluntary Protection Programs sequence, and what does a company need to do to be accepted into the program?
12. What is the most important step when selecting an alternative, safer process or chemical? Why?

REFERENCES

American Industrial Hygiene Association. *Industrial Hygiene Performance Metrics*. Fairfax, VA: AIHA, 2001.

———. *Value Strategy Manual*. Fairfax, VA: AIHA, 2010.

American National Standards Institute. *ANSI/AIHA Z10-2005. Occupational Health and Safety Management Systems*. Fairfax, VA: American Industrial Hygiene Association, 2005.

———. "Overview of the U.S. Standardization System." Accessed April 4, 2015. www.ansi.org/about_ansi/introduction/introduction.aspx?menuid=1.

Basu, R., and J. N. Wright. *Quality Beyond Six Sigma*. New York: Elsevier, 2003.

Deming, W. E. *Out of the Crisis*. Cambridge, MA: MIT Center for Advanced Engineering Study, 1982.

Inc.com. "Total Quality Management." Accessed June 2015. www.inc.com/encyclopedia/total-quality-management-tqm.html.

Montgomery, D. C. *Introduction to Statistical Quality Control*. Hoboken, NJ: John Wiley & Sons, 2004.

W. Edwards Deming Institute. "Teachings." www.deming.org/theman/teachings02.html.

Youngless, J. "Total Quality Misconception." *Quality in Manufacturing* (January 2000).

4
Basic Concepts in Industrial Toxicology

LEARNING OBJECTIVES

After completing this chapter, readers should be able to do the following:

- Define the study of toxicology and explain why it is important to the practice of industrial hygiene.
- List different mechanisms and modes of toxic chemical action and the various methods used to measure and document toxicity.
- Explain various criteria for causation of toxic effects.
- Describe toxicokinetics.
- Identify and list routes of toxic agent exposure in the human body.
- Understand toxic chemical absorption, distribution, metabolism, and elimination.
- Explain the concept of dose response.
- Describe the toxic effects typical in the major organ systems.

Photo credit: tunart/iStock

INTRODUCTION

A traditional definition of toxicology is "the science of poisons." However, this simplistic definition creates more questions than it answers: What is a poison? What does it mean for something to be poisonous? Can something be poisonous to one person but not poisonous to another? Paracelsus, a 16th-century scientist, recognized these limitations and the lack of understanding. He is quoted as saying, "What is not a poison? All things are poison and nothing is without poison. Solely the dose determines that a thing is not poison." Because he brought these issues to the forefront, he is credited with establishing the discipline of toxicology.

Through numerous medical and technological advances since Paracelsus's time, toxicology as a scientific study has become more defined and rigorous because of its multidisciplinary nature. Toxicology is a broad field that studies the effects of chemical, biological, and physical agents on the health of an organism. What once were considered poisons we now call **toxicants**, which are chemical (e.g., alcohol), biological (e.g., bacterial toxins), and physical (e.g., radiation) in nature. The field of toxicology includes studies at the molecular basis of the biological response up to and including evaluations at the cellular, tissue, organ, organ system, and whole body levels. Toxicology integrates several areas of science, including chemistry, biology, physiology, and pathology. Knowledge in all these areas is used to gain an understanding of how a toxicant causes adverse reactions in living organisms.

This chapter focuses on industrial toxicology, with the primary objective of explaining the effects of chemicals on human health, since this aspect of toxicology is most commonly pertinent to the role of the industrial hygienist. Consistent with the concept that "the dose makes the poison," all chemicals can cause adverse health effects at some dose. In this context, "adverse" can be understood as the disruption of normal physiology to the degree that cell or tissue damage occurs or impaired function results. Because the nature of the effects of a chemical exposure and the severity of the damage vary greatly, examining toxicity on an organ systems basis, as reflected in this chapter, is necessary.

WHY THE INDUSTRIAL HYGIENIST NEEDS TO UNDERSTAND TOXICOLOGY

Industrial hygiene focuses on the worker and on the occupational environment in which work is done. Because workers can come in contact with potentially hazardous chemicals in certain environments, toxicology is a core competency for the practice of industrial hygiene. Knowledge of toxicology principles allows the industrial hygienist to determine the potential hazards of a material and to determine and define acceptable exposures. The industrial hygienist also has the crucial responsibility of communicating this information to workers who may come in contact with hazardous chemicals.

One common way that industrial hygienists use their toxicology knowledge and experience is through implementing hazard communication programs. These programs require identifying the chemicals present in the workplace, providing worker access to information about the hazards of the chemicals, labeling containers, and training workers. An important source of toxicology information for chemicals handled in the workplace is the **Safety Data Sheet (SDS)**, formerly Material Safety Data Sheet (MSDS). An SDS is a document that provides information about the hazards of a particular chemical substance. The Occupational Safety and Health Administration (OSHA) requires that the manufacturer, distributor, or importer of dangerous substances provide hazard information on an SDS. Companies that register chemicals for use in Europe have similar requirements.

Two sections of the SDS contain toxicological information:
- Section 2: Hazard(s) Identification—identifies the hazards and warning information of the substance via the following required information:
  - Hazard classification, signal word, hazard statement, pictograms, precautionary statements, description of any hazards not otherwise classified, and statement of unknown toxicity of mixture ingredients
- Section 11: Toxicological Information—provides toxicological and health effects information of the substance via the following required information:
  - Exposure routes
  - Description of the delayed, immediate, or chronic effects from short- and long-term exposures
  - Measures of toxicity (e.g., lethal doses)
  - Symptoms of exposure
  - Other toxicology information such as the potential for carcinogenic activity

In order to understand the SDS content and be able to train and communicate this information to workers, an industrial hygienist must have a thorough understanding of how toxic chemicals cause damage, including their routes of exposure and their degrees of toxicity. Industrial hygienists must also be able to recognize the particular symptoms of different toxic agents when workers are exposed to them.

Similar toxicological expertise is required to make meaningful use of occupational exposure limits (OELs). For example, appropriate judgments about the need for action as exposure approaches the OEL will often depend on the nature and severity of the health effects at the OEL and the industrial hygienist's thorough knowledge of the toxicology of the chemical. (For more information on OELs, see Chapter 5, Occupational Exposure Limits and Assessment of Chemical Risks.)

ESTABLISHING MECHANISMS OF ACTION

The **mechanism of action** is a detailed description of the process involved in an agent's toxic effects, from initial exposure to the development of adverse reactions. However, despite years of study, complete information on all the aspects of a chemical's toxic mechanisms is rarely available. The good news is that it is not necessary to know every aspect of a chemical's effect on physiology to reasonably characterize its hazards. Rather, it is acceptable to rely on knowledge of the chemical's **mode of action** or the key events in the toxic response that can be measured or observed. A **key event** in a mode of action is an empirically observable step that is itself a necessary element of the mode of action or is a biologically based marker for such an element. Key events and toxicological effects are the results of many different types of responses to toxic agents. Figure 4–1 illustrates the concepts of mechanism of action and mode of action. The goal in Figure 4–1 is for a student to get from home to the library. The arrows represent the detailed pathway. This pathway, or process, to the library is akin to the *mechanism of action*. In this process, the student has to stop at the doctor's office and then the store, in that order, before going to the library. The necessary stops are considered the *key events*, and the sequence of these stops represents the *mode of action*.

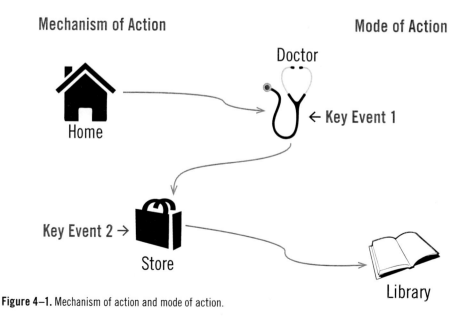

Figure 4–1. Mechanism of action and mode of action.

HOW IS THE MODE OF ACTION ESTABLISHED?

When a toxicologist wants to establish a chemical's mechanism and mode of action, he or she needs to determine the path and the steps, similar to the diagram in Figure 4–1. A toxicologist's first course of action is to evaluate data from **epidemiological studies**, which are human studies; from laboratory studies in **animal models**; and from special molecular studies (which often use cell cultures and other specialized test systems) in which researchers study effects in animals such as rats and extrapolate the results to humans. These types of studies provide data on **biomarkers of effects**, which are observable and quantifiable data on an organism's responses to a chemical that a toxicologist will use to establish the chemical's mechanism and mode of action. Such effects help a toxicologist map out the key steps in the biological response process, from the onset of initial changes in normal physiology to severe effects and overt clinical disease.

BIOMARKERS OF EFFECTS

Clinical Measures. An animal's body weight is measured before and throughout the study as a general indication of well-being. Clinical chemistry can also be monitored during the study. One type of clinical measure is urinalysis, which analyzes the volume and density of the urine as well as the presence of electrolytes, metabolic products of the toxic agent, and protein in the urine. Results can be monitored by evaluating the blood as well (i.e., hematology). Hematology markers for toxic effects include the absolute and relative values of different blood cells, such as white blood cells, red blood cells, and platelets. Also monitored are blood serum biochemical markers such as electrolytes, metabolic products, and blood levels of protein and enzymes, which often indicate the level of organ function or damage. Table 4–1 indicates the observations that can be made during a study and what each measures.

Table 4–1. List of Biomarker Data

Type	Measure	Data
Clinical	Body Weight	Changes in weight from before and throughout the study
	Clinical Chemistry	Urinalysis, blood monitoring, serum biochemistry
Pathological	Gross Pathology	Changes in organs' size, weight, structure
	Histopathology	Changes in tissue such as hyperplasia and hypertrophy
Developmental	Reproductive	Reproductive capacity, fetal toxicity, teratogenicity
	Developmental	Postnatal and developmental effects
Neurological	Behavioral	Changes in appetite, aggressiveness, lethargy, and signs of increased nervousness, tremor, decreased motor skills
Immunological	Response	Increased risk of infection, sensitization, irritation, anaphylaxis

Pathological Measures. At the end of a study, the animals can be euthanized to examine the chemical's pathological effects. When a toxicologist studies the pathology, he or she examines the animal's body at different levels. The largest level involves studying organs and organ systems, which is called **gross pathology**. Gross pathological examination looks at any size or weight changes in an organ and whether the organ appears abnormal in any way. A gross examination could identify potential tumors or cancer, necrosis (death of body tissues), structural changes, hemorrhaging, or other signs of damage to the tissues or organs.

The next level of pathology is **histopathology**, which is the study of the body's tissues and fluids and describes the microscopic changes associated with the gross pathological changes. Histopathological examination will determine, among many other things, the origin of tumor growth, whether there are any changes in the microscopic organization of the cells that make up an affected tissue, and whether changes in organ weight and size are due to increased cellular duplication (hyperplasia) or increased cellular size (hypertrophy).

A robust examination of all major organs and organ systems provides information about the **portal of entry**. The portal of entry, or the site of a chemical's entry into the body, is closely inspected for effects that relate to initial contact with the toxic agent. For example, such an examination would look for corrosive effects and signs of irritation in the lungs after the inhalation of a chemical. To supplement the study of general organ systems, a variety of special studies can be performed to evaluate toxicological effects on both organ structure and physiological system function.

Reproductive and Developmental Measures. It is also possible to measure the effects of a potential toxic agent on the offspring of an animal that was exposed to the chemical. To measure effects on reproduction and development, toxicologists look at **multigenerational studies**, in which both parents are treated with a toxic agent both before and during pregnancy. This assesses the effects of the agent on the reproductive capacity of the animals as well as any effects on the development of the embryo or fetus. A subset of developmental effects that cause structural changes in organs or tissues is called **teratogenicity**. Some toxicology study designs also evaluate effects that occur after the offspring are born (i.e., postnatal effects).

Other Measures. To measure **neurological effects**, assays on motor activity function can be conducted before, during, and after treatment. Studies that measure motor activity function look for changes in gait or the development of tremors and motor coordination deficiencies. Neurological toxicity studies also assess behavioral changes such as increased or decreased food and water consumption, increased aggressiveness, lethargy, and signs of increased nervousness. **Immunological studies** are performed to assay potentially toxic agents' effects on the immune system such as suppression, which decreases the effectiveness of the immune system and increases the risk of infection. Alternatively, some chemicals cause increased immune reactions (sensitization or allergy) that result in an exaggerated immune response to later exposures. Sensitization can result in simple irritation of the exposed tissues or dangerous systemic anaphylactic reactions.

WEIGHT OF EVIDENCE

Once all the data have been measured, the toxicologist evaluates the data to establish the mode of action. In doing so, the toxicologist bases all judgments on relevant information including the **weight of evidence** that a potential toxic agent will cause a potential effect. One method of determining whether a potential toxic agent and a measured effect are related is to compare the data with Hill's criteria for causation. Adapted from epidemiology, **Hill's criteria for causation** help determine the evidence for a causal relationship between two events such as an exposure to a potentially toxic agent and a health effect. There are six types of evidence, or criteria, to consider.

(1) **Strength.** Strength of the effect describes the magnitude of the relationship between events. A strong association between the exposure and a specific effect indicates that the exposure likely caused the effect. In toxicology, a strong association is signified by a high incidence of effect at a given dose or the appearance of severe effects as opposed to mild effects.

(2) **Consistency.** Different studies with different populations producing a correlation between a specific exposure and a specific effect indicate a causal relationship between the exposure and the effect, which can also occur for laboratory and epidemiological data. This criterion also incorporates information on consistency in the range of biological responses within a study. For example, in a toxicology study where an increase in blood markers of liver damage was observed, the researcher would check to see whether the histopathological examination revealed any evidence of damage.

(3) **Specificity.** The third criterion of consideration is specificity. If a rare disease or effect is associated with a group of people who all had a specific exposure and if the effect's incidence in the population is not otherwise explainable, the specificity of the association between the exposure and the effect greatly strengthens the case for the causality of the exposure.

(4) Temporality of Events. A fourth criterion is the temporality of the events. In order to determine that an exposure causes an effect and disturbs a particular mechanism of action, the exposure must occur before the effect.

(5) Coherence and Plausibility. The fifth criterion to examine is the coherence and plausibility of the relationship between the exposure and the potential effect. Does the laboratory biological evidence show that a particular chemical is usually not absorbed into the body at all? Then it is not plausible that the chemical has a systemic effect. The question for this criterion is, "Is the effect reasonable given all the known data on this chemical?"

(6) Dose-Response Relationship. Finally, it is important to establish a dose-response relationship between the exposure and the effect. Exposure to a large amount of the agent of interest should cause a greater response than exposure to a lesser amount. If the agent of interest accumulates in the body, then chronic exposure should cause an increased effect. In addition, the doses needed to initiate early (less severe) key events should be lower than those that initiate later (more severe) effects.

TOXICOKINETICS

To better understand the mode of action for toxicity, toxicologists look at the ways chemicals are taken into the body, move throughout the body, and are removed from the body. This aspect of toxicology, called **toxicokinetics**, helps to answer important questions about a chemical's potential effects:

- Is the chemical absorbed into the body via routes of workplace exposure to a sufficient degree to cause adverse effects?
- Is the potential target effect plausible on the basis of the distribution of the chemical in the body—that is, does the chemical reach the targeted site?
- Are there key similarities in toxicokinetics among species that explain the relevance of animal laboratory studies to human exposures?
- Are interindividual differences in toxicokinetics significant, and do they represent the range of human variability in susceptibility?

The toxicokinetic behavior of a chemical is reflected in a series of processes—including Absorption, Distribution, Metabolism, and Elimination (ADME)—that describe the overall effect of a chemical in the body.

ROUTES OF EXPOSURE

There are three possible ways for a chemical to come in contact with the body: inhalation, dermal contact, or ingestion. These methods are called **routes of exposure.**

The most prevalent route of exposure in the occupational setting is **inhalation**. When aerosols are breathed into the respiratory tract, they interact differently at each anatomical location. For water-soluble gases, the fluids covering the nasal passages remove some of the chemical, which protects the lungs from exposure and decreases absorption into the blood, but can also result in nasal irritation and carcinogenic effects. When gases and vapors that are not highly water soluble are inhaled into the lungs, the gas molecules quickly pass to the thin capillaries lining cells. Because all molecules like a state of **equilibrium,** or a state of balance, the gas molecules divide up between the capillary air and the blood. The ratio of the gas dissolved in the blood to the gas in the air dissolved at equilibrium is the **blood-to-gas partition coefficient**. This ratio is constant, unlike its concentration.

Consider the following example. Gas A has 10 molecules. Three of the 10 molecules are dissolved into the blood the same way that carbonation is dissolved into a can of soda. Seven molecules stay in the capillary lining and are later exhaled out through the lungs. The blood-to-gas partition coefficient for Gas A is thus 3:7. This ratio will stay the same for Gas A, which means that if there were 100 molecules, then blood would dissolve 30 molecules and 70 molecules would be exhaled. Note that the ratio is substance dependent; that is, Gas B could have a different ratio. For gases with a low blood-to-gas partition coefficient such as ethylene, only a very small percentage of the gas is moved from the lungs into the blood. For gases with a high blood-to-gas partition coefficient such as chloroform, most of the agent is transferred from the lungs into the blood.

Absorption of inhaled particles and aerosols primarily depends on particle size and water solubility. Size also determines where particles are deposited in the respiratory tract. Table 4–2 lists the approximate sizes of particles and their deposited locations. The general rule is that larger particles are deposited higher in the respiratory region. Particles smaller than 1 mm are deposited in the alveoli of the lungs. Removing any of these particles from the nose or throat is relatively inefficient. Once particles are deposited, the speed at which they are removed or cleared from the lungs depends on their solubility.

Dermal contact is another important route of exposure in the occupational setting. Although the skin serves as a barrier of protection against the outside environment, some agents can be absorbed through the skin and cause damage

Table 4–2. Particle Size and Respiratory Tract Location

Size	Respiratory Tract Location	How It Exits
≥ 5 µm	Nasopharyngeal region (nose and back of throat)	Wiping or blowing nose, sneezing
2 – 5 µm	Tracheobronchiolar region (throat and top of the lungs)	Trapped in mucous, swallowed, or coughed out
≤ 1 µm	Alveoli of the lungs	Typically dissolved into blood

to internal organ systems (i.e., cause systemic effects). Because chemicals that are absorbed through the skin have to pass through several layers to reach the blood, their ability to pass through the outermost layer, the stratus corneum, affects how quickly they do so. This layer is replenished every 3 to 4 weeks. During this process, the cells that will make up the stratus corneum dehydrate and form a barrier that is less permeable to passive diffusion. Nevertheless, in the stratus corneum, passive diffusion is the way toxicants move across the skin. For this reason, lipophilic, fat, and oil-soluble agents are more likely to diffuse through the skin than are charged chemicals or particles. The rate of this diffusion is proportional to the fat solubility of the agent and its molecular size. Chemicals with a low molecular weight are more able to diffuse across the skin than are larger molecules. The permeability of the skin also depends on the condition of the skin and the where the skin is exposed to the chemical, given that skin thickness varies across the body.

Although **ingestion** might be the least common route of direct exposure to toxic chemicals in an occupational setting, compared to inhalation and dermal exposures, it is still an important route to understand. In relation to occupational toxicology, oral exposures are important for three reasons:

- Many inhaled particles can be ingested following their clearance of the particles in mucous, which is subsequently swallowed.
- Workers can be exposed by the oral route if improper hygiene in the workplace is present.
- In many cases, occupational exposure limits are based on toxicology studies from oral-dosing studies of laboratory animals.

Because different regions of the gastrointestinal (GI) tract have different pH and structural properties, absorption via the GI tract can take place at any region of that tract. Molecules that are lipid soluble will diffuse across the membrane of the GI tract in the areas where the pH renders those molecules most fat soluble. While most toxic agents enter the body via passive diffusion into the GI tract, the mammalian GI tract also contains numerous transporters for the uptake of nutrients, and some toxic agents are taken up via these mechanisms as well. The amount of a chemical that reaches the systemic blood circulation via the oral route depends on the amount of the chemical metabolized by the GI cells or in the liver. The role of the liver in the absorption of chemicals following ingestion is important because blood is first passed through the liver from the GI tract before entering the circulation and reaching other tissues.

ABSORPTION

When a person or animal is exposed to a certain amount of a chemical, that amount is the **total external dose**. The amount that is absorbed into the body is the **internal dose**.

The properties of the chemical affect the internal dose. Chemicals move into the body by passive or active transport. Passive transport is often a matter of simple diffusion, like walking through a door. For example, a lipid-soluble or hydrophobic molecule (meaning it dissolves in lipids and separates from water) often can pass directly through a cell membrane because the membrane is rich in lipids. Hydrophobic molecules have different levels of lipid solubility that in turn affects the rate of diffusion. Alternatively, active transport uses special molecules to help move a chemical into the cell against a concentration gradient or to accommodate movement of large molecules that are too big to diffuse.

The rate at which a chemical is absorbed is best indicated by the time it takes for the chemical to reach peak blood concentrations after exposure. This rate is known as the T_{max}, where T stands for time. The degree to which a chemical is absorbed by the body is determined by measuring the percent of the total dose that is recovered from the excreta or by comparing blood levels after administration by the route of interest with blood levels following intravenous administration. This measurement is called the **percent bioavailability**. Simply put, percent bioavailability is the amount measured divided by the total dose $\left(\frac{\text{Amount Measured}}{\text{Total Dose}}\right)$. Knowledge of percent bioavailability is useful when extrapolating toxicity data from one route of exposure to another.

DISTRIBUTION

Distribution of a chemical is how it moves through the body once it is absorbed. The chemical will partition between the air and the blood and between the blood and the solid tissues of the body. The ratios at equilibrium between the blood and the varying tissues of the body are known as partition coefficients. The partition coefficients for an agent describe the agent's affinity for a particular tissue or organ. This information can be used to establish doses of an agent that are made available to specific organs, which is important to know when an agent has a specific target organ whose function it disrupts. A comprehensive toxicokinetic study will provide data on the amounts of the chemical in different organs, in the blood compartments (plasma versus whole blood), and in the excreta over time. This study will also identify sites where the chemical may accumulate as well lead to establishing the effective dose at targeted organs and tissues.

METABOLISM

Metabolism is the process by which the body transforms a molecule, be it a toxicant, nutrient, or normal physiological (i.e., endogenous) chemical or molecule. For nutrients, metabolism transforms the materials to a more usable or storable form. For toxicants, metabolism often serves to transform the agent into a form that is more easily removed from the body. This chemical reaction is known as **biotransformation,** and the resulting forms of the chemical are known as **metabolites.** Parent compounds are removed from the body either by producing a metabolite that is more water soluble and that can thus be excreted in the urine or by adding bulky molecular groups that lead to biliary excretion into the feces. For each compound, there are often several competing metabolic pathways that perform these functions, leading to many different metabolites. While biotransformation by metabolism can detoxify and remove potentially toxic agents from the body, in some cases, metabolism creates a metabolite that is reactive, or more toxic. In these cases, the metabolite is the toxic form, rather than the original parent compound to which the worker was exposed.

Metabolic biotransformation is facilitated by enzymes. As such, the rates at which enzymes can process a reaction are often vital to how quickly an agent is removed from the body. The concentration at which the enzyme activity is at half its maximal activity is known as **Km** and is an indicator of the enzyme's affinity for the chemical agent it metabolizes. Km is measured in units of mg/L. A low Km means that very low amounts of the chemical are required for the enzyme to become maximally active, and therefore a low Km indicates a high affinity for the molecule of interest. The measure of how quickly an enzyme can process the metabolic reaction is expressed as mg of molecule of interest/min (mg/min). This measurement is known as V_{max} and indicates the metabolic capacity for that reaction. The V_{max}/Km, expressed in L/min, is referred to as **intrinsic clearance**.

This is a measure of how quickly the molecule can be metabolized and cleared from the system (i.e., how many liters of blood per minute can be cleared of the chemical). An enzyme with high affinity for a molecule (i.e., a low Km) and a high processing rate (a high V_{max}) has the greatest capacity to process or "clear" the parent compound from the body.

ELIMINATION

Elimination is how the body removes agents from the body. Elimination of toxic agents and their metabolite compounds occurs through several mechanisms. The most important route of elimination is through the kidneys in the urine, as more chemicals and their metabolites are eliminated in that manner than in any other. Elimination in the urine is primarily of small molecules and for molecules that are hydrophilic. This includes the metabolites of larger and less hydrophilic compounds that have been biotransformed to smaller, more hydrophilic metabolites.

The second most common route of elimination is in the feces. This is the route by which nonabsorbed ingested materials are eliminated as well as the route of biliary excretion. Large molecules are excreted into the bile via the liver and then eliminated in the feces.

Highly volatile chemicals form a state of equilibrium with the air in the lungs, as determined by the previously discussed blood-to-gas partition coefficient. Because of this equilibrium, exhalation is also a type of elimination from the body. For some chemicals, like ethanol, the amount eliminated by exhalation is small, while other highly volatile compounds, such as diethyl ether, are excreted almost exclusively via the lungs. Some chemicals can be detected in the breath weeks after exposure because of their retention in tissues with a high affinity for the molecule and slow release from those tissues.

Significantly less common pathways of elimination are via sweat for small hydrophilic molecules, via the hair and skin for metals and molecules that bind protein, and via milk for lactating females. Determining whether compounds are eliminated via milk is important because this pathway also exposes the nursing infant and can serve as a route for human exposure to chemicals from affected cows. For many lipophilic compounds such as DDT, PCBs, dioxins, and furans, milk is known to be a major route of elimination.

The most basic description given by toxicokinetics for elimination is the time it takes for the concentration of a chemical in the body to decrease by half, or its **biological half-life** (T½) in the body or tissue. The half-life is inversely related to the clearance. That is, a chemical with a small (short) half-life suggests a high (rapid) clearance. Another important description that can be determined from the toxicokinetics of a toxic agent is its **area under the dose versus time curve (AUC)**. The AUC reflects the total mass of the chemical present over a specified amount of time. The AUC is inversely proportional to the clearance; if the clearance is high (rapid), the amount of dose that remains over time is lower.

DOSE/RESPONSE

Determining the dose, or how much of an agent yields an adverse effect, is a critical component of toxicology. For example, sodium chloride (table salt) is essential to biological life. However, too much salt intake can be fatal.

How much is too much? What is the safe dose? The answer is that it depends. There is a wide range of doses required to produce adverse effects. The **median lethal dose (LD_{50})**, the dose of a substance that causes the death of 50% of the exposed population after a single exposure, is one way to compare toxicities of chemicals. Of course, different times—minutes to years—can be associated with the amount of time it takes for a dose to be lethal. A common terminology used in toxicology is $LD_{50/30}$. This is the dose that will kill 50% of those exposed in 30 days. The dose administered to an animal is typically measured in milligrams (mg) of chemical administered per kilogram (kg) of body weight of the animal exposed. It is often presented as LD_{50} (mg/kg). A list of LD_{50} for rats via the oral route is in Table 4–3.

Table 4–3. List of Rat LD_{50} via Oral Route

Chemical	mg/kg	Reference
Ethanol	7,060	Fisher Scientific, 2012a
Salt (NaCl)	3,000	Fisher Scientific, 2013
Aspirin	200	Acros Organics, 2009a
Caffeine	192	Fisher Scientific, 2009
Nicotine	65	Acros Organics, 2010
Potassium Cyanide	5	Fisher Scientific, 2009b

Table 4–3 lists the LD_{50} dose for rats that were treated with the chemical in the table. This means that when the study rats were given the dose mentioned in the table, 50% died.

Death is a very crude measure of toxicity and does not represent the range of the more subtle effects of interest for protecting worker health. Adverse effects that present as altered biological functions or full-blown disease can often be subtle, with a mode of action that includes earlier biological effects that can go completely unnoticed without clinical or pathological evaluation. Toxicologists expend significant effort to evaluate the range of responses and the corresponding exposure levels to or doses of a chemical to assess the potential for adverse effects.

The correlation between the amounts of a chemical a person is exposed to and the occurrence or severity of an adverse effect is the dose-response concept. Basically, as the amount of exposure to a chemical is increased, the likelihood that an adverse event will occur also increases, with the severity of the event increasing as well. As the amount of a chemical exposure dose decreases, the likelihood of adverse effects decreases. This dose could be related to a concentration in inhaled air, an amount ingested, or an amount received across the skin. For most toxic responses, there is a dose under which a particular effect is not observed. That dose is the chemical's toxic threshold dose for that effect. Further, the effective dose could be the result of several repeat exposures or a continuous exposure. Therefore, the exposure scenario also plays an important role in the toxic effects a substance may cause. The LD_{50} is a representation of a serious effect due to an acute, or very-short-duration, exposure. This dose-response concept is presented in Figure 4–2.

Regular, constant, or repeated exposures over a long time are considered chronic exposures. Depending on the physical and biological properties of the chemical and the nature of the exposure, chronic exposures to much lower concentrations of a substance can have similar and as serious effects as large, acute exposures. Small doses, if the chemical is not quickly removed

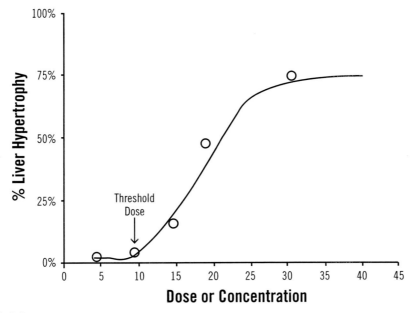

Figure 4–2. Dose response.

from the body, can accumulate over continuous or repeated exposures, and as those doses accumulate, the response grows accordingly. Additionally, if the key effects within a substance's mode of action are not repaired or reversed between exposures, then the effects can continue to progress in response to continuous or repeated small exposures. Thus, the total dose received and the duration of the exposure are both important factors that affect the pattern of toxic response to a chemical.

LIVER TOXICITY

The liver is the primary organ that metabolizes substances, including molecules in foods, endogenous chemicals such as hormones, and molecules that are foreign to the body (also called **xenobiotics**). The liver is where most metabolism after ingestion takes place, which is referred to as **first-pass metabolism.** Blood traveling from the GI tract passes immediately into the liver via the portal vein. That blood then travels through the lobules of the liver along cords of hepatocytes, and enzyme systems in the hepatocytes metabolize the incoming chemicals. Large molecules are excreted into the bile ducts and further excreted in the feces. Metabolites that are small and hydrophilic are released into the blood for filtration by the kidneys, which allows for some toxic chemicals to undergo metabolism before being delivered to the rest of the body. The liver is thus one of the most common sites of toxicity, since it processes many chemicals, including those whose metabolic products are toxic.

Damage to the liver usually takes a fairly predictable course. Early stages of damage from toxic chemicals lead to accumulation of lipids in the hepatocytes. The liver has a paler appearance in radiographs because of the increased density and mass of the lipid globules. Known as **steatosis**, or fatty liver, this condition is reversible and does not necessarily lead to hepatocyte cell death. Many but not all toxic agents lead to steatosis, which can also be caused by other conditions, including obesity, making correlation between exposure and steatosis difficult. Following steatosis, hepatocytes often begin to undergo cell death (necrosis). Where tissue necrosis happens, scar tissues begin to form, a condition known as liver **fibrosis**. As the fibrous scar tissue continues to develop, the amount of blood that can diffuse through the liver becomes limited, as does the amount of functional hepatocytes to filter the blood. The liver becomes deformed, and its ability to function is greatly decreased. When the fibrosis reaches this stage, it is diagnosed as liver cirrhosis, which is irreversible, and can be fatal. In order to monitor damage to the liver, serum levels of liver cell–associated enzymes are checked. The serum levels of many liver enzymes increase during liver damage because of leakage from the dying hepatocytes. The most commonly monitored liver enzymes are alanine aminotransferase (ALT), aspartate aminotransferase (AST), sorbitol dehydrogenase (SDH), glutamate dehydrogenase (GDH), and lactate dehydrogenase (LDH).

KIDNEY TOXICITY

The kidney plays vital roles in removing metabolic waste, regulating body fluid volumes, maintaining electrolyte balance, and ensuring proper body pH. Additionally, the kidney produces many signaling molecules such as renin and erythropoietin, and it transforms vitamin D into its active form. Blood enters the kidney and passes into one of the kidney's nephrons, which are made up of three main elements: the glomerulus, the vascular element, and the tubular element. Blood enters the vascular elements via the renal artery and flows into the glomerulus. The glomerulus comprises a series of specialized capillaries that filter fluids from the proteins and cells, creating an ultrafiltrate that passes into the tubular section of the nephron. These tubular elements are divided into the proximal tubule, the Loop of Henle, and the distal tubule. As blood flows near the proximal tubule, fluid is reabsorbed, concentrating the filtrate from the glomerulus. In the Loop of Henle, many of the ions and electrolytes in the filtrate are reabsorbed into the blood in order to maintain the body's electrolyte balance. Additionally, as fluid passes through the Loop of Henle, urea is initially able to pass freely from blood to tubule filtrate, but in the later parts, the Loop of Henle is no longer permeable to urea, thus regulating the overall levels of urea in the blood and urine (see Figure 4–3). The distal tubule regulates the flow of fluid through the kidney and, as a result, regulates overall blood pressure through mediation of the renin-angiotensin hormone system. The distal tubule ends in the collecting ducts, which further concentrates the filtrate and regulates final urine volume.

Kidney toxicity is evaluated by monitoring the level of serum electrolytes, the total amount of protein, and

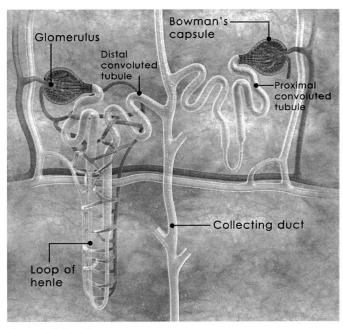

Figure 4–3. Excretion

(Source: 7activestudio/iStock)

the levels of urea and creatinine in the blood. Kidney function can also be assessed by examining the properties of the urine (via urinalysis). An increase or decrease in urine volume, a change in the specific gravity of the urine, or a change in urine osmolality, or an increase in protein concentration in the urine can all indicate kidney dysfunction. An increase in high-molecular-weight proteins may indicate injury to the glomerulus and that early filtration is not functioning properly. An increase in low-molecular-weight proteins suggests tubular injury. The kidneys are very adaptable, and in response to insult or decreased efficiency, they will often increase glomerular fluid flow, increase tubule solute and fluid reabsorption, and undergo compensatory hypertrophy. As a result, damage may not be apparent until compensatory mechanisms are overburdened.

LUNG TOXICITY

In the occupational setting, often the respiratory tract has the most contact with and absorption of toxic agents and is thus an important site of potential toxicity. Air inhaled into the upper respiratory tract passes down the conducting airways, the trachea, and the bronchi. These airways produce mucous, which serves as a protective layer that captures pollutants. Chemicals captured in the mucous are moved via cilia up the respiratory tract, where they are either swallowed or expelled. The airways end in the alveoli, where gas exchange occurs (Figure 4–4). The alveolar wall contains capillaries where the blood is separated from air space by a thin layer of epithelial, interstitial, and endothelial cells. The alveoli interstitial cell population includes fibroblasts, monocytes, lymphocytes, and macrophages. The total volume of air the lung can hold is known as the **total lung volume** (TLV), and the amount of air moved into and out of the lungs when taking the largest possible breath is the **vital capacity** (VC). The amount of air moved into and out of the lungs into a typical resting breath is the **tidal volume** (TV). In diseased and damaged states, measures of changes in these volumes over specific, short periods of time are the bases of pulmonary function testing (spirometry).

The airways are also the primary tissues exposed to gases and vapors. Gases and vapors are often irritants that can lead to inflammation and edema (an accumulation of fluids) in the pulmonary tissues. Inhalation of and overexposure to particles lead to several types of damage in the respiratory airways and lungs, depending on where the irritants are deposited. Dust and nuisance particles can lead to irritation of the upper respiratory tract. Long-term irritation and some particularly reactive particles, such as silicone aerosols and titanium dioxide, can lead to tissue inflammation. Due to the presence of a

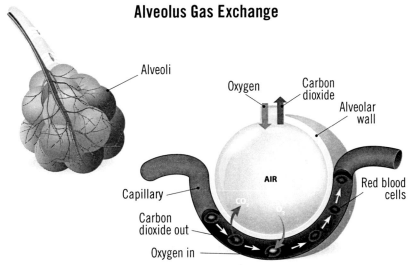

Figure 4–4. Gas exchange.

(Source: ttsz/iStock)

large number of immune cells in the lungs, overexposure to some biological particles (e.g., enzymes) and to certain metals and organic molecules (e.g., diisocyanates) can cause allergic sensitization. Prolonged exposures and the persistent nature of some particles cause damage to the lung that leads to cell death, which in turn leads to fibrosis and scarring. Fibrosis decreases the surface area available for gas exchange and causes the lungs to become less compliant (stiffer), making it harder to draw air into the pulmonary region of the lungs. Another type of injury caused by toxicants in the lung is destruction of the gas exchange surfaces of the alveoli. When these surfaces are severely damaged, a condition known as **emphysema**, which causes inefficient exchange of gases, results and leads to larger, softer lungs that are filled with air. In addition, some chemicals such as asbestos and arsenic are pulmonary carcinogens.

NEUROTOXICITY

The nervous system is composed of nerve cells that interconnect throughout the body. The highest level of classification of the nervous system includes two components: the central and the peripheral nervous systems. The central nervous system consists of the brain and the spinal cord, while the peripheral nervous system consists of the rest of the nerve cells. The nerves are specially adapted receiving information from the environment, from the body, or from other nerve cells and transmitting that information to either a neuron or some other cell that needs to react to the information (e.g., a muscle or gland). A nerve cell has short extensions from its primary cell body known as dendrites and longer extensions from the primary cell body known as axons. Axons extend out to interact with the dendrites of other neurons or with cells that will receive a signal from the neuron. Dendrites are the sites where neurons receive signals. The area where the axon of a neuron interacts with another cell or with another neuron's dendrite is called a synapse. Signals are transferred across the synapse in the form of specialized molecular messengers known as neurotransmitters.

When neurotransmitters are released from the axon into the small space in the synapse, receptors on the cell receiving the signal bind to the neurotransmitter. This initiates an inflow of ions into the receiving cell and creates a change in the electrical charge of the neuron that initiates the release of neurotransmitters from its axons and transmits the signal to the receiving cell. Much of the central nervous system is protected from toxic agents by the blood-brain barrier, in which endothelial cells around the nervous system maintain tight junctions between adjacent cells. This means that, with the exception of molecules that are actively transported into the nervous system, a toxic agent's ability to gain access to the nervous system largely depends on the agent's lipid solubility and its ability to pass through cell membranes. The basic anatomy of a nerve cell is shown in Figure 4–5.

Neurotoxicity affects the transmission of numerous signals in the body that control motion as well as autonomous body functions. Therefore, examining for neuronal damage is usually performed by examining a variety of functions such as pupil dilation and breathing rates. Other signs of autonomic dysfunction include tremors or convulsions, gait abnormalities, decreased grip strength, or a decreased ability to remain upright. Pathological examinations can discover changes in axon length, decreases in the amount of myelin surrounding axons, or deaths of neuronal cells.

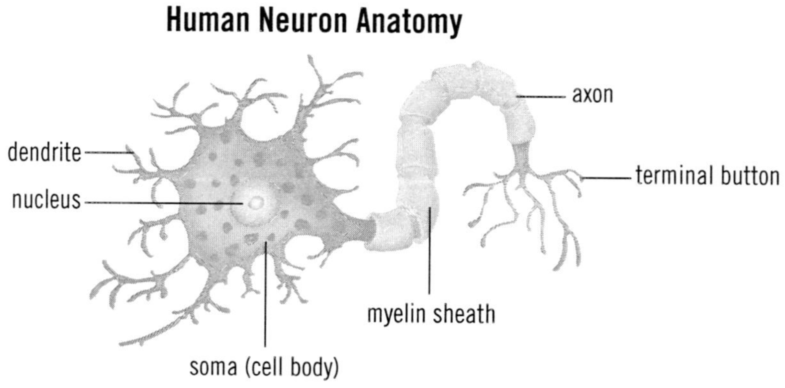

Human Neuron Anatomy

Figure 4–5. Anatomy of a nerve cell.

(Source: blueringmedia/iStock)

REPRODUCTIVE AND DEVELOPMENTAL TOXICITY

Many toxic agents can affect the ability of exposed workers to reproduce or can cause damage to developing offspring. Chemicals that inhibit the ability to reproduce do so via dysregulation of the endocrine hormone systems that control reproduction or by directly damaging the reproductive tissues. The endocrine system consists of glands and organs that release hormones to regulate various body functions. The hypothalamus, pituitary gland, thyroid gland, parathyroid gland, adrenal gland, and ovaries and testes are all important parts of the endocrine system. Changes in the endocrine hormones can lead to decreased libido in both males and females and cause disruption of germ cell development. Chemicals' effects include decreasing sperm count, motility (movement), or morphology (shape) in the male anatomy and abnormal ova and abnormal estrous cycling in the female anatomy.

A full toxicological assessment of reproductive toxicity in animals includes exposing multiple generations of the animals to the toxin. Dosing the animals (usually rats) typically begins 8 weeks before mating is intended to occur and continues throughout pregnancy, after weaning, and until the offspring are mated with nonlittermates. Mating and fertility indices such as mating behavior, ovulation, and the number and quality of sperm are checked. The sex organs are inspected for pathology, and the weights of pups and litters are measured. After the study concludes, uterine, ovary, and testes weights are taken, and histopathological examination is performed on all sex organs. Litters are examined for sex, growth, and any gross abnormalities. The numbers of live and dead offspring are tabulated, and the survival rates to weaning are also monitored. The second generation is then mated, which provides the opportunity to determine whether there were any toxic effects on the eggs or sperm of the second generation or whether there is any tendency for accumulation of damage in subsequent generations.

Modern history shows us the effects of chemical exposures on the development of human embryos and fetuses in utero. In the 1960s, a sudden increase in the number of newborns with serious limb malformations was linked to prescriptions of the apparently safe (in adults) sedative thalidomide, which lessened nausea in pregnant women. In the late 1960s, several young women presented with adenocarcinomas of the vagina and cervix, which were extremely uncommon in women that age. It was eventually discovered that exposure during gestation to diethylstilbestrol (DES), a synthetic estrogen that was widely used to prevent miscarriage, induced genital tract malformations that increased the risk of developing cervical and vaginal adenocarcinomas. In response to these and other tragic developmental toxicity events, studies on determining whether a chemical is a development toxicant were greatly increased. In the 1970s, the description of fetal alcohol syndrome (FAS) detailed the developmental effects of alcohol on developing fetuses.

Typically, developmental studies are performed with rats or rabbits. The mother is exposed to the chemical during **organogenesis**, the period from implantation through gestation. Offspring can be euthanized just prior to birth, just after birth, or after the postnatal period, depending on the type of measurement of interest. Developmental toxicity can be a direct result of the chemical on the growing fetus or embryo, or it may be a secondary effect due to maternal toxicity. Therefore, developmental toxicity studies evaluate both maternal and offspring toxicity. Maternal evaluation fertility indices include gestation length, body weight, food and water consumption, and eventually clinical evaluations such as a gross necropsy, a histopathology, and organ weights. Embryo and fetal toxicity are assessed, and offspring undergo external, skeletal, and visceral checks to determine the presence of any teratogenicity and postnatal survival and development markers, such as bone ossification, a marker of growth-related effects.

IMMUNOTOXICITY

Similarly to red blood cells, many of the cells of the immune system are derived from hematopoietic precursor cells in the bone marrow. Immune system cells include macrophages, monocytes, natural killer (NK) cells, B cells, and T lymphocyte cells. The hematopoietic precursor cells develop into either a myeloid precursor or a lymphoid precursor. Lymphoid precursors further develop into the NK lymphocytes or B lymphocytes, or they travel to the thymus to develop into T lymphocytes. The myeloid precursors develop into macrophages, granulocytes, and monocytes, as well as the red blood cells. There are two types of immunity conferred by the immune system: cell-based immunity and the more specific, antibody-directed immunity known as humoral immunity. Cell-based immunity relies on NK cells, macrophages, and cytotoxic T lymphocytes. B lymphocytes produce the antibodies that target specific molecules (including chemicals) encountered by the immune system and thus enhance the responses of other immune cells. The most common concern with toxic agents that affect the immune system is functional immunosuppression which prevents the immune system from working well or at all to protect the body.

Because of the complexity of the immune system and the immune response, a multitiered approach is necessary to assess whether an agent is immunologic. First, it is essential to establish that all the important cells of the immune system are developing from the hematopoietic precursors. This process begins with a hematology profile that determines the numbers of cells including red blood cells, immune cells, and others. These individual populations of immune cells can be labeled with specific markers to determine the numbers of each subset of cells. Functional tests can also be performed. For cell-based nonhumoral immunity, the ability of macrophages to carry out phagocytosis and the ability of NK cells to kill tumor cells must be established.

To conduct the humoral immunity test, animals are exposed to a foreign antigen, often sheep red blood cells, and then the amount of antibody produced in the spleen and sera to that antigen is assayed. Additionally, isolated B cells from treated test animals must be assayed for their ability to undergo activation when exposed to an antigen. Some of the tests to establish the functionality of the cells involved in acquired immunity used by the National Toxicology Program (NTP) are the cytotoxic T lymphocyte (CTL) assay, the delayed hypersensitivity response (DHR) assay, the macrophage activation factor assay, and the T cell proliferative response to antigens assay.

The immune system may be disrupted in such a way that it damages tissues because of the system's improperly heightened response to a chemical. This response can lead to hypersensitivity or autoimmunity. In hypersensitivity, the antibodies or memory T cells of the immune system respond in an exaggerated way to a chemical they were previously exposed to. In chemically induced autoimmunity, a chemical that interacts with an endogenous protein elicits a response from the immune system. If the immune system later encounters the protein, the result may be an autoimmune disease, which involves the immune system inappropriately attacking the protein in healthy tissues.

SKIN TOXICITY

Another important site of toxicity, especially in the occupational setting, is the skin. Skin toxicity includes nonallergic irritation, allergic contact dermatitis, corrosive chemical burns, and chemical-induced photosensitivity. Nonallergic irritation can be difficult to identify because of the great variety of sensitivities and factors contributing to the irritation. Nevertheless, nonallergic contact dermatitis is generally reversible and has no serious, long-term consequences. Allergic contact dermatitis is common and requires an initial sensitization exposure. Corrosive agents cause damaging necrosis in the top layers of the skin, the dermis, and the epidermis. Symptoms are thus wide ranging and include bleeding ulcers on the skin, areas of hair loss, scabs, and sometimes permanent scars. Some chemicals, most often those stimulated to higher energy levels by ultraviolet (UV) light, induce photosensitivity in the skin. These activated molecules can interact with oxygen or some biological molecules to produce free radicals, which damage cells. When these types of molecules are present in skin cells, exposure to the sun or other UV light sources can begin an inflammatory response and lead to death of the skin cells.

Skin toxicity is typically tested on the skin of lab animals. The skin is usually exposed to a chemical for a short time, often less than 4 hours, and monitored for 14 days for redness, fluid accumulation, necrosis, and scab formation. Increasingly, models to test skin exposure that do not require animals are being developed and utilized. Some commercially available models of human skin use cultured cells to reconstruct a section of human skin in the laboratory.

CARCINOGENESIS AND GENOTOXICITY

Strict regulation of cell growth and differentiation is critical to the survival of multicellular organisms. Precursor cells proliferate and differentiate into mature cell types, and upon differentiation to the terminal, mature cell type, cells enter into stasis, or quiescence, at which point they begin performing their cellular functions. Quiescent cells no longer grow or proliferate. If cells become damaged, are no longer useful, or for some other reason need to be eliminated, they undergo a regulated cell death process called **apoptosis**.

Neoplasia originates from inheritable changes in cell growth that lead to the loss of normal control processes. This autonomous growth is the result of altered gene expression, due either to factors inherent in the neoplastic cells (i.e., gene mutation) or to factors in the environment. Neoplasms are classified using several criteria, the first of which is whether they are benign or malignant. The critical distinction between benign and malignant neoplasms is their ability to engage in successful metastatic growth. **Metastasis** is the movement of neoplastic cells to locations beyond the primary site of origin. Benign neoplasms are unable to form metastatic growths. While any space-occupying growth is referred to as a **tumor**, tumors may or may not be neoplastic, and neoplastic tumors may or may not be malignant. The term *cancer* refers to malignant neoplasms.

The names of neoplasms depend primarily on whether they are malignant or benign and on the tissue type of their origin. Benign neoplasms are named by adding the suffix *-oma* to the tissue of origin. So a benign fibrotic neoplasm is a **fibroma**, a benign growth of adipose tissue is a lipoma, a benign tumor of a gland is an adenoma, and so on. Malignant neoplasms of mesenchymal origin are named by adding the descriptor of *sarcoma* to the tissue of origin, as in *fibrosarcoma*, *adenosarcoma*, and *liposarcoma*. Neoplasms of epithelial origin are given the descriptor *carcinoma*, as in *hepatocellular carcinoma* (liver cell), *epidermoid carcinoma* (skin), and *gastric adenocarcinoma*.

Carcinogenesis progresses through key steps. Although **initiation** of carcinogenesis does not usually have easily detectable morphological or biological changes, mechanistic molecular biology studies have described this stage of development. Three processes are important for initiation: metabolism, DNA repair, and cell proliferation. Anything that alters these pathways can lead to initiation. For example, agents that damage DNA, such as many hydrocarbons and UV radiation, can lead to initiation. Depending on the type of damage, repair processes can lead to increased DNA activity or mutation of the genes related to metabolism and regulation of proliferation.

The next important step in carcinogenesis is **promotion**. Promoting agents control gene expression by altering signal transduction pathways. Many promoting agents disturb the ligand-receptor pathways that regulate cell proliferation. Promotion is typically a reversible process.

The final step in carcinogenesis is **progression**. In progression, cells are often very distinct from the cell type of origin. For example, they may show signs of decreased differentiation, which is typical of progenitor types of cells. Another important hallmark of carcinogenesis progression is genomic instability. As the neoplastic cells divide rapidly, numerous errors in their genome accumulate. Cells can accumulate or lose entire chromosomes, and often the mechanisms for DNA repair or apoptosis to eliminate such abnormalities are suppressed or altered, allowing for propagation of further errors in the neoplasms.

Carcinogenesis is a lengthy process, taking sometimes a large percentage of a person's lifespan before the effects are noticeable. One source of the potential carcinogenicity of a potentially toxic agent is epidemiological studies. By determining whether populations of people with certain exposures are developing particular types of tumors after exposure, correlation can be established. The National Toxicology Program (NTP), under the NIH's National Institute of Environmental Health Sciences, performs testing to establish toxicological profiles of chemicals. These profiles typically include a 2-year exposure study in rats and mice, which is typically a whole lifespan of exposure for rodents. Fifty test animals of each sex in each of the species are included in the test. A range of doses are delivered and the resulting tumors—benign and malignant—are evaluated, with the various tumor types considered independently. These 2-year bioassays provide excellent information on incidences of cancers associated with a lifetime exposure to the agent of interest and are the primary source of information on cancer end points for toxicologists.

Other assays related to carcinogenesis test for mutagenicity and genotoxicity. **Genotoxicity** is the ability of a chemical to cause heritable changes in the DNA or cause other damage to the chromosomes of a cell. Mutagenicity is the ability of an agent to induce changes in the DNA. While all mutagens damage the DNA and are thus genotoxic, some genotoxic agents are not necessarily mutagenic. Genotoxic agents can induce breaks in the DNA, can cause chromosomes to distribute unequally during cell division, can inhibit or alter the mechanisms that and repair damage of the DNA, or can in other ways damage the DNA. While not all genotoxic events are carcinogenic, damage to DNA is known to cause changes that often do lead to the development of cancer. As such, genotoxicity is often, but not always, closely related to carcinogenesis. Several techniques have been developed to determine a molecule's genotoxic capacity, as described in Table 4–4.

Table 4–4. Assays for Genotoxicity

Assay	Description
Ames assay	The Ames assay uses mutation-sensitive bacteria to assay a chemical's potential to cause genetic mutations.
Ames-S9 assay	The Ames-S9 assay modifies the basic Ames assay by introducing metabolic components from the liver to determine whether the metabolites of a chemical are capable of causing genetic mutation.
TK assay	Similar to the Ames assay, the TK assay uses mouse lymphoma cells to determine the mutagenic capability of chemicals in a mammalian cell system.
Comet assay	DNA from mammalian cells that have been exposed to a chemical is run on a gel, and if the DNA has been damaged, the smaller DNA fragments will appear as a "comet tail" on the assay.
Micronucleus test	In the micronucleus test, animals or cells are exposed to a chemical, and if the chemical disrupts the ability of the cell to properly separate its DNA during cell division, micronuclei (small pockets of DNA in the cell cytoplasm) are visible under a microscope.

Several other sources of toxicological information are also available. Some of these are listed in Table 4–5.

Table 4–5. Toxicology Information Sources

Source	Type	URL/Citation
AIHA's Toxicology Principles for the Industrial Hygienist	Textbook	www.aiha.org/marketplace
Casarett & Doull's Toxicology: The Basic Science of Poisons	Textbook	www.mheducation.ca/professional/products/9780071769235
NIH-NIEHS National Toxicology Program	Online database	ntp.niehs.nih.gov
NLM NCBI PubMed	Online literature database	www.ncbi.nlm.nih.gov/pubmed
NLM NIH ToxNet	Online resource database	toxnet.nlm.nih.gov
GESTIS	Online hazardous substance database	www.dguv.de/ifa/Gefahrstoffdatenbanken/GESTIS-Stoffdatenbank/index-2.jsp
European Chemical Substances Information System	Online chemical information database	esis.jrc.ec.europa.eu/
International Program on Chemical Safety INCHEM	Online chemical safety database	www.inchem.org
NLM NCBI PubChem	Online chemical information database	pubchem.ncbi.nlm.nih.gov
NLM NIH Toxicology Tutorials	Online toxicology tutorials	sis.nlm.nih.gov/enviro/toxtutor.html
CDC ATSDR	Online toxicology information database	www.atsdr.cdc.gov/toxprofiles/index.asp
Weight of Evidence: A Review of Concept and Methods	Review article	Weed DL; *Risk Analysis*; Volume 25, Pages 1545–1557, 2005
ECHA Practical Guide 2 : How to report weight of evidence	Guidance document	echa.europa.eu/practical-guides
Using Mode of Action and Life Stage Information to Evaluate the Human Relevance of Animal Toxicity Data	Review article	Seed, J. et al.; *Critical Reviews in Toxicology*; Volume 35, Page 664–672, 2005
U.S. EPA Framework for Determining a Mutagenic Mode of Action for Carcinogenicity	Guidance document	www.epa.gov

A list of agents and common toxic effects is provided in Table 4–6.

Table 4–6. Example Toxic Agents by Target

Organ System	Chemical	Effect
Liver	Acetaminophen	Hepatocyte death
	Arsenic	Cirrhosis
	CCl4	Fatty liver
	Copper	Hepatocyte death
	Ethanol	Fatty liver, hepatocellular death, cirrhosis
	Methylenedianiline	Bile duct damage
	Vinyl Chloride	Carcinogenesis
	Vitamin A	Cirrhosis
Kidney	Acetaminophen	Tubular necrosis/damage, acute renal failure
	Aminoglycosides	Renal vascular constriction, glomerular injury, tubular injury, acute renal failure
	Amphotericin B	Renal vascular constriction, glomerular injury, acute renal failure
	Cadmium	Tubular inflammation
	Cisplatin	Tubular necrosis/damage, tubular inflammation
	Cyclosporine	Tubular inflammation
	Heavy Metals	Tubular necrosis/damage, acute renal failure
	Nonsteroidal Anti-inflammatory Drugs	Renal vascular constriction, glomerular injury, tubular inflammation
Respiratory system	Ammonia	Edema, irritation, chronic bronchitis
	Arsenic	Bronchitis, carcinogenesis
	Asbestos	Fibrosis, carcinogenesis
	Cadmium oxide	Irritation, emphysema
	Chromium (VI)	Nasal irritation, bronchitis, fibrosis, carcinogenesis
	Nickel	Pulmonary edema, nasal and lung carcinogenesis
	Sulfur dioxide	Bronchoconstriction, irritation, chronic bronchitis
	Tungsten carbide	Irritation, hyperplasia, bronchial fibrosis
Neurotoxicity	Aluminum	Dementia
	Arsenic	Encephalopathy, neuropathy, axon degeneration in peripheral nervous system
	Cyanide	Coma, convulsions, demyelination, neuronal loss
	Doxorubicin	Axon degeneration
	Lead	Encephalopathy, learning deficits, peripheral nervous system axonal loss and demyelination

Organ System	Chemical	Effect
Reproductive and developmental toxicity	Ethylene glycol	Male gonadotoxic
	Ethanol	Neural development delays, craniofacial malformations, decreased birth weight
	Ketoconazole	Decreased steroid hormone production
	Methotrexate	Ovarian dysfunction
	n-Hexane	Male gonadotoxic
	Prednisone	Ovarian dysfunction
	Retinoids	Teratogenicity
	Thalidomide	Teratogenicity
Immunotoxicity	Arsenic	Decreased response to sheep red blood cell (SRBC) challenge
	Chromium (VI)	Sensitization
	Cobalt	Sensitization/asthma
	Lead	Decreased proliferation of myeloid progenitors, decreased pathogen resistance
	Nickel	Sensitization
	Polychlorinated Biphenyls	Decreased response to SRBC challenge, immune suppression
	Urethane	Decreased myeloid progenitors, decreased natural killer cell activity, decreased antibody response to SRBC
Skin toxicity	Ammonia	Corrosive
	Anthracene	Photosensitization
	Bisphenol A	Allergic contact sensitization
	Chromium (VI)	Allergic contact sensitization
	Hematoporphyrin	Photosensitization
	Hydrogen chloride	Corrosive
	Neomycin	Allergic contact sensitization
	Nickel	Allergic contact sensitization
	Phenol	Corrosive
	Sodium hydroxide	Corrosive
Carcinogens	Asbestos	Lung carcinogen
	Benzene	Mutagen
	Benzo(a)pyrene	Mutagen
	Cyclophosphamide	Bladder, blood carcinogen
	Ethidium bromide	Mutagen/DNA damaging agent
	Sodium azide	Mutagen
	Ultraviolet radiation	Mutagen
	Vinyl chloride	Liver carcinogen

REVIEW QUESTIONS

1. What is the difference between mechanism of action and mode of action?
2. What are biomarkers, and why are they important?
3. What are some examples of biomarkers?
4. List and describe Hill's criteria of causation
5. What is ADME?
6. What is the blood-gas partition coefficient?
7. What is biotransformation?
8. Define V_{max} and Km. If a chemical has a V_{max} of 150 mg/min and a Km of 3 mg/L, what is that chemical's intrinsic clearance rate?
9. What is the half-life of a chemical that has a volume of distribution of 2 L and that can be cleared from 1,500 mL of blood per hour?
10. What is the difference between an acute and a chronic exposure, and which is more dangerous?

REFERENCES

Acros Organics. *Acetylsalicylic Acid MSDS*. Fair Lawn, NJ: Acros Organics BVBA, 2009.

———. *Nicotine Ditartrate Dihydrate MSDS*. Fair Lawn, NJ: Acros Organics BVBA, 2010.

Fisher Scientific. *Caffeine MSDS*. Fair Lawn, NJ: Fisher Scientific, 2009.

———. *Ethyl Alcohol MSDS*. Fair Lawn, NJ: Fisher Scientific, 2012.

———. *Potassium Cyanide MSDS*. Fair Lawn, NJ: Fisher Scientific, 2009.

———. *Sodium Chloride MSDS*. Fair Lawn, NJ: Fisher Scientific, 2013.

Seed, J., E. W. Carney, R. A. Corley, et al. "Overview: Using Mode of Action and Life Stage Information to Evaluate the Human Relevance of Animal Toxicity Data." *Critical Reviews in Toxicology* 35 (2005): 664–72.

Weed, D. L. "Weight of Evidence: A Review of Concept and Methods." *Risk Analysis* 25 (2005): 1545–57.

5
Occupational Exposure Limits and Assessment of Workplace Chemical Risks

LEARNING OBJECTIVES

After completing this chapter, readers should be able to do the following:

- Describe how Occupational Exposure Limits (OELs) are used in workplace risk assessment.
- Identify the key components of an OEL.
- Identify the OELs created by various organizations.
- Outline the process of OEL derivation.
- Apply available information to working conditions to assess hazardous exposure levels.

Photo credit: baona/iStock

ROLE OF OCCUPATIONAL EXPOSURE LIMITS IN WORKPLACE RISK ASSESSMENT

A traditional description of the industrial hygienist's role includes anticipating, recognizing, evaluating, and controlling occupational health risks. These tasks are consistent with the risk assessment and risk management framework described by the National Academy of Sciences (NAS) (NRC 2009). The industrial hygiene community has continued to emphasize the similarities between traditional occupational risk assessment practices and the approaches outlined in the NAS framework (Dotson et al. 2012). Key elements of the NAS approach include problem formulation, risk assessment, risk management, and risk communication (see Figure 5–1). The risk assessment phase includes several distinct activities that pertain to the development and application of the Occupational Exposure Limits (OELs):

- Hazard characterization—systematic process for weighing the complete data to identify human health effects caused by exposure to a chemical in a specific scenario.
- Dose-response assessment—evaluation of the changes in incidence and severity of effects with increasing exposure (or dose). The assessment identifies a quantitative dose for adverse effects that serves as a starting point for derivation of the OEL. The OEL is intended to provide a quantitative estimate of the maximum air concentration that is believed to be safe for an occupational population that is exposed daily for a working lifetime.
- Exposure assessment—evaluation of the magnitude, duration, and frequency of exposures of a population of interest.
- Risk characterization—integration of the OEL derived from the dose-response assessment and the exposure assessment that supports conclusions about the level of risk.

OELs are important because they provide a scientific basis to evaluate existing exposure control technologies and to identify worker health and medical surveillance needs. By providing more complete health and safety guidance to those exposed to chemicals, OELs can enhance sustainability and stewardship efforts, and promote risk communication by informing workers of potentially adverse health effects of chemical exposure. In the formal risk assessment process, OELs provide a benchmark for exposure assessment comparisons. As the level of exposure reaches and exceeds the limit, the level of risk also increases and may require risk management

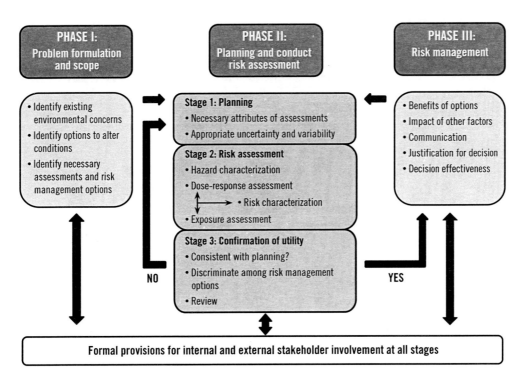

Figure 5–1. Risk assessment and risk management framework.

(Source: Adapted from NAS 2009)

actions to reduce or eliminate the exposure. Thus, in the context of risk assessment methods, OELs are the outcome of the dose-response assessment that is used in combination with the exposure assessment to develop the risk characterization.

KEY COMPONENTS OF AN OCCUPATIONAL EXPOSURE LIMIT

The definition of an OEL and the methods used to derive it differ among the organizations that establish OELs (Waters et al. 2014). The definitions used by these organizations vary primarily in the language regarding the intent of the OEL to protect all versus most workers and the degree to which susceptible populations are accounted for in the OEL (Table 5–1). Several elements of an OEL (in addition to the actual value itself) include the concentration units, the recommended time-weighted average, and the additional notations for hazards not addressed quantitatively in the OEL.

Table 5–1. Full-Shift Airborne Occupational Exposure Limits Developed by Various Organizations

OEL[a]	Definition
ACGIH TLV	Threshold Limit Values (TLVs) refer to airborne concentrations of chemical substances and represent conditions in which *nearly all* workers may be repeatedly exposed, day after day, over a working lifetime, without adverse health effects. TLVs are developed to protect workers who are healthy adults. They do not distinguish between safe and dangerous exposures, nor are they a relative index of toxicology.
AIHA-OARS© WEEL	The Workplace Environmental Exposure Levels (WEELs) are health-based, airborne chemical OELs established to provide guidance where other OELs are not available. This guidance protects most workers from adverse health effects related to occupational chemical exposures. WEELs are expressed as either time-weighted averages or ceiling limits.
DFG MAK	The MAK values are daily, 8-hour time-weighted average values that apply to healthy adults. MAKs give the maximum concentration of a chemical substance in the workplace.
EC SCOEL	The SCOEL time-weighted average exposures represent the levels to which an employee may be exposed via the airborne route for 8 hours per day, 5 days per week, over a working lifetime without adverse effects on the health of the worker or his or her future children.
NIOSH REL	RELs are occupational exposure limits designated by NIOSH as being protective of worker health and safety over a working lifetime. The REL is used in combination with engineering and work practice controls, exposure and medical monitoring, labeling, posting, worker training, and personal protective equipment. This limit is frequently expressed as a time-weighted average (TWA) for up to a 10-hour workday and a 40-hour workweek.
OSHA PEL	OSHA sets enforceable permissible exposure limits (PELs) to protect workers from the exposure to hazardous substances. PELs are regulatory limits on the amount or concentration of a substance in the air. OSHA PELs are based on an 8-hour time-weighted average (TWA) exposure.

[a]ACGIH TLV—American Conference of Governmental Industrial Hygienists (U.S.) Threshold Limit Values; AIHA-OARS WEEL—American Industrial Hygiene Association and Occupational Alliance for Risk Science Workplace Environmental Exposure Levels; NIOSH REL—National Institute for Occupational Safety and Health (U.S.) Recommended Exposure Level; OSHA PEL—Occupational Safety and Health Administration (U.S.) Permissible Exposure Level; DFG MAK—Deutsche Forschungsgemeinschaft (German Research Foundation) Maximale Arbeitsplatz-konzentrationen (Maximum Airborne Concentration); EC SCOEL—European Commission Scientific Committee on Occupational Exposure Limit Values.

Source: Adapted from Waters et al. 2014.

CONCENTRATION UNITS

Most OELs address exposures from inhalation and are presented as concentration of units in the air. For chemicals that are present in the air in particulate form, the OELs' common unit of measure is the chemical's mass per unit of air volume. For most chemicals, the OEL is presented as milligrams per cubic meter of air (mg/m^3). For highly potent chemicals that have very low OELs, units of micrograms or even nanograms per cubic meter are used. The mass per unit volume is the appropriate unit of an OEL for solid particulates and liquid aerosols. When indicating the occupational risks of fibers, units such as fibers per cubic centimeter of air could be used. There is also ongoing debate about the traditional use of mg/m^3 for nanometer-scale particles. Units based on available surface area or density have been suggested as more relevant to biological activity because a chemical's toxicity may relate more to surface area available than to total mass of the chemical.

For gases and vapors (i.e., gases formed when liquids evaporate), the mass per unit volume is also appropriate to indicate the air concentration of the gas or vapor. One representation of an OEL for gases and vapors is based on the **molar ratio,** which is the number of chemical molecules per air molecule. However, the most common representation is parts per million (ppm) or, for very potent chemicals, parts per billion (ppb). Most organizations that set OELs use ppm to represent gases and vapors and provide the equivalent mass per unit air volume (mg/m^3) as well. In some cases, a ppm to mg/m^3 conversion factor is provided. This factor is derived from the ideal gas law equations, which relate mass and molar volume of a gas at a defined temperature and pressure. For occupational situations, the condensed form of this equation is ppm = mg/m^3 × 24.45 / molecular weight. The value of 24.45 is based on the volume of a mole of air, assuming standard pressure and temperature. For example, 1 ppm of benzene is equivalent to 3.2 mg/m^3 of benzene (i.e., 1 ppm = 3.2 mg/m^3 × 24.45 / 78.11 g/mol). See Chapter 6, Gases and Vapors, for more information.

TIME-WEIGHTED AVERAGE

The intent of OELs is to protect workers, assuming daily exposures for a full 40- to 45-year career (i.e., a working lifetime). The appropriate application of an OEL in a given workday depends on the nature of a chemical's toxic effects. Most OELs are given as one of the following: (1) full-shift time-weighted average (TWA), (2) short-term exposure limit (STEL), or (3) ceiling limit. Example definitions are presented in the American Conference of

Governmental Industrial Hygienists' Threshold Limit Values (ACGIH TLV®) *Documentations* (ACGIH 2014).

FULL-SHIFT TWA

The **full-shift TWA** is usually given as an 8- or 10-hour time-weighted average limit. These limits have the underlying assumption that fluctuating exposures during the day will not significantly impact a chemical's potential to cause adverse effects. These limits are developed to prevent chronic health effects that depend on the cumulative dose of the chemical that a worker receives over hours, days, or even months.

SHORT-TERM EXPOSURE LIMIT (STEL)

Many chemicals can also generate health effects following short periods (i.e., minutes) of exposure. A common example is the weakening of the central nervous system associated with intense, short-term exposures to solvent vapors such as from an acetone spill in a research laboratory. For chemicals that create effects over a short time period, a STEL may be established. The STEL is a time-weighted average exposure limit whose averaging time is short. In most organizations, a STEL is a 15-minute TWA exposure limit, although other averaging times can be specified for a given chemical.

CEILING LIMIT

Some chemicals' effects are rapid or nearly immediate, such as the irritation of chlorine gas. Where such data are available, a maximum exposure or **ceiling limit** that should not be exceeded at any time during the workday might be established.

In some cases, the data are not sufficient to develop a STEL or ceiling limit for a chemical that is suspected of generating adverse effects following acute or peak exposure. The **excursion limit approach** addresses this lack of information. For example, the ACGIH TLV approach has an excursion limit of three times the value of full-shift TWA, as long as five times the full-shift TWA limit is never exceeded. These excursion limits are based on well-controlled emission processes and are not biologically based. However, they do provide a method for developing risk management goals when acute limits are not known. These three common limits do not preclude developing OELs specific to task durations. Options for adjusting OELs to match alternative workshift schedules have also been developed (see Paustenbach 2011 for a review of OEL adjustment methods).

HAZARD NOTATIONS

Another common element of OELs is additional **hazard notation**. The criteria for assigning hazard notations are not standardized among organizations. However, in general, most organizations have methods that address the following end points: (1) toxicity via skin absorption, (2) sensitization (i.e., an allergic reaction that can occur when the skin becomes increasingly reactive to subsequent exposures to an irritating chemical) potential, and (3) cancer potential. Each of these end points can affect the interpretation and use of an OEL.

The **skin notation** that accompanies an OEL indicates that dermal exposure to a chemical may be a significant source of a systemic (absorbed) and a toxic dose for workers exposed at or near the OEL, which takes the form of a maximum level of a chemical on work surfaces or of total doses of a chemical on a worker's skin. As traditionally used, the skin notation alerts the industrial hygienist that skin exposure should be limited and that if skin exposure occurs, the inhalation-based OEL might not adequately protect workers. However, in the context of most OELs, the notation does not provide information on damage or toxicity to the skin itself. To address this deficiency, the National Institute for Occupational Safety and Health (NIOSH) developed a skin notation strategy that incorporates into the traditional use of the skin notation the potential for skin damage as well as skin sensitization (Dotson et al. 2011; NIOSH 2009).

Another hazard notation that might be included with the OEL relates to a chemical's ability to be a sensitizer. This notation is necessary when the OEL alone will not protect a worker from developing sensitization via dermal contact. Some organizations differentiate between chemicals that are known or expected dermal sensitizers (DSEN) and those that can cause respiratory tract sensitization (RSEN). Dose-response analysis for sensitization is complex, and in only a few cases are data sufficient to establish an OEL that can protect against sensitization of the respiratory tract. Thus, the RSEN and DSEN notations are valuable additions to OELs that do not fully address a chemical's potential to create sensitization responses.

The approach used to address a chemical's cancer-causing potential diverges among groups that establish OELs. Some organizations do not set traditional OELs for carcinogens but establish an acceptable cancer risk (e.g., 1 in 1,000) based on the theory that, for at least some carcinogens, any dose has some risk. NIOSH RELs include a cancer notation (CA), and in some cases the OEL is established to restrict exposures as low as reasonably achievable (ALARA) using best management practices; however, some other

RELs for carcinogens provide a quantitative OEL. The German *maximale Arbeitsplatz-Konzentration* (MAKs) also provide a notation of carcinogenic potential and refer the user to the regulatory OELs for risk management purposes. The ACGIH TLVs include an evidence-based notation for cancer with multiple levels of evidence represented.

Although the focus of this chapter is traditional OELs for airborne exposures, exposure limits based on levels of chemicals in the body or excreta can be developed to measure the internal dose for specific risk assessment needs. Measures of lead in blood or cadmium in urine are examples of such biological exposure limits. The American Conference of Governmental Industrial Hygienists (ACGIH) Biological Exposure Indices (BEIs) provide such limits, as do some of OSHA's chemical-specific standards. Exposure limits based on measurements of chemicals in biological media are ideal because they assess actual absorbed doses measured concentrations in the environment can only be estimated.

FUNDAMENTALS OF OEL DERIVATION

Just as the definition of an OEL varies among organizations, the specific approaches used for OEL derivation also vary. However, there are common elements used by most organizations that develop OELs. Key steps in OEL derivation (from Deveau et al. 2014) include the following:
1. Define the scenario and develop the problem formulation.
2. Gather and summarize the scientific literature most relevant to the problem formulation (e.g., primary literature and existing reviews on toxicology, epidemiology, pharmacokinetics, and physicochemical properties).
3. Select a dose point that will protect workers from the chemical's toxic effect (i.e., a point of departure such as a no observed adverse effect level [NOAEL], lowest observed adverse effect level [LOAEL], benchmark dose [BMD], or risk-based level) based on factors included in the problem formulation, such as protectiveness, strength of evidence, and human relevance.
4. If necessary, perform extrapolations to increase the relevance of the point of departure:
    a. Adjust for route of exposure and exposure duration/patterns using default assumptions on rates of ingestion/inhalation or physiologically based pharmacokinetic (PBPK) models.
    b. Perform animal-to-human extrapolations and human variability ex-

trapolations using uncertainty factors, chemical-specific adjustment factors, or PBPK modeling.
   c. Apply any additional uncertainty factors (e.g., database deficiency, severity of effect).
5. Submit the value for external review.

The initial step in the process is the critical assessment of all the available literature to identify sensitive adverse effects that are relevant to occupational scenarios and worker health. Key resources for such information are highlighted in Chapter 4, Basic Concepts in Industrial Toxicology.

Dose-response assessment is critical when setting an OEL. The basic premise is that as the dose increases, the likelihood that an effect will occur or that an effect's severity will grow also increases. Likewise, as the dose decreases, the potential for adverse effects decreases. Note: For this discussion we use the generic term *dose* to refer to the amount of the chemical received from the external exposure. For inhalation studies, the term *concentration* is also used. Each chemical has its own dose response, depending on its toxic properties. For most types of toxicity, there is presumed to be a dose below which adverse effects are very unlikely to occur. Identifying this **safe dose** is thus the goal of an OEL. Unfortunately, determining the OEL or safe dose is done using imperfect data from human-effects studies (i.e., epidemiology), which have unclear exposure data, or laboratory animal studies (i.e., toxicology), which have uncertain relevance to human exposure situations. Therefore, the OEL is only an estimate of the concentration or dose that is expected to be safe for workers.

The OEL derivation process often finishes with two key steps—selecting a measure of dose response and considering factors that account for uncertainty when extrapolating from less than ideal data. The quantitative OEL value is calculated by dividing the measure of dose response by the composite of the uncertainty factors, as shown in Figure 5–2. The dose-response curves are evaluated to indicate the highest dose that does not cause an adverse effect (NOAEL). In this case, the minimal amount of inflammation was observed at the NOAEL but not considered toxicologically significant. Dividing the NOAEL by the composite uncertainty factors estimates a dose (or exposure concentration) that is below the threshold for workers who are sensitive to the chemical.

The numerator of the OEL equation is a dose-response measure from a health effects study. Because the goal of the derivation process is to identify the concentration at which a chemical has no adverse effects, the ideal

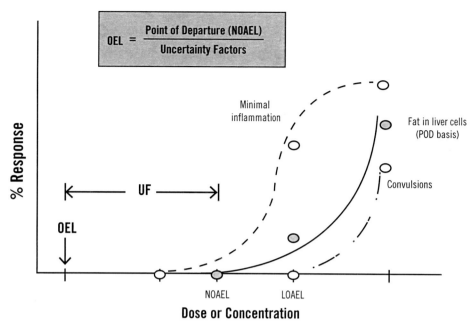

Figure 5–2. Occupational exposure limit derivation.

dose-response measure is a chemical's concentration that has no adverse effects in a representative sample of workers sensitive to the chemical. Because sufficient studies of this kind are rarely available, a compilation of imperfect studies is often sorted through to determine other indications of dose response. In each study, the highest dose that does not cause adverse effects is identified; this dose is often termed the NOAEL. Alternatively, the lowest dose to cause adverse effects might be identified; this dose is often termed the LOAEL. The NOAEL and LOAEL from all the studies are then examined to identify the boundaries of effects.

Typically, the NOAEL or LOAEL that best identifies the boundary of the onset of adverse effects is selected for OEL derivation; this value is often called the **point of departure (POD)**. The POD is the point of the dose-response curve where extrapolation to estimate the safe dose begins. Thus, the procedure extrapolates from the POD to a lower dose that incorporates any remaining uncertainty. Additional methods that are more comprehensive than selecting of a NOAEL to evaluate the chemical dose response and set OELs have been developed (Wheeler et al. 2014).

In most cases, the POD will come from a study that does not directly identify the dose or concentration that has no effects in a representative population of sensitive workers. To compensate for this shortcoming, the typical

Table 5–2. UF Used in Deriving Occupational Exposure Limits

Factor	Area of Uncertainty	Basic Principle
UF_A	Animal to human	Adjusts for differences in sensitivity between animals and the average human, when the POD is based on animal studies
UF_H	Average human to sensitive human	Adjusts the POD for the difference between the average human and the most sensitive applicable worker subpopulation
UF_L	LOAEL to NOAEL	Adjusts for uncertainty in the value of the POD as an estimate of the threshold for the onset of effects, if based on a LOAEL rather than a NOAEL
UF_S	Short-term to long-term exposure	Adjusts for the possibility of identifying a lower POD for chronic toxicity when extrapolating from a study of shorter duration
UF_D	Database insufficiency	Adjusts for the possibility of identifying a lower POD (or more sensitive effect) if additional studies were available
M	Modifying factor	Adjusts for additional considerations, including difference in exposure routes or other toxicokinetic considerations

practice is to divide the POD by a series of factors that consider the possibility that the true safe dose is lower than that identified in the study used for POD selection. The net effect of applying uncertainty factors (UFs) is typically to lower the dose-response estimate (Figure 5–2). As with POD selection, applying uncertainty factors (which are sometimes called safety factors) is not standardized among organizations that derive OELs; and thus the factors that are used, their default values, and the levels of documentation are highly variable. Despite these differences in qualitative terms, the areas of uncertainty considered, and the approaches to applying a composite UF, a well-accepted risk assessment practice is employed by all the major OEL-setting groups. Areas of uncertainty that are traditionally considered are shown in Table 5–2 (Dankovic et al. 2014).

For each area of uncertainty, most organizations have a default value or range of default values to apply based on the nature of the POD used. When the POD addresses a key uncertainty, a factor of 1 would be used. For example, if the POD were derived from a long-term (chronic), lifetime-dosing study in animals, a factor of 1 would be applied as the UF_S in Table 5–2 because the POD already considers the impact of duration of dosing. Each individual UF is multiplied, and their product represents the composite UF, which becomes the denominator in the OEL equation shown in Figure 5–2.

> **Example**
>
> A Typical OEL Data Set Assessment:
>
> - Rats were exposed 6 hours/day, 5 days a week to 0-, 10-, 25-, or 50-ppm solvent for 2 years.
> - No effects were observed at 10 ppm, but signs of liver toxicity were observed at 25 ppm and above.
> - No significant effects data in humans are available.
> - No studies of reproductive effects are available.
> - The chemical has moderately acute toxicity but is not genotoxic or a sensitizer.
>
> An OEL might be derived as follows:
>
> OEL = (Point of Departure)/(Uncertainty Factors) = (10 ppm (NOAEL))/(3 (UF_A) × 3 (UF_H) × 1 (UF_L) × 1 (UF_S) × 3 (UF_D)) = 0.4
>
> OEL = 0.3 ppm (rounded) as a full-shift TWA with no hazard notations assigned

There are additional advanced methods for OEL derivation. In most cases, these methods extend the use of available chemical-specific data and provide better estimates of the chemical's dose-response behavior or address important areas of uncertainty. For example, rather than select single points of departure from existing study doses (i.e., the NOAEL), using statistical dose-response modeling to fit a nonlinear regression curve to the data is becoming standard practice. Physiological modeling approaches are improving the use of data on chemicals' dispositions in the body (toxicokinetics) to refine extrapolations from animals to humans and to better characterize humans' variabilities in chemical sensitivity. Physiological modeling also improves extrapolating from oral and dermal toxicity data to safe doses via the inhalation route. These methods and their use in OEL derivation are reviewed in Haber et al. (2012). The use of reliability and relevance considerations to develop a systematic OEL-selection process and an overview of the OEL-selection process are illustrated in Figure 5–3.

OEL USE AND INTERPRETATION

The OEL is one of the required inputs to support decisions about risk because the OEL considered in the context of the level of exposure enables the characterization of risk. In industrial hygiene practice, the most common approach to characterizing risk is to develop a **hazard quotient** (HQ). The HQ is the ratio of measured or estimated exposure to the OEL derived for a similar exposure

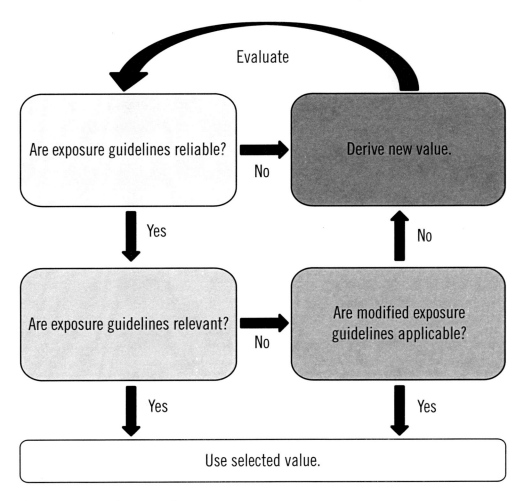

Figure 5–3. Process for OEL selection.

scenario. A value of unity indicates that the exposure estimate is equal to the OEL on the basis of this formula: HQ = exposure / OEL. Because exposure estimates and the OEL both have some degree of imprecision, it is essential to characterize the hazard quotient in qualitative terms—that is, the higher the HQ, the higher the likelihood for adverse effects. In practice, many organizations recommend that exposures above the OEL (HQ above a value of 1) generate management actions to reduce exposures. It is also common to use some fraction of the HQ as an **action level** to initiate controls. The distance from an HQ value of 1 that is selected as an action level depends on the confidence in the exposure measurement and the OEL as well as on the severity of the toxic end point. Thus, if the exposure was developed with a large number of samples and worst-case-scenario assumptions, and the critical adverse effect that served as the basis for the OEL was of minimal severity (e.g., minor eye irritation), then a risk manager might feel comfortable approaching an HQ value of 1 and not taking additional action. On the other hand, if the exposure was highly variable, the data for the OEL were limited, and the critical effect was severe (e.g., developmental toxicity), then an HQ significantly below the value of 1 might be used as an action level. Although there are no specific guidelines on interpreting the hazard quotient and selecting an action level, some OSHA substance-specific standards do include an **action limit** in the regulations to spur medical surveillance and other requirements.

Although the HQ concept is the most common risk characterization method used in occupational risk assessments, a related concept is the margin of safety. In this concept, the OEL and exposure limit are reversed, and the goal is a **high margin of safety** (i.e., the OEL is much greater than the level of exposure). In addition to the imprecision of the HQ and margin-of-safety approaches, a significant limitation to their use is the qualitative nature of the risk characterization. For example, the HQ approach cannot answer questions such as how the probability of an adverse effect changes as the HQ increases, or what the risk reduction would be if controls to reduce the hazard quotient from 1.0 to 0.5 are implemented.

Other, more advanced risk characterization techniques that evaluate risk probability are sometimes used. Of the several techniques available, the most common is to extrapolate from the dose-response curves to estimate the level of risk of a given dose. This method is the foundation of the linear low-dose extrapolation currently used in many risk assessments of carcinogens. Applying similar methods to assess noncancer effects and in occupational scenarios has also been proposed.

The OEL and resulting HQ should be used in the context of their degrees of precision and levels of accuracy. Something that has precision varies minimally from a defined standard. OELs are not definitive limits between safe and dangerous because they are not precise. Instead, OELs are to be used and evaluated in the context of the uncertainties in their derivations. Some key considerations include the following:

- OELs are derived using semiquantitative UF factors that often reflect order-of-magnitude differences in judgment.
- The variability in OELs reflects many parameters: data differences, method differences, and risk tolerance differences.
- After adjusting for risk tolerance and method differences, residual variability can indicate the strength of the data and identify data gaps. Different OEL values for the same chemical can all be accurate.

Users of an OEL also must consider its accuracy, or exact conformity to fact. OELs often are accurate in that they are estimates of a concentration that is safe, but they can be poor estimates of the actual boundary between effect and no effect because of the interplay between risk tolerance and acceptable distance from the effect-versus-no-effect boundary. An OEL may be viewed as the best estimate, the upper-bound estimate, or the lower-bound estimate of the **safe concentration,** depending on the organization or the OEL user. Thus, very different OEL values may all be protective and below the actual human dose-response threshold despite their differences in value.

The OEL and resulting HQ are important quantitative indications of risk. However, an effective **risk characterization** goes beyond the calculations and documents the basis of key decisions (e.g., POD and UF selection), the strengths and weaknesses of alternative approaches, and the key uncertainties that affect the interpretation of the risk. Therefore, reading the OEL documentation is essential to correctly using the OEL value to make risk management decisions.

OEL RESOURCES

Establishing quantitative air concentration limits to protect workers dates back to the 1920s. Since those initial efforts, OEL development has received significant interest from the global industrial hygiene community. (See a library of OEL links at www.tera.org/OARS.) Numerous organizations develop OELs and the detailed documentation that describes the basis of the OEL (see Table 5–1). The activities of these groups provide significant guidance to the industrial hygienist on assessing workplace risks, but all the activities can also lead to a confusing landscape of alternative OELs and resources on well-studied chemicals. On the other hand, no quantitative health-based exposure guidance is available for the majority of chemicals encountered in the workplace. Either extreme can present challenges, but strategies exist to address both situations.

For chemicals that have multiple potential OELs or other exposure guidelines that could be used as the basis of an OEL derivation, a systemic identification and selection process is needed. One element of such a process is developing a go-to list of organizations or websites from which to compile a list of candidate values. The resulting OELs can be evaluated for reliability based on considerations such as level of peer review involved and date of the last revision. A smaller set of values can then be evaluated for their relevance to the scenario or risk assessment of interest. This critical step will help ensure that an OEL relevant to the exposure situation is used. Important questions pertain to the population of concern, the nature of the effects that are used to define risk, and the temporal patterns of the exposure of interest.

If no OELs are available, alternative quantitative health-related benchmarks can be used for screening purposes or as substitutes of the OEL in risk management. Generally, if exposures are low and well controlled, a screening-level benchmark might be satisfactory to guide risk management decisions. As exposure approaches OELs, a refined assessment including derivation of more rigorous OELs might be warranted. This approach enables the industrial hygienist to employ a group of OEL-related tools to support quantitative risk characterizations (e.g., to develop an HQ). The use of such a suite of tools reflects the **Hierarchy of OELs** approach, which is presented in Figure 5–4 (Deveau et al. 2014; Laszcz-Davis et al. 2014). As more toxicological and epidemiological data become available, the industrial hygienist moves up the hierarchy.

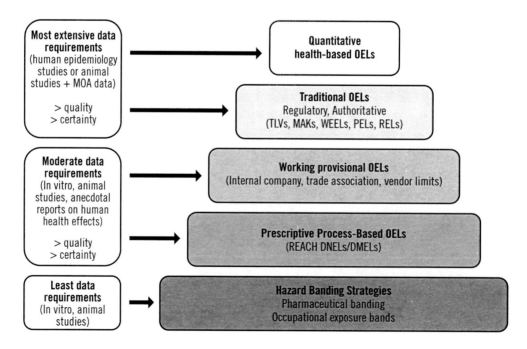

Figure 5–4. A hierarchy of risk-based occupational exposure benchmarks.

(Source: Adapted from a version of the hierarchy developed by Laszcz-Davis, Perkins, and Maier [2014] and reprinted from Deveau et al. [2014].)

SUMMARY

Key roles of industrial hygienists are assessing occupational exposures to hazardous materials and evaluating associated risks. Occupational exposure limits are derived to quantify the levels of risk associated with different chemicals exposures, which in turn helps the industrial hygienist provide appropriate controls and preventive measures to protect workers from chemical hazards.

REVIEW QUESTIONS

1. What are the components of the risk assessment and risk management framework of NAS?
2. What is the general purpose of an OEL?
3. The risk assessment phase includes several distinct activities that pertain to the development and application of the OEL. What are those activities?
4. In addition to its actual value, what are the other elements of an OEL?
5. Identify and describe the three types of OELs.
6. Identify the key steps to OEL derivation.
7. Identify and describe the basis of each of the six common uncertainty factors.
8. How does the industrial hygienist calculate the HQ?
9. True or False: The HQ concept is the most common risk characterization method employed in occupational risk assessments.
10. What concept can be used to provide a quantitative exposure benchmark when an OEL is not available?

REFERENCES

American Conference of Governmental Industrial Hygienists. *2014 TLVs® and BEIs®: Threshold Limit Values for Chemical Substances and Physical Agents and Biological Exposure Indices.* Cincinnati, OH: ACGIH, 2014.

Dankovic, D. A., B. D. Naumann, M. L. Dourson, M. A. Maier, and L. Levy. "The Scientific Basis for Uncertainty, Safety and Modifying Factors in OEL Setting." *Journal of Occupational and Environmental Hygiene* (2014).

Deveau, M., C.-P. Chen, G. Johanson, D. Krewski, A. Maier, K. J. Niven, S. Ripple et al. "The Global Landscape of Occupational Exposure Limits—

Implementation of Harmonization Principles to Guide Limit Selection." *Journal of Occupational and Environmental Hygiene* (2014).

Dotson, G. S., C. Chen, B. Gadagbui, A. Maier, H. W. Ahlers, and T. J. Lentz. "The Evolution of Skin Notations for Occupational Risk Assessment: A New NIOSH Strategy." *Regulatory Toxicology and Pharmacology* 61 (2011): 53–62.

Dotson, G. S., A. Rossner, A. Maier, and F. W. Boelter. "Risk Assessment's New ERA. Part 1: Challenges for Industrial Hygiene." *The Synergist* (April 2012): 24–26.

Haber, L. T., J. Strawson, A. Maier, I. M. Baskerville-Abraham, A. Parker, and M. Dourson. "Noncancer Risk Assessment: Principles and Practice in Environmental and Occupational Settings." In *Patty's Toxicology*. 6th ed. New York: John Wiley and Sons, 2012.

Laszcz-Davis, C., J. Perkins, and A. Maier. "The Hierarchy of OELs: a New Organizing Principle for Occupational Risk Assessment." *The Synergist* (March 2014): 27–30.

National Institute for Occupational Safety and Health. *Current Intelligence Bulletin 61: A Strategy for Assigning New NIOSH Skin Notation. DHHS (NIOSH) Publication No. 2009-147.* Atlanta: NIOSH, 2009.

National Research Council. *Science and Decisions: Advancing Risk Assessment.* Washington, DC: National Academy Press, 2009.

Paustenbach, D. J. "Pharmacokinetics and Unusual Work Schedules." *Patty's Industrial Hygiene* (2011): 957–1046.

Waters, M., L. McKernan, A. Maier, M. Jayjock, L. Brosseau, and V. Schaeffer. "Occupational Risk Probability and Interpretation of Traditional Occupational Exposure Limits: Enhanced Information for the Occupational Risk Manager." *Journal of Occupational and Environmental Hygiene* (2014).

Wheeler, M. W., R. Park, A. J. Bailer, and C. Whittaker. "Historical Context and Recent Advances in Exposure-Response Estimation for Deriving Occupational Exposure Limits." *Journal of Occupational and Environmental Hygiene* (2014).

6
Gases and Vapors

LEARNING OBJECTIVES

After completing this chapter, readers should be able to do the following:
- Describe the various properties of the chemicals that are most important to the practice of industrial hygiene.
- Calculate the concentration of a chemical in the air when provided with the chemical's basic parameters.
- Describe the basic behaviors of gases and vapors in scientific and mathematic terms.
- Identify various methods of collecting and analyzing gas and vapor concentrations in the air.
- Identify the handheld and portable detectors used for direct measurement of gases and vapors in the environment.

Photo credit: U.S. Navy

INTRODUCTION

The ability to accurately anticipate and evaluate hazardous gases and vapors in the air is a core responsibility of an industrial hygienist. Therefore, an industrial hygienist must have a basic understanding of chemistry and be able to apply that understanding in workplace situations. Knowing a chemical's quantity and location will help an industrial hygienist determine the chemical's behavior in the workplace.

PROPERTIES OF CHEMICALS

Normal temperature and pressure (**NTP**) used for gas and vapor calculations are 25°C and 760 mmHg. At NTP, 1 mole of any gas will occupy 24.45 L.

The number of moles of any chemical is the number of grams of the material divided by its molecular weight.

> **Example**
>
> If mercury (Hg) has a molecular weight of 201, how many grams are in 2 moles of Hg?
>
> **Solution:**
> grams = moles × molecular weight
>   = 2 moles × 201 = 402 g

The concentration of gases and vapors in air is typically expressed in terms of parts of contaminant per million parts of air.

$$\text{ppm} = \frac{\text{parts of contaminant}}{\text{million parts of air}}$$

For a gas or vapor of a certain mass at NTP, the concentration in air can be calculated in a given volume using the following equation:

$$\text{ppm} = \frac{(\text{mg/m}^3)(24.45 \text{ L})}{(\text{molecular weight})}$$

> **Example**
>
> If you spill 60 mg of naphthalene (molecular weight = 128) in a room that measures 3 m × 4 m × 4 m, what will be the concentration in ppm after all the liquid has evaporated?
>
> **Solution:**
> V = 3 m × 4 m × 4 m = 48 m³
> ppm = (60 mg)(48 m³)/128 = 22.5 ppm

Occupational exposures to airborne chemicals typically take place in the presence of an open container or a spill when the liquid evaporates. The rate of evaporation, and thus the amount of liquid that becomes concentrated in the air, depend on characteristics such as vapor density and vapor pressure.

For a given volume (such as that of an enclosed room), it is also possible to calculate the ratio of the volume of the contaminant to the volume of the room:

$$\text{ppm} = \frac{V_{contaminant}}{V_{air}} \times 10^6$$

> **Example**
>
> What is the vapor concentration if 1 gram-mole of acetone is spilled in a room that is 8 ft × 11 ft × 15 ft?
>
> (At NTP, 1 mole of any vapor occupies 24.45 L. One cubic foot equals 28.32 L.)
>
> **Solution:**
> V = 8 × 11 × 15 = 1,320 ft³
>
> $$\text{ppm} = \frac{24.45 \text{ L}}{1,320 \text{ ft}^3 (28.32 \text{ L/ft}^3)} \times 10^6$$
>
> ppm = 654 ppm

Vapor pressure is one important indicator of a chemical's ability to evaporate. Vapor pressure is the amount of pressure that a liquid exerts on the inside of a closed container above the surface of the liquid. At equal temperatures, chemicals with higher vapor pressures tend to evaporate more quickly than chemicals with lower vapor pressures. In general, chemicals will evaporate more quickly at higher temperatures.

Vapor density is the measure of how heavy the vapor is in air. Chemicals with higher vapor densities tend to settle near the floor, whereas those with lower vapor densities tend to rise to the ceiling. It is important to consider vapor density when deciding where to take air samples in the workplace. If all of the vapor is near the ceiling but the sample is taken near the floor, the measurements will underestimate the workers' exposure.

It is also important to note that vapors with high densities can displace oxygen from the workers' breathing zones, which can lead to physical/simple asphyxiation.

Density is the ratio of mass to volume. **Specific gravity** (SG) is the mass of a substance compared to the mass of an equal volume of water. Water has an SG of 1.0 gram per milliliter or centimeter cubed. A chemical with an SG of less than 1 g/mL will float on the surface of water. A chemical with a higher SG will sink to the bottom of the container and form a layer beneath the water layer.

Example

If acetone has an SG of 0.79, how many mL are in 1.0 g of acetone?

Solution:

$$SG = \text{mass of material (g)/volume of material (mL)}$$
$$0.79 = 1.0 \text{ (g)}/x \text{ (mL)}$$
$$0.79x = 1.0 \text{ (g)}$$
$$x = 1.3 \text{ mL}$$

SG is important to industrial hygiene because it helps determine the amount of chemical that is released into a given volume of air. For example, for a chemical with an SG of 0.5g/mL, half the amount of the chemical is in the same volume of water. If the chemical is less dense than water, less of the chemical will be able to evaporate into and fill the air space. This fact must always be considered when calculating an evaporated chemical's concentration in the air.

Boiling point is relevant to industrial hygiene applications because chemicals with boiling points at NTP evaporate quickly and are generally considered gases. Boiling points are affected by atmospheric pressure and tend to decrease with lower pressure and higher altitude.

The **flash point** is the lowest temperature at which a material gives off enough vapors to form an ignitable mixture. The **fire point** is the temperature at which the ignitable mixture will continue to burn. The **auto-ignition temperature** is the point at which a material will ignite without an external source of ignition.

Molecular weight is useful in industrial hygiene because it indicates a lot about how a chemical moves and interacts with air or other gases. Chemicals with high molecular weights are larger and move more slowly than chemicals with low molecular weights.

BEHAVIORS OF GASES

Gases move very quickly through the air because of **Brownian motion**. Brownian motion (Figure 6–1) is a measure of how much a gas or vapor molecule moves in the air. Depending on a room's conditions and on each molecule's size and shape, the continuous ricocheting of molecules bouncing off one another allows them to spread out in space as quickly as a couple meters per second.

Another concept that can affect gas and vapor movement is **diffusion** (Figure 6–2). Gases and vapors tend to spread out and move from locations of high density to those of low density. This movement causes them to become equally distributed in a space such as a worker's breathing zone.

GAS LAWS

In order to determine how to conduct appropriate air sampling and properly assess workers' airborne exposures, industrial hygienists must understand how gases behave. Following are some of the basic laws that govern how gases behave.

Ideal Gas Law

The ideal gas law is the basis of the relationships among pressure, volume, and temperature and the number of moles of the chemical being considered. When the amount of the chemical is known and it is entirely evaporated, use the following equation to calculate the volume that the chemical will occupy at a given temperature and pressure:

$$PV = nRT,$$

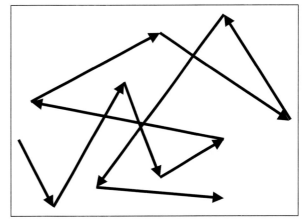

Figure 6–1. Brownian motion of a particle in space

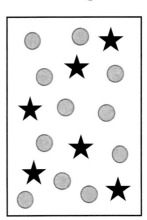

Figure 6–2. Diffusion of a gas in an enclosed space over time

where P = atm, n = moles, R is a constant (0.0821), T is in units of Kelvin (degrees centigrade plus 273), and V is in cubic meters.

Example

If 3 moles of benzene evaporate at an ambient pressure of 760 mmHg and a temperature of 23°C, what volume of space would the benzene fill?

Solution:
PV = nRT
P = 760 mmHg
n = 3 moles
T = 23°C + 273 = 296°K
V = nRT/P
　= 3 moles × 0.0821 × 296°K/760 mmHg
　= 0.096

Generalized Gas Law

In most industrial hygiene applications, the generalized gas law can be used to quickly determine the pressure, volume, or temperature of a system. The relationships among temperature, volume, and pressure are given by the following equation:

$$\frac{P_1 V_1}{T_1} = \frac{P_2 V_2}{T_2}$$

This straightforward equation demonstrates that when the temperature in a closed vessel increases, the gases and vapors inside the vessel expand, causing the volume to increase like a balloon, or if they are constricted and cannot expand, the pressure inside the vessel increases.

Example

If the volume of a balloon of nitrogen is 3.5 L at a room temperature of 24°C and a pressure of 760 mmHg, what is the volume of the balloon at the top of a mountain where the pressure is 700 mmHg and the temperature is 18°C?

Solution:
$P_1 V_1 / T_1 = P_2 V_2 / T_2$
　$V_2 = P_1 V_1 T_2 / T_1 P_2$

　　$V_1 = 3.5$ L
　　$P_1 = 760$ mmHg
　　$P_2 = 700$ mmHg
　　$T_1 = 22°C + 273 = 295°K$
　　$T_2 = 18°C + 273 = 291°K$

　　$V_2 = 760$ mmHg × 3.5 L × 291°K/295°K × 700 mmHg
　　　≈ 3.8 L

If the pressure in two different conditions remains the same and only the volume and temperature change, then the generalized gas law can be simplified, or reduced, to Charles' Law with the following equation:

$$\frac{V_1}{T_1} = \frac{V_2}{T_2}$$

If the volume and pressure in two different conditions change but the temperature remains the same, then Boyle's Law can be used to calculate the new values:

$$P_1 V_1 = P_2 V_2$$

CALCULATING A GAS OR VAPOR CONCENTRATION IN AIR

With an understanding of the information covered in this chapter so far, the industrial hygienist can calculate the concentration of an airborne chemical. For example, if a worker spills 12 mL of acetone and it completely evaporates into a room that is 3 m × 4 m × 3 m = 36 m^3, what is the airborne concentration of the acetone in the room?

Because the SG of acetone is 0.79 g/mL, 1 mL of acetone equals 0.79 g at NTP. So if 12 mL are spilled, 12 × 0.79 g = 9.48 g. Thus, there are now 9.48 grams of acetone in the room, or 9.48 g/36 m^3 = 0.26 g/m^3.

COLLECTING GASES AND VAPORS IN AIR

In addition to calculating the amount of gases or vapors that have evaporated into the air, the industrial hygienist can collect a sample of the air to measure the chemicals directly. There are four important reasons to collect air samples rather than just depend on calculations:
- to obtain an accurate representation of air concentrations
- to identify leaks or releases
- to evaluate the effectiveness of controls
- to assess exposures during particular work processes or activities.

Air is sampled using an air pump that sucks air through tubes and a filtering device that captures the contaminants. What is captured can then be measured and compared with the amount of air drawn through the filter to calculate the concentration of contaminants in the room.

Using the acetone scenario mentioned previously, if a pump is set up to suck air through the filter at a rate of 1.5 L per minute and the pump runs for 480 minutes (an 8-hour workday), 720 L of air will be drawn through the collection filter.

Because there are 1,000 L in a cubic meter, the 720 L of collected air converts to 0.72 m³. Using the room concentration of 0.26 g/m³, the concentration in the collected sample is multiplied by 0.72 m³ to get 0.19 g.

> ### Example
> If an air sample in a room is collected at a rate of 1.2 L per minute for 4 hours, and there is 0.3 g of turpentine (SG of turpentine = 0.87) in the room, what is the actual concentration of turpentine?
>
> **Solution:**
> T = 4 h = 4 h × 60 min/h = 240 min
>
> Sample volume collected = 1.2 l/min × 240 min = 288 L collected
>
> 0.3 g/288 L = 0.001 g/L = 0.001 g/L × 1,000 L/m³ = 1.0 g/m³ as an air concentration in the room where the sample was drawn, independent of the room size

LIMITS OF DETECTION

The limit of detection (LOD) is the point at which the measurement of an agent first becomes possible. The LOD is determined through statistical measures of samples ranging from blank (negative) samples in progressively increasing concentrations to samples in which the agent is detected with some consistency. Similarly, the limit of quantitation (LOQ) is the concentration at which quantitative results can be measured with a high degree of confidence.

For any type of gas or vapor sampling and analytical method, the LOD and LOQ need to be known and compared with the concentration levels of interest. These levels may be of the permissible exposures, the hazardous exposures, or the odor thresholds. Whatever the level of interest, the industrial hygienist needs to be sure that the system of sample collection and analysis is capable of measuring that level.

ERRORS IN SAMPLING AND ANALYSIS

Any sample result is subject to errors. These errors could be in the flow of the pump or in laboratory methods. The total of errors for a collected sample is called the sampling and analytical error (SAE), which can be estimated using the following equation:

$$SAE = \sqrt{(\text{pump error})^2 + (\text{lab error})^2}$$

SAE is typically around 10%.

Example

If the pump error in a collected sample is 2% and the laboratory error is 8%, what is the SAE for the sample?

Solution:
SAE = $\sqrt{(\text{pump error})^2 + (\text{lab error})^2}$

Pump error = 2

Lab error = 8

SAE = $\sqrt{(2)^2 + (8)^2}$ = 8.2%

Accuracy indicates how close a measurement is to the so-called real value.

Precision indicates the reliability or reproducibility of a measurement. These relationships are illustrated in Figure 6–3.

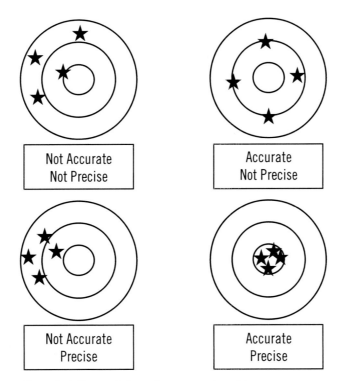

Figure 6–3. Representations of accuracy and precision.

COLLECTION THEORIES, METHODS, AND MEDIA

SAMPLING DURATION

Air samples might be taken at different times of the day to obtain information about airborne concentrations during different worker tasks or operational processes. Samples may also be taken over different lengths of time, depending on what is being measured. For instance, a sample might be taken over an 8-hour period to obtain an average air concentration for an entire workday. Or if work processes and the chemicals used do not change during the day, a 4-hour sampling duration may provide enough information about worker exposure for an entire shift.

Depending on the gas or vapor concentration, it may be necessary to extend the sampling duration or increase the flow rate of the sampling pump to ensure there is enough analyte collected to achieve the necessary LOD and LOQ for analysis. The relationship can be shown by the following equation:

$$\frac{\text{Minimum Required}}{\text{Concentration (mg/m}^3)} = \frac{\text{LOD (mg)}}{\text{Volume Sampled (L)} \times (1\text{m}^3/1{,}000\text{ (L)})}$$

The total quantity of air sampled (Q) is the product of the pump flow rate and the sampling time duration, which is demonstrated in the following equation:

$$Q\text{ (L)} = R\text{ (L/min)} \times T\text{ (min)}$$

Example

If the LOD for the sampling and analysis method is 1.0 mg and the sampling is conducted at a rate of 1.3 L/min for an 8-hour duration, what would be the minimum concentration of the air contaminate required to be detected?

T = 8 hour = 60 min/h × 8 h = 480 min

Q = 1.3 L/min × 480 min = 624 L collected = 0.624 m³

Minimum required concentration = 1.0 mg/0.624 m³ = 1.6 mg/m³

GRAB SAMPLING

Grab samples are taken at a precise moment in time. The grab, which might last only a minute or as long as an hour, is a very accurate measure of the air concentrations during a specific task or process. A grab sample can be useful when it is assumed that the ambient air concentration is constant; the sample is thus a representation of what is in the air space.

Grab Sampling Methods

Summa canisters are sealed and have a vacuum inside. When a canister valve is opened, the canister draws in air from the surrounding space. Once equal pressure inside the canister is reached, the valve is closed and the canister is sent off to a laboratory for analysis. Although Summa canisters are often made of stainless steel, their liners can be composed of materials such as ceramics or glass. Sizes range from as small as one liter up to several liters.

Grab samples can also be collected in a tedlar (plastic) bag and sent to a lab for analysis. In these types of systems, the bag, which has a tube sticking out of it, is placed into an air-proof box, with the plastic tube sticking out of a hole in the side of the box. Another tube inserted through the side of the box is used to suck air out of the closed box. In doing so, the bag inside the box expands because of the suction created in the box, and as the bag expands, it sucks in air through its tube. Once the bag is full of air, the bag's tube is sealed, the box is opened, and the full bag is sent to a lab for analysis. This type of bag is shown in Figure 6–4.

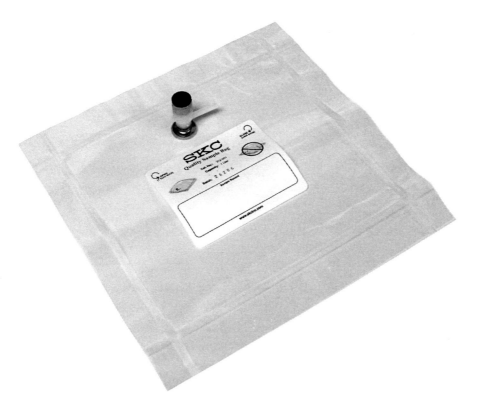

Figure 6–4. Tedlar bag system for collection of gases and vapors.

(Source: SKC Inc.)

DETECTOR TUBES, COLORMETRIC INDICATOR TUBES, AND PUMPS

Detector tubes are the third type of grab sample collection devices. These tubes use a hand pump or bellows to pull air through a small tube (Figure 6–5) that has a series of prefilters and different types of collection media. The level of the chemical agent's ppm concentration in the air will be indicated by a color change in the collection media inside the tube. The indication on the tube thus looks somewhat like the speedometer on a car, with the color change proceeding farther up the tube the more contaminant there is in the air.

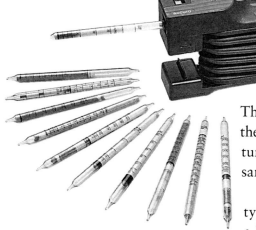

Figure 6–5. Detector tubes and hand pump.

(Source: SKC Inc.)

Detector tubes are made by several different manufacturers, each of which offers different tubes for different chemicals. For example, Draeger Safety Inc. has tubes available for more than 1,000 different chemicals. The amount of air collected and the methods of reading the detector tubes differ among different tube manufacturers and even among different types of tubes from the same manufacturer.

The devices that draw air through the glass tubes typically have a hand pump with either a plunger or a bellows-type device that sucks air into the tube and through the collection medium.

The advantages of detector tubes are that they are relatively easy to use and can indicate the air concentrations of a variety of chemicals. The tubes and collection devices are also relatively inexpensive.

Their disadvantages are that they are typically only accurate to about ±25% of the actual concentration and that each tube can collect and analyze only one chemical or a couple of chemicals. In addition, although more than 1,000 different chemicals can be analyzed, a large number of hazardous chemicals cannot be analyzed using these tubes.

Only qualified industrial hygiene professionals should use these types of devices. Although they might seem easy to use, understanding how gases and vapors behave in the workplace is necessary to accurately draw samples and interpret the results. The LOD, competing chemicals, and other limiting factors must also be understood before using these devices.

CONTINUOUS SAMPLING

Continuous sampling typically involves drawing a sample of air through a collection device over an extended period of time. The air is drawn through a collection filter or tube containing a collection medium that lets the air pass through and at the same time filters the gas or vapor out onto the reagent.

Sample collection tubes may work by adsorption, whereby the contaminant is collected on the surface of the tube. Activated charcoal works well as an adsorption medium because of the large surface area of all the small passageways through the charcoal. Other collecting reagents work by absorbing the contaminant directly.

After a collection period, the collection tube is detached from the sampling train, and the ends of the tube are capped so that the gas or vapor cannot evaporate or escape. The tubes are then sent to a laboratory, where samples are extracted from the collection medium and analyzed.

Because efficiency depends on the volume of air sampled, the type and amount of the absorbing or adsorbing material, and the volatility of the contaminant, agents used in gas and vapor sampling are never 100% efficient.

Activated charcoal adsorbent is created by heating normal charcoal to temperatures in excess of 800°C. This adsorbent is effective and commonly used for most organic vapors. In the laboratory, agents are fairly easy to extract from activated charcoal using carbon disulfide.

AIR-SAMPLING INSTRUMENTATION

A basic sampling train for collecting air samples operates similarly to a pump that draws air through a fish aquarium. A basic pump connects to a tube and sucks air through a filter attached to the end of a collection tube.

The pump can be set to collect air through the tube and filter at any rate desired, usually around 1.0 L to 1.5 L of air per minute. The filter or sample collection device is always on the front end of the system because the pump needs to suck the air through that device.

Collection devices or filters can consist of many designs and materials and vary depending on the specific characteristics of the gases and vapors. Because gases and vapors can be difficult to catch or pull out of an air stream passing through a filter, filters are not 100% efficient, and a lot of the air stream that is pulled through the filter is not collected.

Activated charcoal is often used as a gas or vapor collection filter medium. This type of charcoal has millions of nooks and crannies that vapors get caught in when they adsorb onto the charcoal's surface.

When contaminants are not easily adsorbed onto activated charcoal, inorganic sorbents can be put on the surface of silica gel beads that are packed into a collection tube. As the contaminant of interest is drawn through the tube, the large surface area of all the packed beads in combination with the attractive sorbents effectively absorb the gases and vapors at they pass through the tube.

These sorbent-covered beads are also good at collecting polar contaminants, including amines and some inorganic chemicals. One disadvantage of these beads is that they also easily absorb water molecules, which then take up space on the sorbent and reduce the efficiency of the collection process.

When gases or vapors are highly reactive and unstable, it can be difficult to get them to stick to any absorbing medium. In these cases, sorbents with chemical characteristics that specifically attract the highly reactive contaminants can be developed for use in the sorbent tubes. The only remaining problem is getting the contaminant out of the sorbent for analysis in the laboratory.

PERSONAL SAMPLING

It is often useful and necessary to take air samples that represent the workers' exposures throughout the day or at different periods. Air-sampling devices can be attached to the workers. The pump and battery pack are typically attached to the belt, and a collection device is placed near the worker's breathing zone around the face. A tube connects the sampling device to the pump. The sampler itself is specially designed to collect the agent of interest and may have activated charcoal or gel-covered beads. Other personal samplers and filters can be used to collect aerosols such as dust or fibers.

The pump is calibrated before and after each use and is typically set at a flow rate between 1.0 and 2.5 L per minute.

REAL-TIME DETECTION INSTRUMENTS

A wide variety of direct-reading instruments has been developed to provide instantaneous readings of gases and vapors in the work environment. These instruments combine air sampling and analysis to provide an instantaneous indication of contaminant concentrations. Some devices have data-logging systems that record data over longer periods.

One advantage of these devices is that they quickly indicate the presence of potentially dangerous levels of gases and vapors. They are also able to point out variations in effluents that might result from different operational processes or work activities. Because the instruments can be used to pinpoint leaks in pipes or ducts, timely repairs can be made. These devices can be attached to system or area alarms and should be placed in work areas to identify leaks that may approach hazardous levels.

Direct-reading devices are often used as indicators of where to conduct extensive workplace air monitoring, which uses more specific methods or for longer durations. Readings from these devices can also be used to determine which workers need personal dosimetry to analyze their exposures in more detail. Direct-reading instruments can also be used to confirm air samples collected by other methods.

Some direct-reading instruments, such as indicator tubes, can be simple to use and provide workers with personalized information in real time; however, they may identify the concentrations of only one or two chemicals. Personal dosimetry devices usually are equipped with alarms that warn workers to leave an area or don respiratory protective equipment.

Direct-reading devices such as multigas analyzers are extremely complex systems that can differentiate and quantify numerous unknown chemicals in the air at the same time. Other devices can provide nonspecific information about a group of chemicals.

Many real-time instruments are lightweight and portable. They can easily be carried into workspaces, including confined spaces. Some of the more sophisticated instruments are too heavy to be considered handheld, but they are still transportable, such as by cart.

Case: EtO in Hospital Sterilization Center

Not long ago, hospitals used large cylinders of compressed ethylene oxide (EtO) to sterilize surgical instruments. The containers were stored in rooms adjacent to the sterilizer room, rolled in when needed, and connected to the sterilizer. This process was quite hazardous because of the size and weight of the containers and the volume of gas that could escape if there was a leak.

EtO is extremely toxic, and at 800 ppm it is immediately dangerous to one's life and health. It is also a carcinogen, embryotoxin, and teratogen. In addition, EtO is highly flammable and self-reactive.

The Occupational Safety and Health Administration (OSHA) has a permissible exposure limit (PEL) of 1.0 ppm for EtO. The OSHA excursion level for EtO is 5 ppm, and the action level is 0.5 ppm.

Most hospital sterilization systems now use specially designed containers that are slightly smaller than a can of hairspray and contain about 100 g of EtO under pressure. When the canisters are placed into the specially designed sterilizer, they are punctured and the gas is released into the instrument space. The high concentration of the gas can kill all of the microorganisms on the surgical devices in about 30 minutes.

Various air monitoring instruments are used to determine the EtO levels in the room with the sterilizer. Because EtO is heavier than air, any leaked EtO will accumulate on the floor. If there is a spark or other source of ignition, the EtO concentration could burst into flames. Therefore, at least one area detector should be located near the floor to monitor for leaks from the system. Another detector is typically located in the breathing zone of a worker inside the EtO sterilizer room. Sometimes a third area detector is located in the room adjacent to the EtO sterilization room to monitor whether toxic gases are escaping or building up. Handheld detectors may also be used to monitoring for leaks.

COMBUSTIBLE-GAS/MULTIPLE-GAS MONITORS

Combustible-gas monitors are relatively simple, handheld devices that can quickly and accurately detect oxygen levels and three to five additional gases or vapors. For example, the Dräger X-am 5000 personal gas detector measures combustible gases and vapors, O_2, and up to three toxic gases including CO, H_2S, CO_2, Cl_2, HCN, NH_3, NO_2, NO, PH_3, SO_2, organic vapors, and ozone. The detector is often lightweight and useful for personal monitoring in hazardous work areas, and has a digital readout and sometimes audible alarm set points. These devices are able to withstand abrasive environmental conditions, shock, and vibration.

THERMOCHEMICAL DETECTORS

Thermochemical detectors are the largest class of direct-reading instruments used to detect the presence of a wide range of airborne gases and vapors. There are two different types of measurement methods.

In the first method, which involves **heat of combustion**, the sample gas or vapor enters the detector sample cell, comes into contact with a heated source, and ignites. The resulting heat changes the resistance of the filament, which in turn changes the current of the detector in proportion to the concentration of the contaminant in the sampled air.

The second method, which involves **thermochemical heat**, employs heated filament or oxidation catalysts. Thermal conductivity detectors work because of their ability to conduct heat away from the hot source contained within the detector. They measure resistance or temperature changes using thermocouples. This type of detector should be used with care near flammable vapors or gases.

Because thermochemical detectors are nonspecific, they cannot identify the chemical or vapor being measured. But more specific heat-of-combustion detectors that measure carbon monoxide, ethylene oxide, hydrogen sulfide, methane, and oxygen deficiency are available. Most of these detectors give measurements in terms of percent of lower explosive limit or ppm.

ELECTROCHEMICAL DETECTORS

Electrochemical sensors consist of a coarse filter, a cell with two electrodes and an electrolyte (a liquid, gel, or solid matrix), and a porous membrane. (See Figure 6–6.) Gas diffuses across the membrane and reacts with the electrolyte. This reaction produces ions that diffuse across the electrolyte to the counting electrode. The concentration of the gas is proportional to the increase in current in the electrode.

Like most handheld detectors, these sensors must be calibrated before and after each use, at the same altitude and ambient temperature, to ensure accuracy. This calibration is particularly important for oxygen sensors.

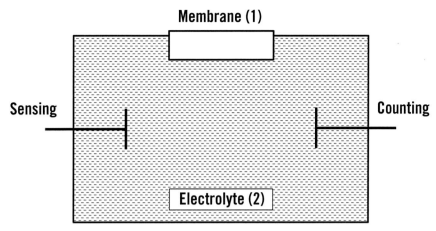

Figure 6–6. A schematic for a typical electrochemical sensor

Electrochemical sensors are available for approximately 50 different gases and vapors. These sensors tend to have low power requirements and can operate for up to four months without battery replacement.

However, there are also drawbacks to using these devices. For example, chemicals can interfere with accurately sensing the agents of interest, membranes can become clogged, the electrolytes can have short shelf lives, and the sensors can be hazards themselves because some contain corrosive electrolytes. In addition, some devices need new sensors as frequently as every six months, making them fairly expensive to maintain.

COULOMETRIC DETECTORS

Another electrochemical detector is the coulometric direct-reading instrument, such as the Conductimetric DRI, which measures the conductivity of ions in proportion to the gas concentration of the sample agent and is based on the electrolysis of a substance. Such detectors use a gold film mercury vapor to measure oxygen, carbon monoxide, oxides of nitrogen, chlorine, hydrogen sulfide, corrosive gases, ammonia, and sulfur dioxide.

IONIZATION DETECTORS

Ionization detectors rely on the energy source of the device to ionize the sample agent and then correlate the ionization to the agent's concentration in the air. (See Figure 6–7.)

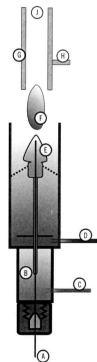

Figure 6–7. Schematic of a flame ionization detector. (A) GC column exit; (B) detector oven; (C) hydrogen fuel enters; (D) oxidant enters; (E) positive bias voltage; (F) flame; (G) collector plates; (H) signal transmitter; (J) exhaust port.

(Source: Mattj63 (Own work) [Public domain], via Wikimedia Commons)

In flame ionization detectors, a hydrogen flame burns the sample agent to produce ions. Organic compounds produce positive ions, which are then collected in an anode, and the detector measures the current produced by these ions. Although these devices cannot differentiate among agents, they can be used for most organic compounds.

Advantages of flame ionization detectors are that they are very sensitive and can detect a wide range, down to nanograms, of organic compounds. Their disadvantages are that they need a hydrogen supply to operate and that they need to be located in a secure environment to ensure safe operation of the flame. They also cannot be used to detect materials in the presence of large amounts of water vapor.

Photo ionization detectors use ultraviolet light to ionize aliphatic, aromatic, and halogenated hydrocarbons and other organic compounds. (See Figure 6–8.) They can also detect some inorganic chemicals such as arsine, phosphine, and hydrogen sulfide.

Electron capture detectors use a radioactive source to ionize agent samples. These detectors are useful with halogenated organic compounds, nitrates, pesticides, and some organo-metallic materials. For more information on ionization, see Chapter 12, Radiation.

Figure 6–8. The MiniRAE 3000 PID can detect hundreds of different volatile organic compounds, with concentrations ranging from 0 to 15,000 ppm, with a response time of 3 seconds.

(Source: SKC Inc.)

SPECTROCHEMICAL DIRECT-READING INSTRUMENTS

The measurements of spectrochemical instruments are based on the principles of Beer's Law. They use infrared (IR), ultraviolet, or visible light and photometers to measure the transmittance of light through the sample chamber. The greater the concentration of the agent in the air or sample chamber, the greater the absorbance of the light passing through. The relationship between absorbance and transmission in Beer's Law is represented by the following equation:

$$A = 2 - \log \%T$$

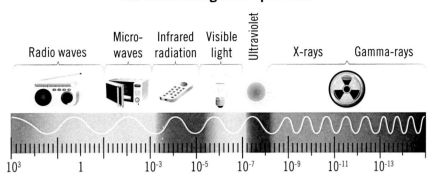

Figure 6–9. Infrared portion of the electromagnetic spectrum.

(Source: ttsz/iStock)

In IR spectrophotometers, the IR radiation is effectively absorbed by organic compounds that absorb energy in the IR wavelength range. (See Figure 6–9.) Meters contain an IR source, wavelength selector, sample cell, optics and mirrors, a detector, and a processor for analysis and readout.

IR spectrophotometry methods are useful for a wide variety of agents because nearly all molecules absorb IR radiation. For example, these methods can be used for general organics and hydrocarbons and for specific chemicals including methane, ethylene, ethane, propane, carbon monoxide, carbon dioxide, and Freon. However, IR spectrophotometry methods are not very good at differentiating among agents or characterizing a mixture because the spectrophotometric peaks may overlap. Also, because water strongly absorbs IR, care must be taken to avoid carrying out measurements at or near these absorbances.

The absorbance in these detectors is related to the molar absorptivity of the agent, the path length of the IR in the detector, and the concentration of the agent. This relationship is demonstrated by the following equations:

$$A = ebc, \text{ where}$$

A = absorbance
e = molar absorptivity
b = path length
c = concentration

$$I/I_0 = e^{-\mu CL}, \text{ where}$$

I = intensity of IR light with molecular absorption

I_0 = intensity of IR light without molecular absorption
μ = molecular absorption coefficient of the chemical
C = concentration of the chemical
L = path length of the IR radiation through the chemical

Some IR sampling techniques use prisms or gratings to separate radiation into component wavelengths to obtain a complete spectrum for a qualitative identification of agents. (See Figure 6–10.) Other nondispersive techniques use filter photometers to identify specific chemicals. Slit width methods define the window of energy used by the detector to identify different samples.

Ultraviolet and visible-light photometers operate on the principle of absorption of radiation. Ultraviolet devices operate from 10 to 35 nm; visible-light devices operate from 350 to 770 nm. These photometers measure the intensity of the light getting through the sample chamber and compare that intensity to a reference quantity. These devices can detect light intensity in the ppm range. (See Figure 6–11.)

Chemicals commonly analyzed by ultraviolet photometers are ammonia, mercury, oxides of nitrogen, sulfur dioxide, and ozone. These devices can analyze a wide variety of substances and can detect chemicals at very low levels.

PORTABLE GAS CHROMATOGRAPHY

In recent years, portable gas chromatographs have been developed that provide detailed separation and quantification of a wide range of chemicals. Substances are carried to the analytical chamber via an inert gas, and their retention times are analyzed via flame ionization, photoionization, thermal conductivity, or electron capture methods. Portable gas chromatographs are good at measuring volatile organic compounds.

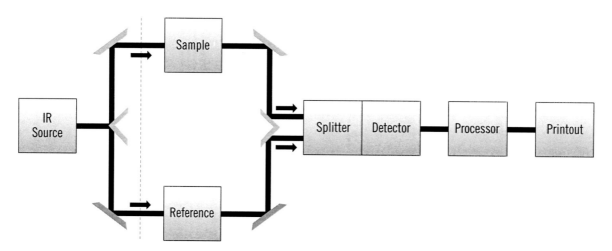

Figure 6–10. Schematic of an IR detection system.

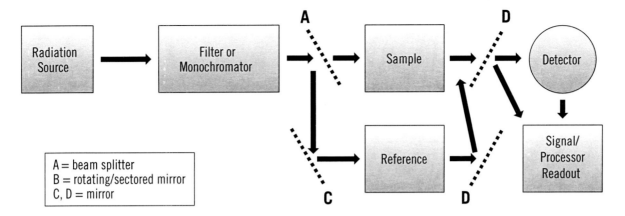

Figure 6–11. Schematic of a visible-light system.

SUMMARY

A significant portion of many industrial hygienists' careers involves analyzing worker exposures to gases and vapors. Therefore, the ability to use basic physical and chemical principles about the behaviors of gases and vapors when monitoring and controlling those materials is essential. In order to make safe and accurate decisions in the workplace, the industrial hygienist must also be familiar with the processes and instruments used for calculating the concentrations of gases and vapors in the air.

REVIEW QUESTIONS

1. If acetone has an SG of 0.79, how many mL are in 2.0 g of acetone?
    a. 2.53 (mL)
    b. 5.0 (mg)
    c. 3.9 (mL)
    d. 7.5 (mg/L)
2. What is the term for the temperature at which a fire can be maintained?
3. Fill in the blanks in the table below for ppm to mg/m³ conversions (assume NTP):

Chemical	Molecular Weight	ppm	mg/m3
a. methanol	32	1,000	_____
b. naphthalene	128	_____	10.0
c. phenol	94	10	_____
d. perchloroethylene	166	_____	100

4. What is the name of the direct-reading meter that uses an ultraviolet lamp to ionize organic compounds to detect and quantify them?
   a. Passive dosimeter
   b. Photoionization detector
   c. Flame ionization detector
   d. Electron capture detector
5. If you sampled for 4 hours and obtained a sample volume of 960 L, what was the flow rate?
   a. 2.5 L per min
   b. 8.5 L
   c. 4.0 L per min
   d. 2.5 (mg/h)
6. A tank car with a tight-fitting lid is reported to contain pure acrylonitrile monomer. You have been asked to verify the tank car's contents. Estimate the concentration of ethyl alcohol in ppm in the air space above the liquid. Acrylonitrile monomer has a vapor pressure of 83 mmHg.
7. According to a NIOSH analytical method, the minimum level of detection (LOD) for a chemical is 50 µg, the recommended exposure limit for this chemical is 1.0 mg/m³, and the fraction is set at 0.1. How many liters of air must be collected to obtain the minimum sample volume?
   a. 25.5 (mL)
   b. 300 (L)
   c. 85 (L)
   d. 500 (L)
8. What are five useful features of direct-reading industrial hygiene air monitoring instruments?
9. What is the principle of operation of electrochemical sensing portable instruments?
10. What is the principle of operation of a flame ionization detector?
11. What environmental conditions can affect the accuracy and usefulness of direct-reading instruments?
12. What type of direct-reading instrument would likely be used to monitor for carbon monoxide or oxygen?
13. If you have a 1-L balloon of nitrogen gas at an initial pressure of 760 mmHg and you take that balloon to the top of a mountain where the pressure is only 610 mmHg, what is the new volume of nitrogen in the balloon?
14. What is the main purpose of collecting a grab sample when analyzing air contaminants?

  a. Collect a relatively instantaneous sample to be compared with (ceiling limits)
  b. Cover the entire period of worker exposure
  c. Collect samples in worker breathing zones
  d. Obtain an 8-hour average
15. What is the main advantage of passive dosimeters?
  a. They provide a relatively instantaneous indication of worker exposures.
  b. They are lightweight and don't interfere with the workers during their activities.
  c. The user doesn't need to know very much about chemistry to use them effectively.
  d. Laboratory analysis does not degrade the sample.
16. Which of the following is NOT a principle of operating a direct-reading instrument?
  a. Electrochemical
  b. Gravimetric
  c. Spectrochemical
  d. Thermochemical
17. What is the basic principle of operating a passive air-sampling device?
  a. Gravity
  b. Electrostatic attraction
  c. Diffusion
  d. Ventilation

REFERENCES

Centers for Disease Control and Prevention. *Guideline for Disinfection and Sterilization in Healthcare Facilities, 2008.* Atlanta: CDC, 2009. www.cdc.gov/hicpac/Disinfection_Sterilization/13_02sterilization.html.

Environment Canada. *June 2003 Consultations on the Risk Management of Ethylene Oxide Used for Sterilization in the Healthcare Sector. Degradation/Decomposition Products of Ethylene Oxide in the Atmosphere.* Gatineau QC, Canada: EC, 2013. www.ec.gc.ca/toxiques-toxics/Default.asp?lang=En&n=C5039DE5-1&xml=9C17E21A-B548-4B04-B643-7D8B6CA6EB06.

SKC Inc. *X-am 5000 Personal Multi-gas Detector/Data logger.* Eighty Four, PA: SKC Inc., 2014. www.skcinc.com/prod/805-43748.asp.

7
Aerosols

LEARNING OBJECTIVES

After completing this chapter, readers should be able to do the following:
- Describe the different types of aerosols.
- Understand the basic effects on the lungs associated with aerosol inhalation.
- Assemble the appropriate equipment and perform calibration using primary and secondary standards on a general total nonfibrous contaminant sampling train.
- List the basic steps to conduct aerosol monitoring on a worker.
- Propose a strategy that can monitor the majority of nonfibrous contaminants based on sampling objectives.
- Conduct nonfibrous aerosol monitoring, properly ship samples to the lab, and request pertinent analysis on the basis of sampling objectives.

Photo credit: GlenJ/iStock

INTRODUCTION

The National Institute for Occupational Safety and Health (NIOSH) defines an aerosol as an airborne suspension of tiny particles, fibers, or droplets—such as dusts, mists, fogs, fibers, smokes, or fumes. Inhalation of, skin contact with, or ingestion of aerosols can adversely affect workers' health because aerosols are a significant source of hazardous agents. Inhalation is considered the main route of entry, and aerosol monitoring combined with laboratory analysis are the most common ways to assess workers' exposure to aerosols.

Aerosols often lead to complex toxicological interactions in the human body. Because aerosols can be difficult to control and measure, their roles in workplace illnesses have been consistently underestimated. Because of their various sizes and characteristics, aerosols tend to affect different areas of the lungs in different ways. Therefore, it is the industrial hygienist's responsibility to understand the different types of aerosols, the various ways they can affect workers, and the different procedures to limit workers' exposures to them.

TYPES OF AEROSOLS

Aerosols can be produced by many workplace activities. The type of aerosol generated depends on the work process (DiNardi 2003).

Dusts are composed of suspended particles in the air, ranging in size from less than 1 μm to around 1 mm, that are produced by mechanical processes such as crushing, grinding, milling, breaking, and pulverizing or that are resuspended in the air by processes such as conveyor transporting, loading, unloading, cleaning, sieving, and bagging. Although the chemical composition of dusts does not change from that of their original parent material, their physical characteristics do change since dusts are smaller in size and have a greater surface area per unit of mass than their parent material. These characteristics increase the possibility that dusts will be inhaled and will penetrate the respiratory tract, which affects their toxicity, rate of solubility, and explosion potential.

Mists are liquid droplets, ranging in size from around 10 μm to 100 μm, produced in liquid processes such as splashing, agitating, electroplating, bubbling, painting, and other spraying applications. Similar to dust, a mist's chemical composition does not change from that of its parent solution, but its physical characteristics do enhance its ability to be inhaled and to penetrate the respiratory tract.

Fogs are liquid droplets, ranging in size from 1 μm to 10 μm, produced by the condensation of the vapor phase of the parent material. Given their small size, fogs tend to remain suspended longer in the air than mists. Fogs are also more likely than mists to be inhaled and penetrate the respiratory tract.

Fibers are particles that have a length-to-width ratio greater than 3:1 and can naturally occur in materials such as asbestiform silicate minerals or in manmade materials such as graphite. The physical and biological behaviors of fibers from a single material can change depending on the length-to-width ratio of the fiber particles, which makes the characterization of fiber aerosols more complicated than that of other aerosols.

Smokes are complex mixtures of solid and liquid aerosols, together with gases and vapors, that range in size from 0.01 µm to 1 µm and result from incomplete combustion of carbon-containing materials.

Fumes are solid aerosol particles, around 1 µm in size, produced from the condensation of a vaporized solid material, generally metal, during processes such as welding and smelting. Metal fumes may have different chemical properties (i.e., metal oxides) than those of their parent metal, rendering them more reactive inside a worker's body. Similar to smokes, their small size enhances their ability to penetrate deeply into a worker's lungs, increasing their hazards.

HEALTH EFFECTS FROM INHALATION

The health effects of inhaling an aerosol depend on several factors, including the chemical and biological composition of the agent's particles. An aerosol agent may be toxic, infectious, radioactive, water soluble, or non–water soluble. For example, the crystalline structure of silica can affect the toxicity of inhaled silica particles because free crystalline silica is more toxic than silica chemically bound with another atom. The size and shape of an aerosol can also affect where its particles are deposited and how they interact with the respiratory tract. Large particles tend to stop in the nasal region; smaller particles can end up in the lower reaches of the lungs and the alveoli. Water-reactive aerosols interact with the mucous on the lining of the tracheobronchial system, whereas non–water soluble aerosols are more likely to make it down into the alveoli. Short fibers are easier to exhale from the lower parts of the lung, while larger fibers can get stuck in the lung. Each of these different characteristics plays a role in whether and where a disease may occur.

The total amount of aerosol inhaled or ingested and absorbed into the body affects the aerosol's toxicity. The dose, or total amount absorbed, is proportionally related to the concentration of and the duration of exposure to the aerosol. The longer a worker breathes in an aerosol, the worse the health effects. The severity of the effects also relates to how much air workers can take into their lungs per hour, which in turn depends on the work rate or metabolism of a worker at the time of exposure. Concurrent exposure to other agents in the workplace can also increase the effects of aerosol exposures.

Pneumoconiosis (major, minor, or benign) is the generic name for the group of lung disorders associated with inhaling inorganic dust. Major pneumoconioses such as silicosis, asbestosis, and coal worker's pneumoconiosis are related to severe fibrosis of the lungs that results in a significant impediment in lung function that can be observed in lung function tests. In minor pneumoconiosis, there is minimal fibrosis of the lungs, resulting in no interference in lung function (Salama 2014). Some of the particulate sources associated with minor pneumoconiosis are clay, kaolin, mica, and nonfibrous silicates. In benign pneumoconiosis, particle deposition in the lungs is not associated with fibrosis or any other disturbance of lung function. Disorders such as siderosis (from iron dust), stannosis (from tin dust), and chalicosis (from calcium dust) are types of benign pneumoconiosis. Inhalation of aerosols can also be associated with chronic obstructive pulmonary diseases such as asthma, bronchitis, emphysema, and byssinosis. The effects of such disorders can range from slightly debilitating to fatal (Rushton 2007). Some organic aerosols, such as red cedar dust, can cause allergic responses and asthma (WorksafeBC 2014).

Many infectious agents are aerosols and can enter workers' lungs via inhalation, which can then lead to diseases such as bronchitis or pneumonia. Tuberculosis is a bacterial disease that becomes airborne when an infected person coughs or sneezes. The particles then float in the air for several minutes before they fall to the floor or are removed by a ventilation system, which is common in a hospital setting. Anyone who enters a room where airborne particles are present might breathe in the bacteria, which then lodge in the lung and transmit the disease to the person.

Some aerosols, such as asbestos and silica, can become lodged in the lower parts of the lung because of their size and shape. As an immune response, the body sends macrophages to dislodge or break down the foreign material. However, because the material is inanimate, the macrophage defenses do not work, and the normal ciliary functions of the lung are unsuccessful at removing the fibers. The result is a scarring of the lung called **fibrosis**.

Fumes have particular actions in the body. For instance, exposure to welding fumes increases a worker's exposure to zinc oxide fumes, which have been associated with **metal fume fever**, a complex acute allergic condition. Other metals such as copper lead to similar health effects (NIOSH 1975).

Lung cancer is associated with inhaling aerosols such as hexavalent chromium, asbestos, beryllium, cadmium, and nickel (Steenland et al. 1996).

AEROSOL DEPOSITION

Even though the human respiratory system is generally efficient at removing inhaled particles and fibers from the body, particles of a certain size, concentration, or toxicity can be hazardous.

The respiratory system is divided into three major regions: the nasopharyngeal region, the tracheobronchial region, and the pulmonary or alveolar region. Figure 7–1 shows the human respiratory system separated into these regions.

In general, the smaller a particle's size, the deeper inside the lungs it can go. Moreover, it is more likely to penetrate into cells and through organic tissues; its surface area's per-unit mass is large, thus increasing its reaction potential, and it dissolves quickly, which increases the bioavailability of its solubilized compounds.

Inhaled aerosols can be deposited in the human respiratory track through different mechanisms: sedimentation, impaction, interception, and Brownian diffusion. Figure 7–2 shows the relationship between particles' **aerodynamic diameter** (AD) and how particles deposit into the human respiratory track.

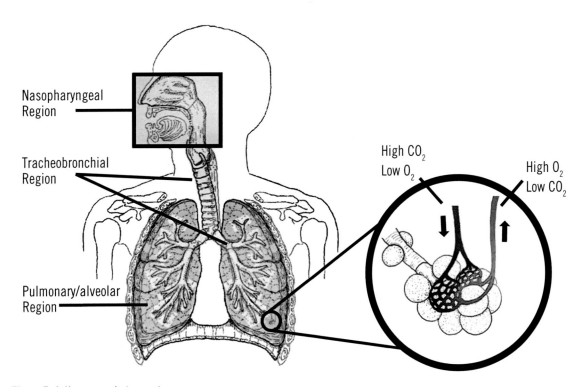

Figure 7–1. Human respiratory system.

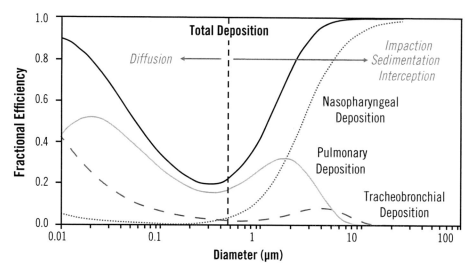

Nasal Breathing and Selected Parameters: Q = 500 cm³/s; V = 1,500 cm³; FRC = 3,300 cm³ and f = 10

Figure 7–2. Particle deposition in human respiratory tract as a function of particle aerodynamic diameter.

(Source: Modified from Bugarski and Gautam 2014.)

Aerodynamic diameter is defined as the diameter of a unit density sphere that settles at the same velocity as the particle in question. As can be seen in Figure 7–2, the AD can be differentiated by three main sizes or categories and where the particles interact with the respiratory tract. In general, the sizes are related to the following types of particles:

- Inhalable particulate mass—100 mm
- Thoracic particulate mass—10 mm
- Respirable particulate mass—4 mm

Inhaled aerosol deposition mechanisms vary according to particle size and the region of the respiratory track where such deposition occurs. In the nasopharyngeal and tracheobronchial regions, particle deposition occurs through inertial impaction and Brownian diffusion. These particles, typically less than 10 mm, are referred to as *inhalable*. In the tracheobronchial airways, deposition occurs primarily through sedimentation and Brownian diffusion. Particles that settle in this region are called *thoracic*. In the pulmonary or alveolar region, deposition occurs primarily through gravitational settling (sedimentation in still air) and Brownian diffusion. Particles smaller than 4 mm fall in this region and are known as *respirable*.

Aerosols can also deposit on workers' skin or eyes and cause irritation or toxic effects. Moreover, very small particles may pass through the skin and enter the body, and soluble particles may dissolve and enter workers' bodies through their skin (Bugarski and Gautam 2014).

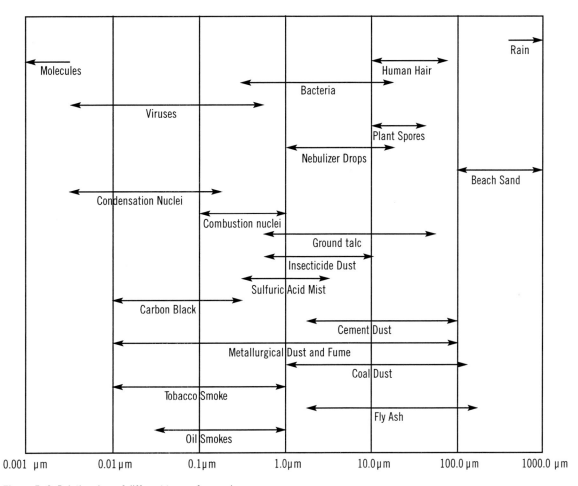

Figure 7–3. Relative sizes of different types of aerosols.

Sedimentation is the process by which an aerosol is deposited under the effect of gravity. The rate of aerosol sedimentation in a given space depends on the aerosol's size, shape, mass, orientation (if the aerosol is not spherical), air density, and viscosity and on the rate at which air is moving in the space. (See Figure 9–3 in Chapter 9, Respiratory Protection.)

Impaction occurs when an aerosol touches and deposits on a given surface, which generally occurs when an aerosol is forced to change direction and its inertial forces carry it across the bending air flow long enough to touch a surface. Generally, impaction increases with the increasing mass and velocity of the aerosol and with the sharpness of changes in the air's movement. (See Figure 9–4 in Chapter 9, Respiratory Protection.)

Interception takes place when an aerosol comes close to a surface and is deposited because of its size. Deposition by interception is directly proportional to the ratio between the aerosol size and the size of the collecting object. In this sense, fibers have a much greater probability of being collected by interception because their size is determined by their length. (See Figure 9–2 in Chapter 9, Respiratory Protection.)

In **Brownian diffusion**, an aerosol is deposited on a surface via random displacement caused by air motion associated with thermal microscopic behavior. This deposition process is more prominent for small-size aerosols, where gravitational forces are not dominant. (See Figure 6–1 in Chapter 6, Gases and Vapors.)

AEROSOL MONITORING

Monitoring aerosols requires a thorough understanding of their physical and chemical characteristics. In addition, industrial hygienists need to define in advance the purpose for which such monitoring is to be conducted (i.e., compliance, hazard control assessment, area sampling, or exposure group assessment). With the purpose of the monitoring clearly known upfront, industrial hygienists can select the appropriate monitoring technique and strategy to assess workers' exposure. Regardless of the aerosol monitoring technique used, the objective is to acquire information about the aerosol concentration and its properties over a specific period of time. In order to achieve this objective, a sample of the air containing the aerosol of interest is drawn by a pump into a sampling container. The air sample is then deposited on a filter or a liquid medium for subsequent analysis, or it is passed through a sensing device for optical detection (DiNardi 2003). In general, air monitoring can be divided into two major categories: **instantaneous** or **real-time monitoring** and **continuous** or **integrated monitoring**.

INSTANTANEOUS OR REAL-TIME MONITORING

Instantaneous or real-time monitoring consists of collecting an aerosol sample for a period of time ranging from a few seconds to less than 10 minutes. One of the advantages of instantaneous monitoring is that after a sample is collected, analysis is available immediately from the monitoring device. Instantaneous monitoring is commonly used to identify and quantify contaminants during short-term operations and to measure levels for specific isolated processes. It is generally also used during hazard control evaluations (Bisesi and Kohn 1995).

CONTINUOUS OR INTEGRATED MONITORING

Continuous or integrated monitoring consists of collecting an aerosol sample continuously over a period of time ranging from 10 to 15 minutes to several hours (typically 8 hours). Most workshifts last for 8 hours, and available occupational exposure limits are usually based on an 8-hour workshift. One of the advantages of conducting integrated monitoring over an 8-hour shift is that the results are representative of workers' exposures during that period of time. However, peak exposures during the workshift will not be identified if the integrated monitoring renders a single-value result that represents the average time-weighted exposure during the sampling period. Also unlike instantaneous monitoring, most samples obtained from continuous monitoring need to be submitted to a laboratory for analysis (Bisesi and Kohn 1995).

Traditionally, the most common industrial hygiene approach to aerosol monitoring is to conduct integrated monitoring using a sampling pump with a filtering collection medium and then send the sample for both gravimetric and chemical analyses in a laboratory.

Integrated monitoring of the great majority of aerosols consists of using battery-operated air-sampling pumps as air-moving devices that create an active flow of the aerosol through a filter or liquid sorbent. Air-sampling pumps are normally rated in three different air-flow ranges: low flow (< 1 L/min), high flow (1 to 5 L/min), and ultra-high flow (> 5 L/min). There are also multiflow pumps available that can be used to monitor aerosol flows from around 50 mL/min to 5 L/min. Industrial hygiene aerosol monitoring is generally conducted with air flows at or above 1 L/min and below 5 L/min, whereas volatile organic compounds, vapors, and gases are sampled at lower flow rates and bioaerosols are sampled at a flow rate between 10 and 15 L/min.

The **sampling train** consists of an air-sampling pump that is connected to the sampling medium by means of flexible and transparent tubing. During the sampling, the aerosol is drawn by suction through the sampling medium, where the airborne contaminant is separated from the air stream by filtration or absorption. From the collection medium, the air continues to be drawn through the flexible hose and finally into and through the air-sampling pump. After the sampling period ends, the collection medium is submitted for laboratory analysis (Bisesi and Kohn 1995).

AIR-SAMPLING PUMP CALIBRATION

Air-sampling pumps should be calibrated both before and after the sampling period to ensure that the sampling flow rate of the air is accurate. Such accuracy is needed in order for the industrial hygienist to calculate the total air volume that was collected throughout the sampling period. Equations 1 and 2 are used to make this calculation. Equation 1 is used to calculate the average flow rate, and Equation 2 is used to calculate the total sampled volume.

$$\text{Flow Rate ``}Q\text{''} \left[\frac{L}{min}\right] = \frac{\text{Pre-Calibration Flow Rate} \left[\frac{L}{min}\right] + \text{Post-Calibration Flow Rate} \left[\frac{L}{min}\right]}{2} \quad \text{Equation 1}$$

$$\text{Total Sampled Volume } V \; [m^3] = \text{Flow Rate } Q \left[\frac{L}{min}\right] \times \frac{1 min^3}{1,000 \; L} \times \text{Sampling Time } [min] \quad \text{Equation 2}$$

Air-sampling pump calibration consists of adjusting the device to a specific, desired flow rate and measuring that rate against a known standard. In general, flow rates are expressed in liters per minute (L/min) or milliliters per minute (mL/min). It is important to note that the flow rates from pre- and post-calibrations must be within 10% of each other. Standards are primarily of two types: primary calibration standards and secondary calibration standards. Primary calibration standards are more accurate because they are based on actual volumetric physical measurements. Secondary calibration standards for air flow are created by using primary calibration standards (Bisesi and Kohn 1995).

The most basic and reliable primary calibration technique for practicing industrial hygienists is the use of the frictionless bubble tube, also known as the soap-film flowmeter. The bubble tube consists of an inverted burette that is wetted on the interior with a soap solution and has its volume displayed and marked with at least two conspicuous lines that indicate the start and finish of the displayed volume. Figure 7–4 shows a typical bubble tube. As shown in the picture, the sampling pump is connected to the outlet of the bubble tube by a flexible and transparent hose. As the pump draws air through the bubble tube, the industrial hygienist creates a bubble or soap film in the bottom of the burette that travels nearly frictionlessly through both marked lines from the bottom to the top of the burette. The flow rate in liters per minute is determined in two steps: (1) by measuring the time (in seconds) required for the above-mentioned bubble or soap film to pass between the markings enclosing the bubble tube's known volume; and (2) by dividing the volume traversed by the soap film or bubble in the bubble tube (in mL) and applying a factor that converts seconds to minutes, as depicted in Equation 3.

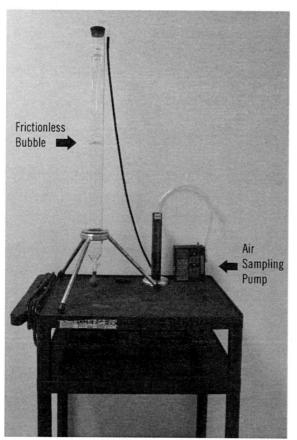

Figure 7–4. Frictionless bubble tube connected in line with an air-sampling pump.

$$Q\left[\frac{L}{min}\right] = \frac{Bubble\ Tube\ Volume\ [ml]}{Time\ it\ took\ the\ bubble\ to\ traverse\ markings\ [sec]} \times \frac{1\ L}{1{,}000\ mL} \times \frac{60\ sec}{1\ min} \qquad \text{Equation 3}$$

Another important aspect of air-sampling pump calibration is that this calibration is performed in the same environment where the air monitoring will take place. In the field, however, such calibration is not always possible, especially when different areas are to be monitored at the same time. Under this circumstance, calibration is often performed in an area that is central to all the places where workers will be monitored. Whenever a sampling pump is calibrated in an area where air temperature and atmospheric pressure differ from those where the monitoring will take place, air temperature and atmospheric pressure in both places should also be measured so that the industrial hygienist can correct the calibrated pump flow rate to the actual conditions in which sampling will take place. Equation 4 shows how to correct the calibration flow rate for actual sampling conditions.

$$Q_{actual} = Q_{cal} \times \frac{T_{actual}[°K]}{T_{cal}[°K]} \times \frac{P_{cal}[mmHg]}{T_{actual}} \qquad \text{Equation 4}$$

In Equation 4, Q_{actual} represents the sampling conditions' calibration flow rate, and Q_{cal} represents the flow rate measured during calibration. After the industrial hygienist measures the air temperature and pressure at the monitoring site, Q_{actual} can be calculated.

The parameter Q_{actual} should be used in Equation 2 to calculate total sampled volume.

Example

In the field, you set the pump's flow rate to 2.0 L/min using a precision rotameter and sampled for 400 min. You note and record the field temperature as 95°F (35°C) and the barometric pressure as 600 mmHg. Calculate the corrected volume of air in liters for normal temperature and pressure (25°C and 760 mmHg).

Solution:

$$Q_{actual} = Q_{cal} \times \frac{T_{actual}[°K]}{T_{cal}[°K]} \times \frac{P_{cal}[mmHg]}{T_{actual}}$$

R = Q/T
1.5 L/min = Q/400 min
Q = 600 L

There are other primary standards besides the manual bubble tube just described. Many of them use the same basic principle as the manual bubble tube but have small variations in the way time is measured or in the type of frictionless device used (soap film in the manual bubble tube). All of these other primary calibration standards have electronic components and periodically need to be factory calibrated or checked against a manual bubble tube. Users should conduct their own research to suit their needs concerning primary calibration standards for air flow.

A rotameter is the main secondary calibration standard for air flow (Figure 7–5). The rotameter consists of a tapered, precision-bored tube made of transparent glass or plastic, inside of which is a solid spherical float. As Figures 7–5 and 7–6 show, a pump connected to the rotameter draws in air that enters the bottom of the marked tube and carries the solid sphere upward until the force exerted by air flow on the surface of the solid float equals the float's weight. At that moment, the flow rate can be read from the marking that represents the height of the float. The reading should be taken at the widest point of the float (Bisesi and Kohn 1995).

As with any secondary standard, rotameters are calibrated by comparing them with a primary calibrator, preferably a manual bubble tube. An example of such a comparison is shown in Figure 7–6, where an air-sampling pump, a rotameter, and a bubble tube are connected in line in that order by means of two flexible hoses. In order to calibrate the rotameter, the air-sampling pump is turned on, which draws air first through the bubble tube, then through the rotameter, and finally into the sampling pump.

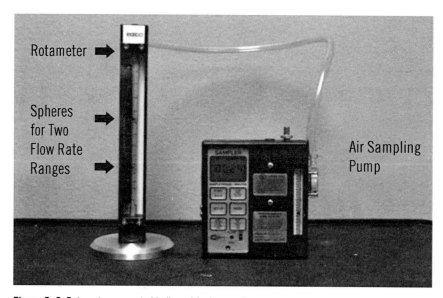

Figure 7–5. Rotameter connected in line with air-sampling pump.

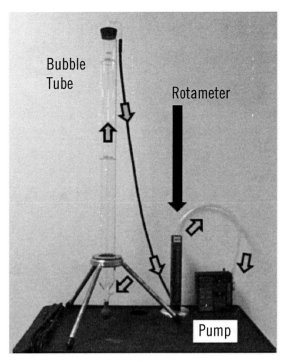 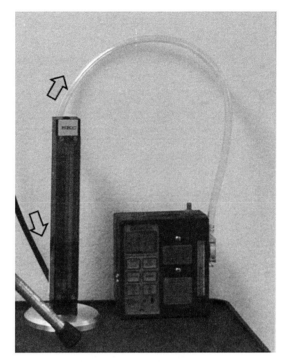

Figure 7–6. Here, the bubble tube is the primary standard and provides the flow rate by measuring the flow through the bubbles moving up the tube on the right. This flow rate is then compared with the flow rate indicated on the rotameter on the right.

The industrial hygienist should select five different flow rates across the work range of the air-sampling pump. For each flow rate, the flow rate measured using the bubble tube and the corresponding rotameter reading should be recorded and then plotted on linear graph paper, with rotameter readings on the vertical axis and bubble-tube flow-rate measurements on the horizontal axis. Figure 7–7 on the next page shows a calibration curve derived from the data shown in Table 7–1.

Table 7–1. Example Rotameter Calibration Data

Rotameter Readings (mL/min)	Bubble-Tube Flow Rate (L/min)
3,060	3
2,540	2.5
2,030	2
1,630	1.5
1,130	1
510	0.5
20	0

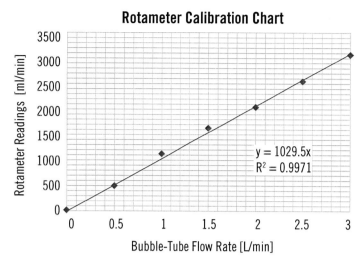

Figure 7–7. Rotameter calibration chart.

As indicated in Figure 7–7, once data from Table 7–1 are plotted in the graph, a linear regression line is adjusted to the plotted data, setting the intercept at 0. For this particular example, the calibration curve displayed in Figure 7–7 explains 99.71% of the variability in the data and indicates that each L/min of actual flow rate represents 1,029.5 mL/min in rotameter readings. Conversely for this example, each 1,000 units in rotameter reading represents approximately 0.971 L/min of actual flow rate in the pump.

Note that a rotameter measures the mass flow, rather than the volume flow, of the gas and therefore is affected by changes in gas density. When a rotameter is not calibrated in the same air temperature and pressure conditions in which it will be used in the field, air temperature and pressure in both conditions (calibration and field sampling) must be recorded and pump flow rate in the field adjusted according to Equation 4. Once properly calibrated, rotameters can be used directly in the field to calibrate the air-sampling pumps without the need of a bubble tube. Overall, the advantages rotameters have over other calibration methods are simplicity, portability, and low price.

ASSESSING TOTAL PARTICULATE AND MIST CONCENTRATION IN THE AIR

The concentrations of particulates and mists are generally expressed as mass per unit of air volume (mass concentration). Most current standards for aerosol exposures express limits using mass concentration in milligrams or micrograms per cubic meter. The sampling equipment used to assess total particulate or mist concentration in the air is basically the same regardless of the contaminant under study. However, the type of filter, pumps' flow rates, number of field blanks, and laboratory analysis are specific for each contaminant. The *NIOSH Manual of Analytical Methods* and the *OSHA Technical Manual* provide details on the sampling parameters and analytical techniques appropriate for assessing exposure to each contaminant. As an example of the sampling equipment and approach to assess total particulate or mist concentration, we will look at the approach used to sample for total dust, nuisance dust, or particulate not otherwise classified. In this approach, the objective is to find out the concentration of particulate, both toxic and nontoxic, in a given area or to determine what an individual's exposure is to the total amount of particulate rather than the exposure to a specific type of particulate.

The filtering medium for total particulate sampling is a filter with a diameter of 37 mm made of polyvinyl chloride and with a pore size of 5.0 µm. Filters that are 37 mm in diameter provide an adequate cross-sectional area for particulate deposition and collection through electrostatic attraction, interception, and impaction. Particulates that are larger than the filter's pore size are collected on the surface of the filter; those with smaller diameters are collected within the pores via impaction and interception and the electrostatic attraction between the particles and the filter. As with any air filter, collection efficiency increases as particulates accumulate on the filter.

Figure 7–8 depicts a 37-mm, three-stage cassette used as the collection medium for total particulate sampling. Contaminated air enters the cassette through the inlet in the third stage, expands throughout the length of the second stage, deposits the contaminants on the filter (placed on top of the support pad that is held in place by the first stage), passes through the support pad, passes through the outlet in the center of the first stage, and then exits the cassette.

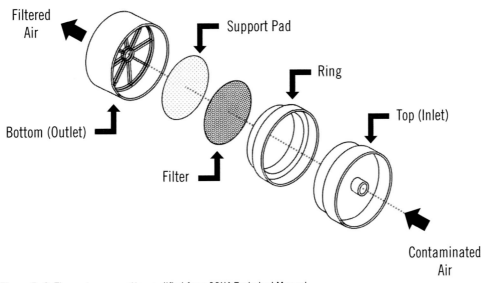

Figure 7–8. Three-stage cassette, modified from OSHA Technical Manual.

(Source: Occupational Safety and Health Administration. "Section II: Chapter 1: Personal Sampling for Air Contaminants." In OSHA Technical Manual. Washington DC: OSHA, 2014. https://www.osha.gov/dts/osta/otm/otm_ii/otm_ii_1.html.)

A sample collection medium is generally prepared and provided by an independent, accredited laboratory. The total particulate medium is prepared by preweighting a filter for each field sample and field blank that is going to be collected. Immediately prior to preweighting, the desiccated filter is passed across an ionization strip to eliminate any static charge that may be present on the filter and cause contamination because of the unwanted attraction of airborne particulate during the weighing process. Preweighted filters are then placed on the support pads, and each three-stage cassette is assembled and closed with plugs in its inlet and outlet. Once cassettes are assembled, they are wrapped with cellulose shrinking bands to ensure that their ring connections are airtight. Finally, cassettes are labeled, coded, and sent to the industrial hygienist for sampling.

The basic sampling train for total particulate consists of an air-sampling pump connected to a three-stage cassette by means of a flexible hose. Sampling trains must be pre- and post-calibrated for air sampling to be properly conducted. Figure 7–9 shows a calibration train consisting of the sampling train connected in line with a primary calibrator by means of a flexible hose. It is important to note that the resistance imposed by the pump is directly proportional to the length of the flexible hose, and therefore industrial hygienists (1) should use the same length of hose in their sampling as they used to calibrate their sampling train and (2) should use the shortest possible flexible hose to connect the sampling train to the calibrator.

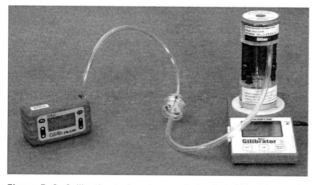

Figure 7–9. Calibration train using an electronic bubble tube as the primary calibrator.

(Source: Occupational Safety and Health Administration. Section II: Chapter 1: Personal Sampling for Air Contaminants. In OSHA Technical Manual. Washington DC: OSHA, 2014. https:// www.osha.gov/dts/osta/otm/otm_ii/otm_ii_1.html.)

ASSESSING RESPIRABLE PARTICULATE CONCENTRATION IN THE AIR

Sometimes industrial hygienists are interested in assessing workers' exposures to the respirable fraction of a particular contaminant in the air, such as when the respirable fraction is more toxic than the nonrespirable particles. The respirable fraction of an aerosol is the subfraction of inhaled particles that have an aerodynamic diameter smaller than 10 μm. For these types of assessments, the sampling train incorporates a cyclone that precedes the collection medium. A cyclone is a gas-solid separation device, as shown in Figure 7–10, that is commonly used in processes across the industry spectrum. As can be observed in Figure 7–10, the contaminated air enters tangentially (black arrow on the left) at the top of the cyclone and immediately starts spinning down the cyclone's body.

The spinning air loses its ability to carry the larger particles (spinning black arrows), which start precipitating to the bottom outlet of the cyclone, while the smaller particles continue to be carried by the spinning air stream downward to the tapered section of the cyclone body. At this point, static pressure starts increasing and creates a pressure differential in the center of the cyclone that is sufficient to carry the air with the smaller particles up and through the center outlet of the cyclone (gray spinning arrows). In a production process, the cyclone is generally used to recover desirable larger particles from the air stream. In industrial hygiene, the cyclone is used to separate the undesirable larger particles from the air stream so that industrial hygienists can determine the respirable fraction of the contaminated air that exits the cyclone through its top outlet toward a collection medium. Because cyclones' separation efficiencies depend on air-flow rate, industrial hygienists must follow manufacturers' flow-rate recommendations in order to ensure a respirable fraction collection (aerodynamic diameter < 10 μm). In general, the higher the flow rate, the larger the particle size fraction that will be carried through the upper outlet of the cyclone by the air stream. Figure 7–11 shows the Dorr-Oliver cyclone, which has been used extensively in industrial hygiene practice. Dorr-Oliver cyclones must be operated at 1.7 L/min. In respirable dust sampling, as with the respirable fraction of any other contaminant, the collection medium is a two-stage cassette that goes in line, right at the outlet of the cyclone.

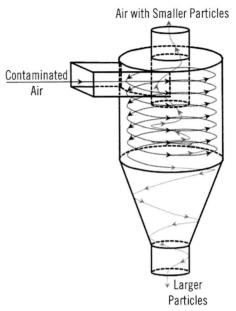

Figure 7–10. Cyclone particle size selection process.

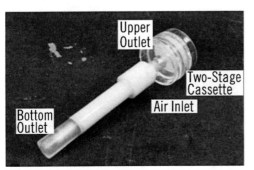
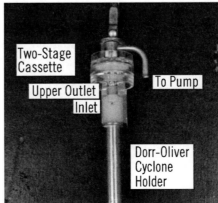

Figure 7–11. Dorr-Oliver cyclone.

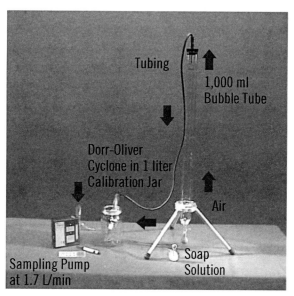

Figure 7–12. Dorr-Oliver Cyclone Calibration, extracted from OSHA Technical Manual.

(Source: Occupational Safety and Health Administration. "Section II: Chapter 1: Personal Sampling for Air Contaminants." In OSHA Technical Manual. Washington DC: OSHA, 2014. https://www.osha.gov/dts/osta/otm/otm_ii/otm_ii_1.html.)

CYCLONE CALIBRATION

When conducting respirable dust sampling, industrial hygienists must follow recommended guidelines on how to calibrate a sampling train with a cyclone in line. Figure 7–12 is a schematic drawing on how the calibration train for a respirable dust sample using a Dorr-Oliver cyclone should be assembled in order to properly calibrate the sampling train.

INSTANTANEOUS OR REAL-TIME TOTAL AND RESPIRABLE DUST SAMPLING

As mentioned previously, instantaneous or real-time monitoring is used to identify contaminants and measure their relative levels during short-term operations; during specific, isolated processes with anticipated peak levels; and during hazard control evaluations. In aerosol monitoring, the only direct readings that can be obtained are of total or respirable dust samples while using a real-time aerosol monitor in the respective sampling train instead of using the traditional air-sampling pump. Real-time aerosol monitors use optical technology (infrared and light scattering) to detect the real-time concentrations of aerosols in the air and display those concentrations to the user.

AEROSOL ANALYSIS

In industrial hygiene practice, unless sampling is conducted solely for research purposes, all laboratory analyses performed on air samples must be conducted by an independent and accredited laboratory. In the United States, the American Industrial Hygiene Association is the accrediting body of laboratories that conduct analyses on industrial hygiene samples. Industrial hygienists are again encouraged to refer to the *NIOSH Manual of Analytical Methods* and the *OSHA Technical Manual* in order to plan the aerosol monitoring study they need to conduct and to learn how to ship their samples for analysis after the monitoring has been completed. Figure 7–13 shows the first page of NIOSH Analytical Method 0500 for Particulates Not Otherwise Regulated, Total.

As indicated in the Sampling section, on the basis of a theoretical contaminant concentration in the air (15 mg/m^3 in the case of total dust), the method provides recommendations on a range of sampling rates and total volumes to

PARTICULATES NOT OTHERWISE REGULATED, TOTAL   0500

DEFINITION: total aerosol mass   CAS: NONE   RTECS: NONE

METHOD: 0500, Issue 2	EVALUATION: FULL	Issue 1: 15 February 1984 Issue 2: 15 August 1994

OSHA: 15 mg/m³   **PROPERTIES:** contains no asbestos and quartz less than 1%
NIOSH: no REL
ACGIH: 10 mg/m³, total dust less than 1% quartz

SYNONYMS: nuisance dusts; particulates not otherwise classified

SAMPLING		MEASUREMENT	
SAMPLER:	FILTER (tared 37-mm, 5-µm PVC filter)	TECHNIQUE:	GRAVIMETRIC (FILTER WEIGHT)
FLOW RATE:	1 to 2 L/min	ANALYTE:	airborne particulate material
VOL-MIN: -MAX:	7 L @ 15 mg/m³ 133 L @ 15 mg/m³	BALANCE:	0.001 mg sensitivity; use same balance before and after sample collection
SHIPMENT:	routine	CALIBRATION:	National Institute of Standards and Technology Class S-1.1 weights or ASTM Class 1 weights
SAMPLE STABILITY:	indefinitely	RANGE:	0.1 to 2 mg per sample
BLANKS:	2 to 10 field blanks per set t	ESTIMATED LOD:	0.03 mg per sample
BULK SAMPLE:	none required	PRECISION ($\bar{S}_r$):	0.026 [2]

ACCURACY	
RANGE STUDIED:	8 to 28 mg/m³
BIAS:	0.01%
OVERALL PRECISION ($\hat{S}_{rT}$):	0.056 [1]
ACCURACY:	±11.04%

APPLICABILITY: The working range is 1 to 20 mg/m³ for a 100-L air sample. This method is nonspecific and determines the black. This method replaces Method S349 [5]. Impingers and direct-reading instruments may be used to collect total dust [3] in addition to the other ACGIH particulates not otherwise regulated [4].

INTERFERENCES: Organic and volatile particulate matter may be removed by dry ashing [3].

OTHER METHODS: This method is similar to the criteria document method for fibrous glass [3] and Method 5000 for carbon black. This method replaces Method S349 [5]. Impingers and direct-reading instruments may be used to collect total dust samples, but these have limitations for personal sampling.

Figure 7–13. NIOSH analytical method for total particulate not otherwise regulated.

(Source: NIOSH Manual of Analytical Methods, 4th Edition.)

be collected, as well as the sampling media and the number of field blanks to be used and any shipment restrictions that should be followed or care that should be taken if shipment is necessary. Using the method as a reference, industrial hygienists can plan their monitoring studies on the basis of their sampling objectives, understanding the overall precision this method should have. Moreover, in the Measurement section, the document indicates the type of analysis to conduct, the analyte to study, and the technique to use as well as the method's precision and limit of detection. In addition, analytical methods also provide information on any possible interference between the contaminant of interest and other substances in the air. The rest of the method is a detailed description of how the laboratory analysis must be conducted, including calibration of the necessary analysis equipment.

CONTROL OF AEROSOLS

SUBSTITUTION AND ELIMINATION

Substitution and elimination of aerosol sources should be considered the first step of control whenever possible. In the construction industry, the most dangerous exposures to crystalline silica aerosols occur during sandblasting to remove rust and paint from old structures. New products for sandblasting have been developed that use alternatives such as plastic or glass beads or other abrasive materials that do not have silica's toxic properties.

Workers' exposures to aerosols can also be reduced by reducing the creation of aerosols during work processes. Wet mopping floors during stone crushing or handling operation facilities can greatly keep materials from becoming airborne while sweeping the floors. Spraying the demolition of concrete or masonry structures with water is another way to keep materials from becoming aerosolized, from affecting workers, and from escaping.

REDUCTION IN THE ENVIRONMENT

Particularly in the case of infectious aerosol agents, it is possible to reduce their airborne concentrations in the environment. Germicidal ultraviolet radiation is one method used in health care and research to reduce the concentration of living materials inside ventilation ducts and recirculating air.

LOCAL EXHAUST VENTILATION

Localized exhaust systems can also minimize the concentrations of materials in the air and reduce worker exposures through inhalation. Specially designed capture hoods collect rock fragments from grinding operations before they reach the worker's breathing zone.

GENERAL EXHAUST VENTILATION

General exhaust ventilation of aerosols can reduce their overall air concentrations in many work settings.

SUMMARY

Inhalation of aerosols is the most common route of occupational exposure to toxic agents. Therefore, a thorough understanding of aerosol characteristics is essential to the effective practice of industrial hygiene. It is important to know how aerosols move in the environment and enter the body via the respiratory system. The ability to measure the concentration of aerosols in the air is a key factor in determining when controls are necessary.

REVIEW QUESTIONS

1. In what ways are mists different from fogs?
2. Why is aerosol size important when determining where aerosols will interact in the lungs?
3. In the field, you set the air-sampling pump's flow rate to 1.7 L/min using a precision rotameter and sample the air for 350 min. You note and record the field temperature as 80°F (26.6°C) and the barometric pressure as 650 mmHg. Calculate the corrected volume of air in liters for normal temperature and pressure (25°C and 760 mmHg). Show all work.

REFERENCES

Bisesi, M., and J. P. Kohn. *Industrial Hygiene Evaluation Methods*. Boca Raton, FL: CRC Press, 1995.

Bugarski, A., and M. Gautam. *Size Distribution and Deposition in Human Respiratory Tract: Particle Mass and Number*. Morgantown, WV: West Virginia University, 2014. www.cdc.gov/niosh/mining/UserFiles/works/pdfs/sdadi.pdf.

DiNardi, S. *The Occupational Environment: Its Evaluation, Control and Management.* 2nd ed. Fairfax, VA: American Industrial Hygiene Association, 2003.

National Institute for Occupational Safety and Health. *Criteria for a Recommended Standard: Occupational Exposure to Zinc Oxide, Publication No. 76–104.* Atlanta: NIOSH, 1975.

———."Particulates Not Otherwise Recognized." In *NIOSH Manual of Analytical Methods.* Atlanta: NIOSH, CDC, 2003. www.cdc.gov/niosh/docs/2003-154/pdfs/0500.pdf.

Occupational Safety and Health Administration. "Section II: Chapter 1: Personal Sampling for Air Contaminants." In *OSHA Technical Manual.* Washington DC: OSHA, 2014. www.osha.gov/dts/osta/otm/otm_ii/otm_ii_1.html.

Rushton, L. "Occupational Causes of Chronic Obstructive Pulmonary Disease." *Reviews on Environmental Health* 22 (2007): 195–212.

Salama, R. *Pneumoconiosis.* Ismaïlia, Egypt: Suez Canal University, 2014. www.pitt.edu/~super7/32011-33001/32551.ppt.

Steenland, K., D. Loomis, C. Shy, et al. "Review of Occupational Lung Carcinogens." *American Journal of Industrial Medicine* 29 (1996): 474–90.

WorkSafeBC. *Western Red Cedar Asthma.* Vancouver, BC, Canada: WorkSafeBC, 2014. www.worksafebc.com/publications/health_and_safety/by_topic/assets/pdf/cedar_asthma_ph51.pdf.

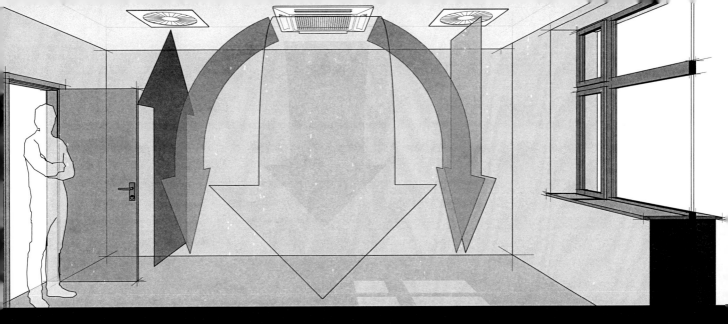

8
Ventilation

LEARNING OBJECTIVES

After completing this chapter, readers should be able to do the following:
- Identify the main types and purposes of different types of workplace ventilation systems.
- List the environmental factors that influence ventilation system operations.
- Make ventilation calculations using standard formulas.
- Explain the relationships among different types of pressures within a duct that are related to air movement.
- Recognize various parts of a ventilation system.
- Identify different types of air capture device hoods and their appropriate uses and limitations.
- Differentiate among the various types of ventilation fans.
- Describe standard air cleaning devices and filters.
- Calculate air exchange rates using provided parameters.

Photo credit: valigursky/iStock

INTRODUCTION

Ventilation in industry has two main purposes. One is to provide a comfortable work environment for building occupants by controlling temperature, odors, humidity, and other undesirable conditions. Second, ventilation can be used to protect workers from hazardous airborne contaminants (dusts, gases, vapors, etc.).

The two main components of general building ventilation are supply and exhaust. The supply brought into the work space provides filtered, tempered air at a rate and quantity that are comfortable for occupants. In addition, an exhaust system draws air out of the work space and into the outdoors or a recirculation system.

Industrial ventilation involves the appropriate design and application of equipment to provide the conditions necessary to efficiently maintain the health and safety of workers. Ventilation is used to control hazardous emissions and conditions and to control worker exposures to toxic agents. By controlling the air flow in the work environment, contaminated and hazardous air is continuously replaced with fresh, clean air. The main objectives of ventilation are as follows:

- to maintain an adequate oxygen supply in the work area
- to prevent hazardous concentrations of toxic materials in the air
- to remove any undesirable odors from a given area
- to control temperature and humidity
- to remove undesirable contaminants at their source before they enter the workplace air

It is important to balance the amounts of supply and exhaust air to meet the various needs of the building occupants and processes. In many cases, the amount of supply air is about 10% to 30% higher than the amount of exhaust air. This balance keeps the building at a slightly positive pressure and keeps dust and unfiltered outdoor air from entering through cracks, leaks, and the actions of opening and closing doors.

Sometimes it may be useful for an area of a building to have negative pressure compared to the outdoors or adjacent areas. For example, the pressure in an area where toxic materials are used, such as laboratories, may need to be negative to ensure that the vapors do not escape. By balancing the flow and pressure of air throughout a building, the contaminated areas will be under control.

An out-of-balance building ventilation system can lead to leakage and other inefficiencies. Rooms or larger areas may be drafty, and opening or closing doors may be difficult if pressure balances between rooms are too great. Well-designed and well-balanced building ventilation leads to the optimization of efficiencies and the minimization of energy costs.

Several operational and environmental factors affect the overall ventilation system design of a building. Some of these factors are the following:
- manufacturing processes
- exhaust air system and local air extraction
- climate requirements (tightness, plant aerodynamics, etc.)
- cleanliness requirements
- ambient air conditions
- heat emissions
- terrain around the plant
- contaminant emissions
- regulations

All of these factors must be considered when developing and designing a building ventilation system.

COMPONENTS OF SUPPLY SYSTEMS

Ventilation air systems typically consist of the following components:
- outdoor air intake
- filters
- heating and cooling units
- fans
- ducts
- registers/grilles for distribution

If air is being recirculated, then a plenum or ducted return system should also be used. A simple air-supply system is shown in Figure 8–1.

In general building ventilation, dilution is used to maintain the temperatures and concentrations of hazardous chemicals. In some situations, however, it is more efficient and safer to capture the hazardous aerosols at their source using local exhaust ventilation (LEV). These types of systems capture the contaminated air and exhaust it from the general space before it can mix with the supply air. An exhaust ventilation system draws off the contaminated air, which is replaced by the supply air, and may exhaust the entire work area (general exhaust) or may be located at the source to remove the contaminant (local exhaust).

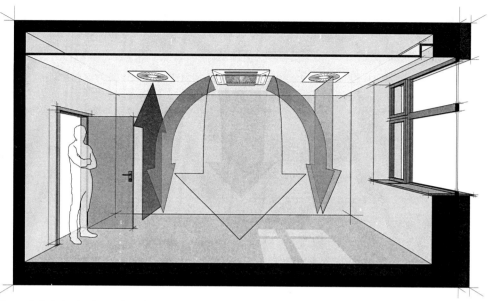

Figure 8–1. A simple air-supply system.

(Source: valigursky/iStock)

General ventilation systems are also used for heat control in an area by introducing large quantities of air, which may be tempered and recycled. This principle is used to dilute contaminants by mixing enough outdoor air with the contaminant so that the average air concentration is safe to breathe. General ventilation is most often used in these circumstances:
- When more cost-effective options are not available
- When gases, vapors, or small aerosols with low toxicity are present
- When emissions occur uniformly in time and space
- When emissions are not near the workers' breathing zones
- When the supply air is clean

The objective of a local exhaust system is to remove the contaminant as it is generated at the source. The advantages of local exhaust systems are that they are more effective than general systems at removing contaminants and that the lower overall flow rates result in lower heating and cooling costs. Additionally, the smaller flow rates lead to lower costs for maintaining the air cleaning equipment.

LEV systems are most effective when the following conditions exist:
- More cost-effective methods do not work.
- The contaminant is toxic.
- The worker is near the emissions source.
- Emission rates vary with time.

- Sources of emissions are large and few.
- Sources are stationary.
- Regulations require sources to be exhausted.

While most sciences use metric units of measure, ventilation engineers still cling to English units. This is slowly changing, as all federal projects now require the use of metric units. In this chapter, a combination of metric and English units will be used and presented. Standard temperature and pressure are given as 70°F, 29.92 in. of mercury (Hg), and dry air.

The amount of air being supplied to a building can be quantified in terms of its volumetric flow rate (Q). The volumetric flow rate is the actual volume of air passing through a given location per unit of time and can be represented by the following formula:

$$Q = VA$$

Where:
Q = Volumetric air-flow rate, in cubic feet per minute
V = Average velocity, in feet per minute
A = Duct cross-sectional area, in square feet

For example, if a square duct is 1 ft high and 1 ft wide, its area is 1 ft^2. Now visualize a 1-ft-long box of air inside the duct. The box would be 1 ft high × 1 ft wide × 1 ft long, giving it a volume of 1 ft^3. If that 1-ft^3 box of air moves down the duct at a rate of 5 ft/min, then the total volumetric amount of air flowing down the duct is 5 ft^3/min.

Example

What is the volumetric flow rate of air in a duct 0.5 ft wide, 1.25 ft high, and moving at a speed of 7.1 ft per min?

Q (ft^3/min) = V (ft/min) × A (ft^2)

Where:
Q = Volumetric air-flow rate (ft^3/min)
V = 7.1 (ft/min)
A = 0.5 (ft) × 1.25 (ft) = 0.625 (ft^2)

Solution:
Q (ft^3/min) = 7.1 (ft/min) × 0.625 (ft^2) = 4.4 (ft^3/min) = 4.4 (ft^3/min)

Because of their versatility and strength, round ducts are used often in ventilation systems. However, because of their shape, it is necessary to understand how to calculate the area of a circle before determining the volumetric flow rate. (See Figure 8–2.) The equation for the area of a circle is as follows:

$$A = \pi r^2 \text{ (r is in ft)}$$
$$Q = V \times A$$

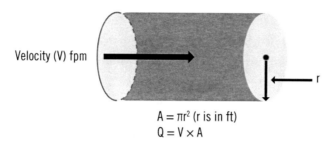

Figure 8–2. Calculating air flow in a duct.

When a round duct is said to be 10 inches, it typically means that its diameter is 10 in. So before calculating the area of a duct, it is necessary to divide the diameter in half to get the radius (r) and then convert that result to feet (divide by 12). Thus, for a 10-in-diameter duct, the radius is 5 in., which converts to 0.42 ft.

Example

What is the volumetric flow rate of air in a 10-in-diameter duct if the air moves at 80 ft per min?

Solution:

$$A = \pi r^2$$
$$= 3.14 \times [0.42 \text{ (ft)}]^2 = 0.55 \text{ (ft}^2)$$

$Q \text{ (ft}^3/\text{min)} = V \text{ (ft/min)} \times A \text{ (ft}^2)$

$Q \text{ (ft}^3/\text{min)} = 80 \text{ (ft/min)} \times 0.55 \text{ (ft}^2) = 44.3 \text{ (ft}^3/\text{min)}$

Because air in a duct is generally considered incompressible, the relationship between the volumetric air flow in a duct tends to be continuous and constant, regardless of the changes in velocity and duct area. This relationship is known as the continuity equation; because air is incompressible, the quantity of air flowing in a duct always stays the same. Therefore, if the area of the duct gets smaller but maintains the same quantity of air, the air has to move more quickly, which means that its velocity increases. (See Figure 8–3.)

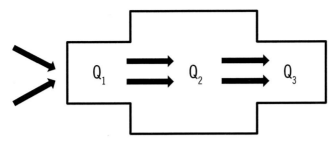

Figure 8–3. The volumetric air flow into a hood and duct stays the same.

Example

If the air in a 14-in. duct with an air flow of 2,000 ft³ per min and a velocity of 13.1 ft per min flows into a smaller, 8-in.- (0.66-ft-) diameter duct, what are the new flow rate and velocity?

Solution:

Q always stays the same; therefore, $Q_2 = Q_1 = 2,000$ ft³/min.

$A = \pi r^2 = 3.14(0.33 \text{ ft})^2 = 0.34 \text{ ft}^2$

$V = Q/A = 2,000 \text{ ft}^3/\text{min}/0.34 \text{ ft}^2 = 5,882 \text{ ft/min}$

MOVING AIR IN DUCTS

The air inside a duct is not likely to move unless force is applied to the air. Typically, that force is created by a fan placed inside the duct, which causes the pressures inside the duct to change, in turn causing the air to move.

When a fan draws air along the duct in a given direction, it creates different pressures at different locations throughout the duct. Three basic types of pressures describe fan-generated movement of air in a duct.

Static pressure in a duct consists of the forces exerted in all directions within the duct. This pressure may be on the sides of the walls of the duct, in the direction of air flow, or against the direction of air flow. Static pressure may be positive or negative, depending on which side of the fan the pressure is measured (Figure 8–4). On the side of the duct past the fan and in the direction of air flow (downstream), the static pressure is positive (Figure 8–5). Conversely, in the area of the duct before the fan (upstream), the static pressure is negative (Figure 8–6). Imagine sucking on a straw with a cherry stuck on the other end—you are the fan, and the straw is upstream of you. The straw will tend to collapse on itself. This indicates negative static pressure acting on the walls of the straw and sucking them in.

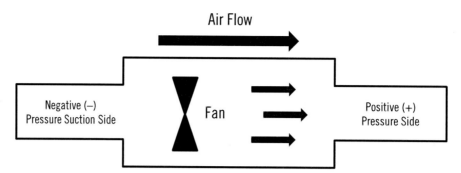

Figure 8–4. The static pressure on either side of a fan.

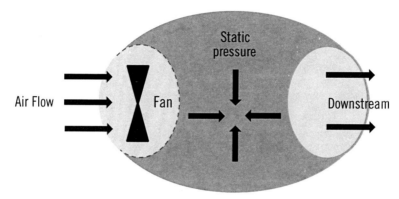

Figure 8–5. Downstream of the fan, the static pressure in a duct is positive, and the duct tends to expand.

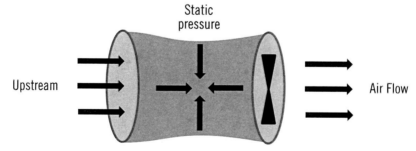

Figure 8–6. Upstream of the fan, the static pressure in a duct is negative, and the duct tends to collapse.

Another representation of static pressure is shown in Figure 8–7. Upstream of the fan, on the left side on the diagram, the static pressure is negative; because it is drawing on the water in the tube, the water level in the tube is pulled upward. On the downstream side of the fan, on the right side of the diagram, the static pressure is positive. So it exerts a force on the water in the tube, and the water level is pushed down.

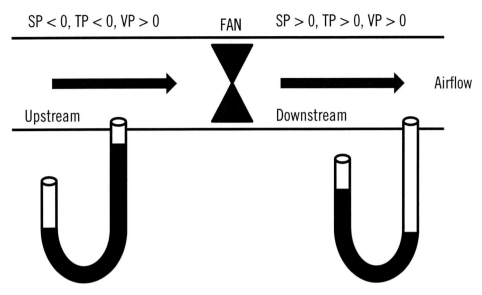

Figure 8-7. Duct pressure upstream and downstream of a fan.

Velocity pressure in a duct is the pressure exerted in the direction of air flow. Velocity pressure for a moving air stream is positive in the direction of travel, both upstream and downstream of the fan.

The total pressure in a duct is the algebraic sum of the static pressure (SP) and the velocity pressure (VP). Total pressure (TP) can be either positive or negative.

$$TP = VP + SP$$

Example

If the static pressure in a duct is −2.5 inches of water (in. wg) and the total pressure is −1.5 in. wg, what is the velocity pressure?

Solution:

VP = TP − SP

  = (−1.5 in. wg) − (−2.5 in. wg)

  = +1.0 in. wg

In general, total pressure falls in the direction of air flow because as the air travels down the duct, energy is lost to turbulence and friction with the sides of the duct.

$$TP_1 = TP_2 + h_L$$

Where:

TP_1 = a point upstream

TP_2 = a point downstream

h_L = energy losses

The effect of energy losses is illustrated in Figure 8–8, where a blockage has been inserted into the duct. As shown in the figure, the total pressure decreases within the duct as energy is lost.

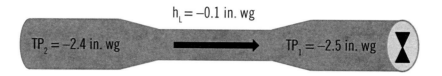

Figure 8–8. Hypothetical energy losses in a duct.

Note in Figure 8–7 that as the air moves downstream toward and through the fan, the pressures change. On the upstream side before the fan, the static and total pressures become increasingly negative the closer to the fan they get. Conversely, the velocity pressure increases as it moves closer to the fan. (This is similar to water going over a waterfall in that the water tends to speed up as it gets closer to the edge.) On the other side of the fan, the static and total pressures become positive and peak. As the air moves farther from the fan, all three pressures tend to continue to decrease as the fan exerts less influence and the duct exerts energy losses. Figure 8–9 is another way to represent the pressures as air moves through a duct before and after the fan.

It is possible to measure the three different pressures inside a duct by attaching two air tubes to the duct and attaching their ends to an inclined manometer. One tube is attached so that its face is on the side of the tube. With this orientation, the tube can measure the static pressure, or the pressure exerted on the side of the duct. The other tube is inserted and positioned so that its opening faces the direction of air flow. With this orientation, the tube can measure the velocity pressure in addition to the static pressure.

If the ends of both tubes—one measuring the static pressure and the other measuring the sum of the static and velocity pressures (that is, the total pressure)—are connected to opposite ends of the inclined manometer, the force of the static pressure against the total pressure will cause the liquid in the tube to subside slightly. The velocity pressure in the duct can thus be measured as $VP = TP - SP$.

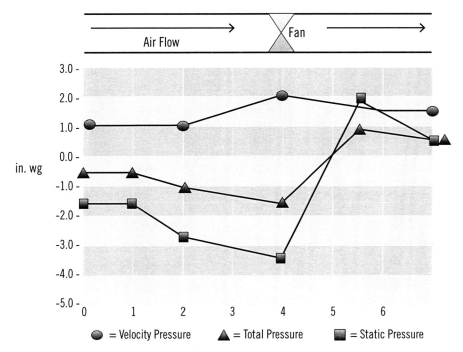

Figure 8–9. Changes in total, static, and velocity pressures in a duct.

In practice, the velocity pressure in the duct can be used to calculate the actual velocity of the air flow inside the duct, using the following equation:

$$V = 4{,}005\sqrt{VP}$$

where the velocity pressure, in inches of water, is the pressure required to accelerate the air from velocity zero to some velocity (V) in feet per minute.

Example
If the velocity pressure in a duct is 0.8 in. wg, what is the velocity of the air?
Solution:
$= 4{,}005\sqrt{VP}$
$= 4{,}005\sqrt{0.8 \text{ in. wg}}$
$= 3{,}583$ ft/min

LEV is used to capture air that is contaminated with toxic or noxious chemicals. In order to do so, the end of the duct is placed as close to the source as possible to suction off the vapors or gases and blow them out of the building or otherwise away from the building occupants.

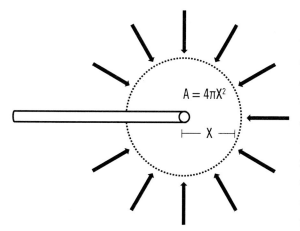

Figure 8–10. Point source suction into a duct.

Unfortunately, contaminated air is difficult to collect into a duct, and a lot of energy is consumed trying to draw the air toward the duct. In addition, because the opening of a duct, or hood, is not effective at separating the contaminated air from the clean air around the opening, trying to collect a toxic dust particle from the air floating in front of the duct opening leads to gathering all the surrounding clean air into the hood. This is illustrated by the schematic in Figure 8–10.

In the figure, if the particle in front of the hood opening is at distance X, then the duct will also draw in air from all of the other directions, which is represented by the area of a sphere, 4. The other air thus competes with the particle at point X. The capture velocity is the speed that the air needs to move toward the duct at any given distance in order to gather the toxic vapors into the duct. The capture velocity can be determined using the following equation for a round duct:

$$Q = VA$$
$$A = 4$$
$$V = Q/(4)$$

Example

What velocity of air would be needed to draw toxic vapors into a duct 12 in. away from a round duct with an air-flow rate of 1,200 ft³ per min?

Solution:

$V = Q/(4\pi X^2)$

$X = 12$ in. $= 1$ ft

$V = 1,200/(4p \times 1^2)$

  $= 1,200$ ft³ per min/12.56 ft²

  $= 95.5$ (ft/min)

By redesigning the opening of a duct, we can reduce the available surrounding air supply and improve the capture efficiency of the system. By adding a circular flange around the duct, we can significantly limit the amount of air that is drawn into the duct.

By adding a flange around the duct, more air is drawn in from farther distances. This improves the efficiency of the hood in capturing noxious aerosols in the spaces nearby.

In general, capture velocities need to be between 50 and 100 ft per min, at a distance of 1 duct diameter, in order to capture gases or vapors released into still air. When moderate, competing air currents are in the area, the capture velocities needed to suck in the vapors must be increased to 100 to 200 ft per min. In areas of rapid air movement where generation of the vapors involves movement of the source, as in spray painting, the capture velocity must be between 200 and 500 ft per min in order to adequately collect the aerosols. When aerosols have considerable mass, as in dust from abrasive blasting or grinding, velocities from 500 to 2,000 ft per min are needed to collect the particles.

Hood Type	Airflow into Hood	Hood Entry Loss Factor (F)
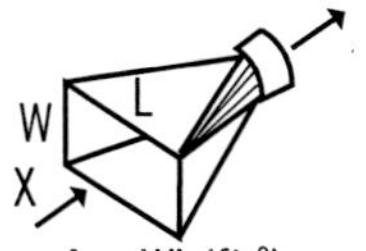 Plain Opening $A = WL$ (ft.^{2})	$Q = V_x(10X^2 + A)$	$0.93\ VP_{duct}$
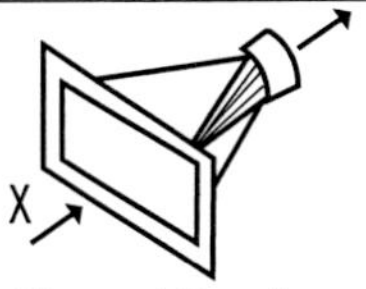 Flanged Opening	$Q = 0.75 V_x(10X^2 + A)$	$0.49\ VP_{duct}$
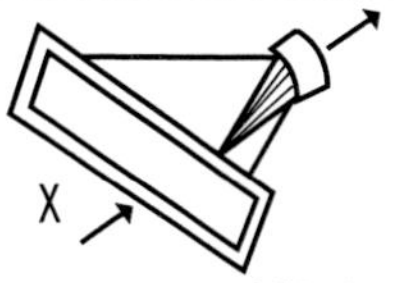 Flanged Slot (L = Length of Slot)	$Q = 2.6\ LVX$	$1.78\ VP_{slot} + 0.25\ VP_{duct}$

Figure 8–11. Hood data for selected capturing hood types, the air flow needed to generate a specified velocity, and the hood entry loss factor.

Other types of hoods can improve capture velocities further. Figure 8–11 shows different types of hood designs and their corresponding capture equations, which represent their quantitative efficiencies.

The following empirical equations can be used to calculate the air flow needed to capture the air at any given distance from a duct opening or hood. For a round or rectangular hood, the air flow needed to capture a contaminant is given by the following equation:

$$Q = V(10x^2 + A)$$

Where:
Q = air flow in cubic feet per minute
V = centerline velocity at x distance from hood in feet per minute
x = distance outward along axis in feet (Note: The equation is accurate only to within 1.5 diameters of the hood opening.)
A = area of hood opening in square feet

> ### Example
>
> Compare the flow rates into a hood necessary to capture a contaminant at 12 in. versus 18 in. Assume a minimum capture velocity of 100 ft per min and a 12-in. round hood opening.
>
> **Solution:**
>
> $A = \pi r^2$
>
> $A = 3.14 \times 0.5 \text{ ft}^2 = 0.785 \text{ ft}^2$
>
> $Q = V(10x^2 + A)$
>
> At a capture distance of 12 in.,
>
> $Q = 100 \text{ ft/min}[10 \times (1 \text{ ft})^2 + 0.785 \text{ ft}^2]$
>
> $Q = 1,079 \text{ ft}^3/\text{min}$
>
> At a capture distance of 18 in.,
>
> $Q = 100 \text{ ft/min}[10 \times (1.5)^2 + 0.785 \text{ ft}^2]$
>
> $Q = 2,329 \text{ ft}^3/\text{min}$

> ### Example
>
> What flow rate would be required to capture a contaminant at a distance of 20 in. given a capture velocity of 110 ft per min and an 18-in. round hood opening?
>
> **Solution:**
>
> $A = \pi r^2 = 3.14(0.75)^2 = 1.766 \text{ ft}^2$
>
> $x = 20 \text{ in.}/12 \text{ in./ft}$
>
> $= 1.66 \text{ ft}$
>
> $Q = V(10x^2 + A) =$
>
> $110 \text{ ft/min}[10(1.66 \text{ ft})^2 + 1.77)] =$
>
> $3,225 \text{ ft}^3/\text{min}$

If we have the same operating parameters as in the example above but put a flange on the end of the duct to better capture the air, the equation to calculate the flow needed changes to $Q = 0.75V(10x^2 + A)$. To capture the contaminants, 25% less air flow is required, but the same distance from the face of the hood is required. Replacing the numbers in the flanged hood equations gives the following:

$$Q = 0.75V(10x^2 + A)$$
$$= (0.75)110 \text{ ft/min}[10(1.66 \text{ ft})^2 + 1.77)]$$
$$= 2,417 \text{ ft}^3/\text{min}$$

We can see that with a flange added to the end of the duct, it takes less air to gather the contaminants into the hood.

By using LEV, it is possible to situate the hood of the capture system as close as possible to the source of contamination to draw off the noxious particles, gases, or vapors before they have a chance to disperse throughout the work space. LEV is the most efficient way to protect workers in the following conditions:

- More cost-effective methods don't work.
- The contaminant is toxic.
- The worker is near the emission source.
- Emissions vary with time.
- Sources of emissions are large and few.

In certain industries and conditions, specific LEV systems are required by law.

The basic components of an LEV system are the hood, which collects nearby air, and the ducts, which then transport the contaminants. Sometimes air cleaning devices are attached in line with the duct to filter aerosols from the air stream. The fan is placed downstream, after the air filter, to pull the air through the filter. Cleaning the air before it enters the fan reduces particle buildup on the fans themselves. After leaving the fan, the air is either recirculated to other parts of the building or ejected to the outdoors through an exhaust stack. A very basic LEV system is shown in Figure 8–12.

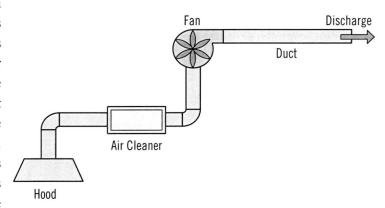

Figure 8–12. A basic local exhaust system.

Source characterization evaluates the workplace processes, chemicals, temperatures, and other factors to determine the type of ventilation that should be used in a work space. Several factors need to be considered when designing a ventilation system:

- location
- relative contribution of each source to the exposure
- characterization of each contributor
- characterization of ambient air
- worker interaction with emission source
- work practices

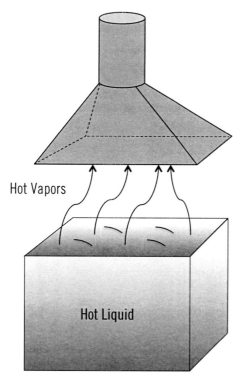

Figure 8–13. Canopy hood.

A wide variety of system designs can meet many different contaminant collection objectives. The effectiveness of a hood is determined by its shape, the extent to which the hood encloses the source, and the quantity of air flowing into the hood. For hot vapors, which tend to rise easily, canopy hoods situated above the source are an efficient means of collecting the vapors, as shown in Figure 8–13. For other sources, it may be advantageous to design slots that run horizontally across the back of a vat in order to draw the vapors or gases away from the workers' breathing zones.

By adding sides to a capture hood, collection efficiencies can be improved, and noise and dust levels from an operation can be minimized. In fact, a chemical laboratory fume hood is designed so that only a small area in the front is open for the worker to stick his or her hands into the hood and do the work. The rest of the fume hood is totally enclosed, with slots in the rear panel drawing the vapors and gases away from the opening (Figure 8–14).

Once the air is captured in the duct, the ventilation system must keep the air suspended and moving. A minimum velocity is needed to ensure that the vapor or dust does not settle in the duct. On

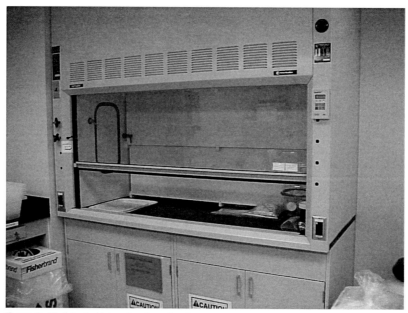

Figure 8–14. A typical laboratory fume hood.

the other hand, excessively high velocity would be wasteful and could cause abrasion to the inside of the duct. Some examples of minimum duct velocities are presented in Table 8–1.

Table 8–1. Recommended Duct Velocities for Various Contaminants

Type of contaminant	Design velocity (ft/min)
vapors, gas, smoke	1,000–2,000
fumes	2,000–2,500
fine dust	2,500–3,000
average dust	3,500–4,000
heavy dust	4,000–4,500
heavy, moist dust	4,500+

FAN LAWS

Fans are the workhorses of ventilation systems. They physically move the air in a room or a duct and create the pressure variations that draw the air one way or the other. A fans' ability to function also depends on the characteristics of the contaminants in the system (mass, density, etc.) and the air temperature.

Energy is provided to the fan so that it can overcome system losses and move the air. Fans are generally categorized as axial, centrifugal, or special.

Axial fans are the most common and resemble a propeller. These fans are also used in general ventilation and can be used to move air in open spaces or to pull air through an open window. Tubeaxial fans are propeller-type fans placed directly inside the duct in line with the direction of air flow. Vaneaxial fans are a specialized type of axial fan that is very efficient but best suited to clean-air situations.

Centrifugal fans have an unusual design. The air is drawn into a sort of cage that contains fan blades, and the blades then spin and eject the air out the other side (the front) of the fan.

The blades on centrifugal fans have different orientations inside the cage depending on the work the fan needs to do and on the air the fan needs to move. Forward-curved blades tend to scoop the air and send it in the direction the blades are traveling. Fans with these blades are good when low to moderate static pressures are needed, the fans need to be quiet, and limited space is available.

If the centrifugal fan blades bend backward and away from the direction of air travel, they are capable of very high speeds and high efficiencies. These

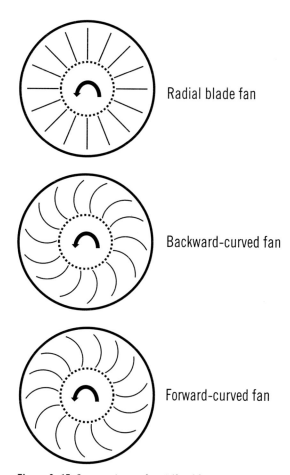

Figure 8–15. Common types of centrifugal fans.

backward-blade fans are quiet and can move a lot of air. But dusts and materials tend to build up on the blade tips over time, so they are best used when the air is relatively clean and has little dust.

The third type of centrifugal fan has radial blades that are flat and straight, with neither a forward nor a backward curve. These are called radial fans, and they have high impact resistance, which makes them effective for moving dirty air with a heavy particulate load. The three most common centrifugal fan types are shown in Figure 8–15.

Special fans include the in-line centrifugal, which is a backward-curved fan placed directly into ductwork. Power exhausters or power roof ventilators can be either axial or centrifugal fans. Other special fans incorporate dust-collection devices into the fan unit.

The amount of air that a fan can move is related to the fan's speed, its ability to create static pressure, and the fan motor's horsepower. The ability to move air through a duct depends on the fan's size, number of blades, and other design factors. A fan's ability to move air, independent of the design, is determined by what is called the fan total pressure, which is a measure of the fan's ability to change the pressure in the duct between the upstream and downstream sides of the fan.

For fans of the same design and size, it is possible to compare some of their significant operating parameters. For a fan operating at a given number of revolutions per minute (rpm), there is a given air-flow rate in the duct. The relationship between flow rate and fan speed is directly proportional and can be shown by using the following equation:

$$Q_2 = Q_1 \times (RPM_2)/(RPM_1)$$

Where
Q_2 = new or final air flow (ft³/min)
Q_1 = initial air flow (ft³/min)
RPM = fan rpm

Example

If a fan initially operating at a speed of 1,000 rpm can move air in the duct at a rate of 10,000 ft³ per min, how fast would the fan have to move to increase the flow rate to 12,500 ft³ per min?

Solution:

$Q_2 = Q_1 \times (RPM_2)/(RPM_1)$

$Q_1 = 10,000$ ft³/min

$Q_2 = 12,500$ ft³/min

$RPM_1 = 1,000$ RPM

$RPM_2 = Q_2/Q_1 \times (RPM_1)$

$RPM_2 = 12,500$ (ft³/min)/10,000 (ft³/min) $\times 1,000$ (RPM)

$= 1,250$ RPM

The relationship between fan static pressure and fan speed is squared and given by this equation:

$$FSP_2 = FSP_1 \times [(RPM_2)/(RPM_1)]^2$$

Where:

FSP_2 = new fan static pressure (in. wg)
FSP_1 = initial fan static pressure (in. wg)
RPM = fan RPM

Example

By increasing the RPM from 1,000 to 1,250, how will our initial FSP of –3.0 in. wg be affected?

Solution:

$FSP_2 = FSP_1 \times [(RPM_2)/(RPM_1)]^2$

$FSP_1 = -3.0$ in. wg

$RPM_2 = 1,250$ RPM

$RPM_1 = 1,000$ RPM

$FSP_2 = -3.0$ in. wg $\times [(1,250$ RPM$)/(1,000$ RPM$)]^2$

$FSP_2 = -4.7$ in. wg

Finally, the relationship between fan speeds has a cubed effect on the horsepower (hp) of a fan system, as shown by the following equation:

$$HP_2 = HP_1 \times [(RPM_2)/(RPM_1)]^3$$

Where:

HP_2 = new or final bhp
HP_1 = initial bhp
RPM = fan rpm

Example

When increasing the RPM from 1,000 to 1,250, how much horsepower is needed? (Initial = 2.0 hp.)

Solution:

$HP_2 = HP_1 \times [(RPM_2)/(RPM_1)]^3$

$HP_1 = 2.0$
$RPM_2 = 1,250$ RPM
$RPM_1 = 1,000$ RPM

$HP_2 = 2.0$ hp $\times [(1,250)/(1,000)]^3$

$HP_2 = 3.9$ hp

By using the equations presented above, it is possible to design and purchase the fans and fan systems that are necessary to meet the design objectives discussed earlier in the chapter. If a certain flow rate is needed to achieve a particular capture velocity for a hood, select a fan that will ensure enough flow in the system. If a system needs just a bit more flow, it may be possible to identify the necessary fan speed to meet the goal. If the horsepower can be increased in the system or if a fan with greater horsepower can be used, the rpm can achieve the necessary air flow.

AIR CLEANING DEVICES

Air cleaning devices are used to filter and remove contaminants from the air stream. The type used depends on the contaminant (particulate or gas/vapor), the possible efficiencies and energy required, and the characteristics of the gas stream (e.g., hot, corrosive, wet).

Fabric filters are one of the simplest devices to clean air. They have a mesh filter that traps particles or aerosols when they interact with the structure of the filter. As the contaminated air stream is pulled through the filter, the larger particles collect on the fabric.

Fabric-type filters are commonly used in baghouse designs, which collect dust in large bags contained in a support structure, such as that shown in Figure 8–16. After dust loads onto the internal bags, the suction is turned off periodically and the bags are shaken to knock the dust off and into a collection hopper placed at the bottom of the structure. Although fabric systems do not have 100% collection efficiency, their efficiency improves with repeated dust loading and accumulation over time. When the air resistance through the clogged filter becomes too great, the bags are shaken to remove the dust collected.

Electrostatic precipitator dust-collection systems collect effluent particles by means of their electrical charge. Effluent streams move through the precipitator between metal plates that have a large electrical charge. Charged particles are attracted to the metal plates and adhere to their surfaces. Meanwhile, the cleaned exhaust stream is sent out the effluent stack. Periodically, the voltage to the metal plates is shut off, and the plates are shaken to release the dust, which falls to hoppers beneath the collection system.

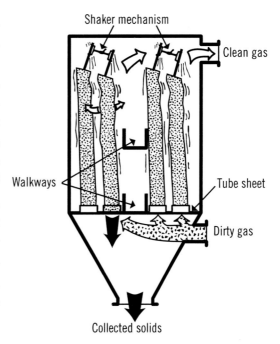

Figure 8–16. A baghouse-type air-filtration system.

(Source: © User:Goran tek-en/Wikimedia Commons/ CC-BY-SA-3.0)

Compared to media filtration devices, electrostatic precipitators require less energy to remove particles from the air stream because of their lower pressure drop.

In another type of effluent cleaning system, the contaminated effluent enters a scrubber, and water or another liquid rains down and washes the dust out of the air stream. These aptly named scrubbers can handle a wide variety of air stream conditions and are often used to satisfy air pollution control requirements. A drawing of a wet scrubbing system is shown in Figure 8–17.

Dry centrifugal air cleaning systems remove particles using centrifugal, inertial, or gravitational forces. These types of devices are also known as cyclone collectors. Because they have low pressure drop requirements, they are relatively inexpensive to operate. In addition, they have low maintenance requirements.

As effluent air enters the cyclone collector, the air begins to spin around, and centrifugal forces cause the larger particles in the effluent stream to move toward the outer edges of the enclosure. As these large particles hit the walls of the collector, they slow down and fall down the side of the collector into a hopper at the bottom of the device. The lighter, cleaner portions of the effluent stream avoid the walls of the device, are sucked upward and out the top of the cyclone, and are ejected from the exhaust stack. (See Figure 8–18.)

In addition to particulate and aerosol collection devices, some air cleaning systems can clean gases and vapors. Charcoal cartridges and filters can be used to adsorb gases and vapors from effluent streams. Wet scrubbers can also be effective in cleaning some gases and vapors from effluents.

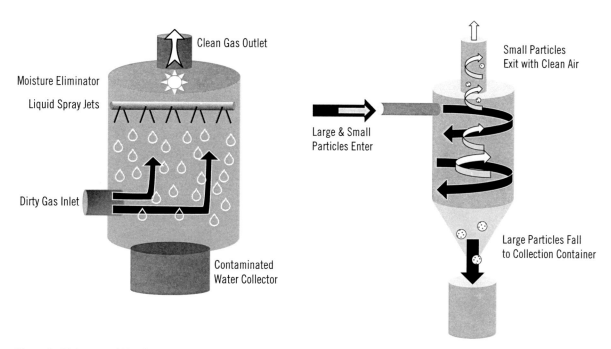

Figure 8–17. A wet scrubbing filtration system.

Figure 8–18. A cyclone filtration system.

GENERAL VENTILATION SYSTEMS

General ventilation might be better called dilution ventilation. It brings new, outdoor air into a space to push out older, dirty air. The new air dilutes the contaminated air and eliminates odors, controls humidity and temperature in the space, reduces fire hazards, and reduces the levels of toxic contaminants in the space. In more specific spaces (e.g., laundries, bakeries, and foundries), general ventilation is used to prevent discomfort and illness.

The value of general ventilation is greatest under the following circumstances:
- More cost-effective options are not available.
- The gases, vapors, or small aerosols present have low toxicity.
- Emissions occur uniformly in time and space.
- Emissions are not near workers' breathing zones.
- The supply air is clean.

In special situations, the principles of general ventilation are used to control the quality of the air. In confined spaces, fresh air can be blown into the space with fans and flexible ducts to increase the oxygen levels, force out toxic gases or vapors, and eliminate explosive environments. Whenever this is done, however, industrial hygienists must be sure not to short-circuit the air supply and must be sure the supply air is fresh.

In occupied spaces of buildings, it is useful to quantify the levels of general ventilation using the unit of air changes per hour (ACH). This unit simply indicates the relative effectiveness of the general ventilation supplied to a space, taking into account the size of the space, using the following relationships:

$$ACH = (Q \times 60)/Vol$$

Where:
ACH = number of ACH
Q = air flow (ft^3/min)
60 = the number of minutes in an hour
Vol = room volume in cubic feet

Example

If a 10 × 10 × 8-ft room is ventilated with both a supply and a return at 100 ft^3 per min, how many ACH are occurring?

Solution:

ACH = (Q × 60)/Vol

Vol = 10 × 10 × 8 = 800 ft^3
ACH = (100 ft^3/min × 60)/800 ft^3
ACH = 7.5 ACH

The recommended ACH levels vary depending on the space. The table on the next page shows some ACH level recommendations for various types of spaces.

ACH	Location
4	Office building
2–10	Warehouse
4–6	Hospital room
2–15	Factory floor
20	Bakery

The air change equation provided earlier has one limitation that we must account for in our analysis. The air in a room is never mixed perfectly because a room will always have corners, pockets, equipment, or furniture that interferes with the flow of the supply and exhaust air. In order to adjust the ACH that is calculated, a mixing factor (K_{mix}) is applied to the equation in the following situations:

K_{mix}	Conditions in Space
1.0	Perfect mixing
1.5	Open spaces, good supply and return locations
3.0	Crowded spaces, poor supply and return locations

Because the primary effect of poor mixing is a reduction in the effectiveness of the air exchange rate, we expect the ACH to be decreased in relation to the mixing factor. By dividing the ACH by the mixing factor, we can calculate the ACH that accounts for less-than-perfect air mixing.

Example

If the room in the preceding example has poor mixing ($K_{mix} = 3$), what is the true ACH?

Solution:

$ACH_{actual} = (Q \times 60)/(K_{mix} \times Vol)$

$Vol = 10 \times 10 \times 8 = 800 \text{ ft}^3$

$ACH = (100 \text{ ft}^3/\text{min} \times 60)/(3 \times 800 \text{ ft}^3)$

$ACH_{actual} = 2.5$

TESTING VENTILATION SYSTEMS

Ventilation systems should be tested at various times and for different purposes. New systems should always undergo commissioning to ensure that the installed systems can operate according to the design parameters and that the installed designs provide adequate air supply and exhaust. Subsequently, periodic testing should be performed to ensure that the systems continue to operate according to design standards. Systems may also need to be reevaluated when building occupants are uncomfortable or when processes or internal architecture is changed.

Three primary methods are used to analyze building air flow. Measurements can be made within ducts by means of an inclined manometer and a pitot tube to assess duct velocity and flow. Air velocity instruments can also be used to measure air currents in ducts and open areas. Third, smoke tubes or tracer gases can be used to measure air flow qualitatively and quantitatively.

Case: Troubleshooting a Chemistry Building Ventilation System

A university chemistry building was designed and built without a mechanical-systems commissioning of the building. After a couple years, when hazardous chemical operations were in full swing, occupants of the building began to notice noxious odors. Many of the problem areas, including offices and classrooms, were nowhere near the toxic chemical laboratories but on other floors and wings of the building.

Numerous experts and consultants were hired to try to find the problems with the ventilation system and where the odors were coming from. After many thousands of dollars were spent, it was determined that toxic chemicals exhausted from the exhaust stacks on the top of the seven-floor building were being drawn back into the air intakes on the side of the second floor of the building. But that still didn't explain why this was occurring or indicate how to correct the problem.

More money was spent on having an engineering firm figure out how the exhaust was getting back into the building. The engineers eventually discovered that as the building was being constructed, the owners decided the exhaust stacks were too tall and detracted from the looks of the building. All the exhaust stacks were then lowered by 6 feet, which resulted in the exhaust not having enough momentum and height to escape the air-supply system that drew air into the second floor of the building.

Needless to say, the underlying problem was that the exhaust stacks hadn't been constructed as designed. An appropriate and thorough ventilation commissioning would have caught the mistake before the building was occupied and people were exposed to toxic chemicals.

AIR SUPPLY AND EXHAUST

Air supply and exhaust can be measured directly using a balometer placed carefully over air-supply or -exhaust ducts. The air moving through the balometer is measured to provide the rate of air flow in cubic feet per minute. These results, in addition to accurate measurement of room size, can provide a good estimate of room air exchange rates.

Pressure gauges measure and indicate building ventilation and air flow, and pressure differentials throughout the building can be used to indicate directions of air flow.

Tracer gases, using measurements of intentionally released contaminants, identify air flow in a building and can provide information about fume hood flow, the dispersion of contaminants in a space, and patterns of building exhaust. The tracer gases used should be easy to collect and analyze, nonreactive, nontoxic, nonexplosive, and not normally present in the air stream.

Smoke tubes release smoke and allow for the observation of air-flow patterns, velocities, and directions. In general, air velocities above 150 ft per min are difficult to see. Smoke can be a quick way to determine room pressurization and air directions.

MEASURING DUCT VELOCITY AND FLOW

The most accurate way to calculate air flow in a duct is to measure the air's velocity and the duct's diameter. An inclined manometer with a pitot tube can be used to calculate the velocity pressure in the duct.

A pitot tube consists of a tube inside another tube. The opening to the internal tube points in the direction of the oncoming air flow; the air pushing into the opening represents both the pressure from the velocity of the air flow and the static pressure in the duct, which is pushing in all directions. In other words, the hole in the center tube represents the total pressure in the duct. The outer tube has holes only in its sides. These holes can't measure the velocity pressure but can measure the static pressure in the duct.

When the outer tube (with the static pressure) is connected to the inclined manometer on one end and the inner tube (representing the total pressure) is connected to the other end of the manometer, the resulting movement of the liquid in the manometer represents the difference between the total pressure and the static pressure, which is the velocity pressure. Using this measurement in the velocity calculation equation from earlier in the chapter, it is possible to calculate the air's actual velocity in the duct.

Because air flow in a duct differs depending on its location in the duct, it is necessary to make several measurements throughout the duct to obtain the average air flow. For example, the flow in the center of the duct tends to be higher than the flow on the duct's sides, where there is more turbulence and friction. Pitot traverse measurements are thus made at the locations shown in Figure 8–19. As a general rule of thumb, the larger the duct, the larger the number of traverse measurements required:

- for round 6-in. or smaller ducts: ≥ 6 traverses
- for round > 6-in. ducts: ≥ 10 traverse points
- for very large ducts: approximately 20 traverse points

After the duct pressure measurements are collected, each one is converted to velocity using the equation. Then all the velocities are averaged to get the final duct velocity.

Pitot tubes are considered primary standards and need no calibration. In order to remain accurate, the pitot tube diameter should be no larger than 1/30th of the duct diameter. Using pitot tubes is not recommended for ducts with velocities greater than 600 to 800 ft per min, and they should not be used in air streams with heavy particulate concentrations because the hole will clog easily.

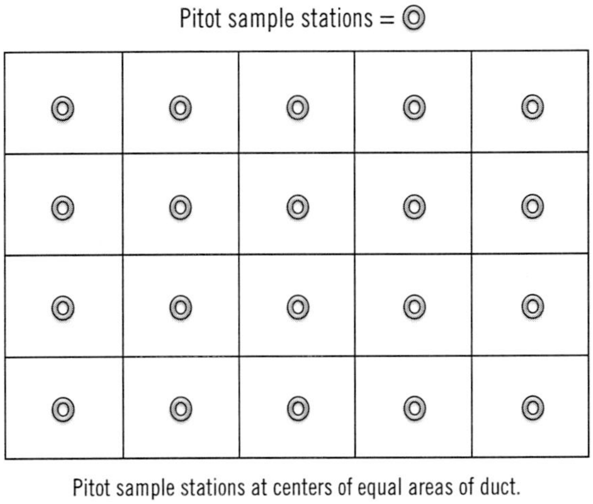

Duct velocities can also be measured using a thermal anemometer, which correlates air speed with the loss of heat from the tip of a hot wire probe that is inserted into the duct. These devices are not as accurate as pitot tubes, however. In addition, the devices need periodic calibration, and they can be dangerous if used near flammable chemicals.

Figure 8–19. A pitot tube traverse plan for round and square ducts.

SUMMARY

An industrial hygienist must understand the basic methods associated with the supply of ambient air and the exhaust of contaminated air from buildings and work spaces. The chapter describes general ventilation systems for dilution, local exhaust systems, fans, stacks, and air filtration systems. Familiarity with the methods of evaluating air-handling systems can help an industrial hygienist make educated decisions that result in a safe work environment.

REVIEW QUESTIONS

1. When is a canopy hood useful?
2. If the mechanical drawings of a ventilation system indicate that the flow rate will be 1,000 ft^3 per min and the duct velocity will be 3,000 ft per min, what diameter of round duct is being used?

3. What do we call the air movement at a given distance from the front of a hood that is necessary to overcome opposing air currents and cause a contaminant to flow into the hood?
   a. Face velocity
   b. Duct velocity
   c. Capture velocity
   d. Slot velocity
4. What type of ventilation system would you recommend for a hazardous-waste storage tank into which toxic chemicals are transferred once a day?
   a. LEV
   b. Supplied-air breathing systems
   c. Secondary spill containment
   d. General ventilation
5. There are certain conditions for which we could use dilution ventilation to control a hazard. List at least three of them.
6. If you have a canopy hood with dimensions 3 ft by 4 ft hanging from the ceiling in the center of a room and located 4 ft above the surface of a solvent tank, and you need a capture velocity of at least 60 ft per min, what flow rate (Q) is required for this system?
7. What are the three types of centrifugal fans?
   a. Forward curve, backward curve, radial
   b. Squirrel cage, forward curve, backward curve
   c. Radial, impeller, backward curve
   d. Tubeaxial, propeller, radial
8. For the duct below, indicate whether the pressure is positive or negative upstream (before) and downstream (after) from the fan.

Upstream	Downstream
VP = ____	VP = ____
SP = ____	SP = ____
TP = ____	TP = ____

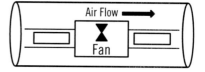

9. What is a thermal anemometer, what does it do, and how does it work?
10. If you measure the average face velocity of a lab hood with the sash in the fully open position to be 45 ft per min and the dimensions of the hood are 2.5 ft wide by 2.0 ft high, at what height should the sash be placed to increase the face velocity to 100 ft per min?
11. A solvent dip tank is equipped with a 2.5-ft-long flanged slot hood to exhaust vapors, and the hood volumetric flow rate is 1,500 ft^3 per min. If this hood needs a minimum capture velocity of 100 ft per min to protect the workers from exposure, how wide can the solvent tank be to maintain

the minimum capture velocity across the tank? (Assume the tank is 30 in. long and the slot opening runs along the entire length of the tank.)

12. The blower you purchased for a local exhaust system is currently operating at 10,000 ft³ per min at –3.0 in. wg FSP using a 2.0-bhp motor running at 500 rpm. If you now need to increase the exhaust to 20,000 ft³ per min, how fast must the fan turn in rpm to achieve the new Q of 20,000 ft³ per min? In addition, calculate the effects of this increase in Q on the FSP and bhp.

13. If $V = 1,096\sqrt{VP/r}$, why do we typically use $V = 4,005\sqrt{VP}$ to calculate velocity flow in a ventilation system?
    a. Fan speed is inversely proportional to area.
    b. The density of air generally stays the same.
    c. Q always stays the same.
    d. All of the above

14. What should be the approximate face velocity of a typical chemical safety fume hood?
    a. 20–50 ft per min
    b. 80–120 ft per min
    c. 200 ft per min
    d. 120–180 ft per min

15. What is the main disadvantage of a canopy hood?
    a. It can't take advantage of the natural direction of the effluent.
    b. It is extremely expensive.
    c. It gets clogged very easily.
    d. The plume may enter the worker's breathing zone.

16. What type of hood would you recommend for a grinding operation?
    a. Receiving
    b. Canopy
    c. Fume
    d. Slot

17. What is the minimum recommended duct velocity for moving heavy dust?
    a. 500–1,500 ft per min
    b. 2,000–2,500 ft per min
    c. 4,000–4,500 ft per min
    d. > 8,000 ft per min

18. If you want to move relatively dirty air, what type of fan would you use?
    a. Backward centrifugal
    b. Forward centrifugal
    c. Radial
    d. Vaneaxial

19. What is the basic characteristic, or special feature, of a scrubber air-filtering system?
20. Why is it necessary to take measurements at numerous places throughout an exhaust duct to estimate velocity?
21. Which is the most accurate method of measuring exhaust duct velocities?
    a. Inclined manometer and pitot tube
    b. Thermal anemometer
    c. Smoke tube
    d. Wind vane anemometer
22. The total pressure of a local exhaust system is −2.4 in. wg, and the static pressure is −2.8 in. wg inside a duct. What is the velocity pressure in inches of water, and what is the velocity in feet per minute for this system?
23. Given a static pressure of −1.7 in. wg and a total pressure of −1.1 in. wg, what is the velocity pressure? On which side of the fan were these readings taken?
24. A duct's diameter is 16 in. at point A, and the velocity is 2,800 ft per min. The duct increases to 28 in. at point B.
    a. What is the Q at point A and the approximate Q at point B?
    b. What is the new velocity at point B?
25. If it is recommended that a patient's hospital room have a minimum of 12 ACH and the room is 15 × 25 × 12 ft, what quantity of air needs to be supplied to (and exhausted from) the room?
26. The velocity of air in a duct is 3,582 ft per min. What is the velocity pressure?
    a. 0.2
    b. 0.8
    c. 54.1
    d. 1,800
27. Which type of exhaust cleaning system requires the least pressure drop or differential in the effluent stream but is relatively expensive to install and operate?
    a. Electrostatic precipitator
    b. HEPA
    c. Baghouse filters
    d. Charcoal
28. What volumetric flow rate is needed to capture contaminant vapors in front of a 6-ft-long slot-type hood in order to maintain a capture velocity of 150 ft per min 2 ft from the slot opening?

REFERENCES

American Conference of Governmental Industrial Hygienists. *Industrial Ventilation—A Manual of Recommended Practice*. 28th ed. Cincinnati, OH: ACGIH, 2013.

———. *TLVs® and BEIs®*. Current ed. Cincinnati, OH: ACGIH.

American National Standards Institute/American Industrial Hygiene Association (ANSI/AIHA). Ventilation-Related Standards. New York: ANSI.
    Z9.1–2006, *Ventilation and Operation of Open Surface Tanks*.
    Z9.2–2012, *Fundamentals Governing the Design and Operation of Local Exhaust Ventilation Systems*.
    Z9.3–2007, *Spray Finishing Operations—Design, Construction and Operation*.
    Z9.4, *Abrasive Blasting Operations—Ventilation and Safe Practices for Fixed Locations*.
    Z9.5–2012, *Laboratory Ventilation*.
    Z9.6–2008, *Exhaust Systems for Grinding, Polishing and Buffing*.
    Z9.7–2007, *Recirculation of Air from Industrial Process Exhaust Systems*.
    Z9.9–2011, *Portable Ventilation Systems*.

American Society of Heating, Refrigerating, and Air-Conditioning Engineers, Inc. *Handbook of Fundamentals*. Atlanta: ASHRAE, 2013.

Harris, M. K., L. E. Booher, and S. Carter. *Field Guidelines for Temporary Ventilation of Confined Spaces*. Fairfax, VA: AIHA, 1996.

McDermott, H. "Dilution Ventilation of Industrial Workplaces." In *Fundamentals of Industrial Hygiene*, edited by Barbara Plog. Itasca, IL: National Safety Council, 2012.

———. *Handbook of Ventilation for Contaminant Control*. 3rd ed. Cincinnati, OH: ACGIH, 2000.

———. "Local Exhaust Ventilation." In *Fundamentals of Industrial Hygiene*, edited by Barbara Plog. Itasca, IL: National Safety Council, 2012.

National Research Council. *Prudent Practices for Handling Hazardous Chemicals in Laboratories*. Washington DC: National Academy Press, 1981.

U.S. Occupational Safety and Health Administration. *Ventilation*. General Industry Safety and Health Regulations, Code of Federal Regulations, Title 29, Part 1910.94, subpart G.

9
Respiratory Protection

LEARNING OBJECTIVES

After completing this chapter, readers should be able to do the following:

- List Occupational Safety and Health Administration (OSHA) requirements for respiratory protection.
- Describe basic human anatomy and function of the lungs.
- Outline the requirements for medical clearance needed prior to wearing a respirator.
- Understand how respirators work and the different types available.
- Review the respirator fit-testing purpose and process.

Photo credit: Used with permission from MSA.

INTRODUCTION

The most common way American workers are exposed to harmful substances in the workplace is through the inhalation of toxic particles, gases, vapors, and fumes. Although employers design and implement engineering controls to reduce or eliminate these airborne contaminants, respiratory protection is needed to protect workers from levels that controls cannot eliminate or as a backup in the event controls fail. In many occupational workplace processes or procedures, engineering controls may not be feasible to implement. In these cases, the employer relies solely on the use of respirators for the protection of workers from hazardous airborne contaminants. Yet using respirators to protect workers may not be effective if the respirator is not selected properly, fitted or used correctly, or if the employee is physically unable to use the respirator. Simply putting a respirator on a worker does not guarantee protection. Having an understanding of the anatomy of the lungs and how a respirator functions will help the safety professional, industrial hygienist, or occupational health nurse make good decisions on using respiratory protection.

This chapter focuses on basic fundamentals of respirator use, a basic understanding of how the lungs function, and basic respirator function. Much of our implementation of respiratory protection in the workplace is driven by OSHA standards, particularly *29 CFR 1910.134 Respiratory Protection*. Although some discussion of the standards is presented in this chapter, the intent is to give a general overview of the regulations, not to explain the specifics of *29 CFR 1910.134*.

REVIEW OF OSHA RESPIRATORY PROTECTION STANDARDS

29 CFR 1910.134 Respiratory Protection is the set of standards used in general industry. Additionally, several *29 CFR 1926 Construction Industry* standards refer to standards in *29 CFR 1910.134* to be followed in construction-related work where respirators are required.

When effective engineering controls are not feasible for any inhalation hazard, the employer must provide appropriate personal protective equipment (PPE). For airborne contaminants, the standards require that the employer provide appropriate respirators for the purpose intended. The employer must also develop a written respiratory protection program that must be administered by a trained program administrator. The written program must include the following:

- procedures for selecting respirators
- medical evaluation of employees using respirators

- fit-testing of respirators
- procedures for proper use of respirators
- procedures for cleaning and disinfecting respirators
- procedures for air quality in using air-supplied respirators
- training and training procedures
- evaluating the effectiveness of the respiratory protection program
- procedures where respirators are not required

ANATOMY AND FUNCTION OF THE LUNGS

The human respiratory tract is made up of the nose, pharynx, larynx, trachea, bronchi, and lungs (see Figure 9–1). These parts are fundamental to external respiration (taking in oxygen from the air, moving it through the body, and then expelling carbon dioxide from the body).

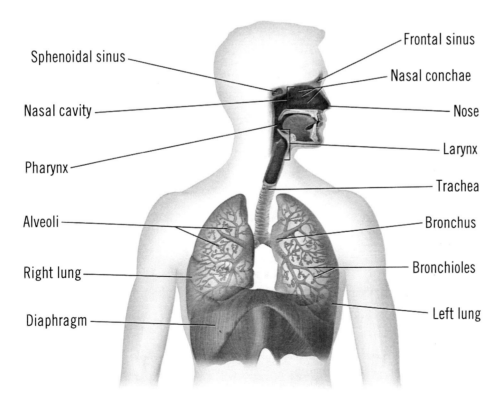

The Respiratory System

Figure 9–1. Schematic drawing of the respiratory system.

(Source: Blausen.com staff. Blausen gallery 2014. Wikiversity Journal of Medicine.)

There are two parts to the nose: the external portion on your face and the internal nasal cavity located in the skull. The nasal septum divides the nasal cavity into two passageways. The nasal cavities open to the outside through the nares or nostrils. The nares contain hairs that trap particles breathed in.

Turbinates are curved projections of tissue found near the middle of the nasal cavity and both sides of the septum. Turbinates help condition the air (warming) as it passes into the lungs. A mucous membrane lines the turbinates and the interior walls of the nose. Mucous traps bacteria and dust as air enters the nasal cavity. It also warms and adds moisture to incoming air.

Cilia are hairlike filaments that line the mucous membrane. Cilia aid the mucous membrane in cleaning incoming air.

The pharynx (throat) is a tubelike structure that connects the mouth and the nasal cavity. The walls of the pharynx are lined with a mucous membrane and cilia. From the pharynx, the larynx (a tube-shaped organ) connects the trachea and the esophagus to the pharynx (Revoir and Bien 1997). The larynx is also known as the voice box. A flap of tissue known as the epiglottis automatically covers the opening of the larynx when food is swallowed, preventing entry into the trachea and then the respiratory tract. Food instead flows to the esophagus and into the digestive system.

The trachea, or windpipe, is a tube lined with cartilage rings in connective and muscle tissue to prevent its collapse. The trachea divides into the right and left bronchi. Each bronchus is similar in structure to the trachea and leads to a separate lung, and further division occurs in smaller tube-shaped passageways called bronchioles. The right lung is divided into three lobes and the left into two lobes. This makes the right lung slightly larger.

The bronchioles do not contain cartilage but are lined with circular muscles (Revoir and Bien 1997). Bronchioles further branch out into atria and then lead into ducts that end with small air sacs called alveoli. Walls of the alveoli are very thin, only two cells thick, which allows oxygen to pass freely. Very small particles of contaminant (< 5 µm) that enter the lungs eventually pass into the bloodstream in this manner as well. The human respiratory tract can contain 300 million alveoli (Balmes 2012). A lubricated membrane, called the pleura, covers the lungs and chest cavity.

RESPIRATION

Respiration is subdivided into four stages: breathing, external respiration, internal respiration, and intercellular respiration. This chapter focuses on just external respiration, which is the transfer of oxygen from the lung air in the alveoli to the blood in the lung cells.

AIRBORNE CONTAMINANTS

Because inhalation is the most common way for hazardous chemicals to enter workers' bodies, safety and health professionals need to understand the process. Airborne toxic substances in the form of gases, vapors, dusts, smoke, fumes, and mists can be inhaled through the nose, throat, bronchial tubes, and lungs into the bloodstream (Goetsch 2005). Toxic materials can be absorbed very quickly in the lungs and passed throughout the body and brain in the blood. As explained in Chapter 7, aerosols are extremely small liquid or solid particles that remain suspended in the air for long periods of time. Fibers, fumes, mists, dusts, fog, and smoke are all considered types of aerosols.

RESPIRATOR FILTRATION

Filtration mechanisms are designed for different types of aerosols, use times, and filter efficiency levels (Bollinger and Schutz 1987). The user needs to be knowledgeable of the contaminant involved in the work task to properly select a filter. The various filtration mechanisms are discussed in the following sections. For details on classes of filters, cartridges, and canisters for respiratory protection, see the *NIOSH Guide to Industrial Respiratory Protection* (Bollinger and Schutz 1987).

INTERCEPTION

As depicted in Figure 9–2, with the interception filtration mechanism, airstreams containing particles approach a fiber in the filter media. The airstream splits as it strikes the fiber and then comes together on the other side of the fiber. If a particle in the airstream is within one particle-size distance of the fiber, it is captured by the fiber. Interception capture increases as particle size increases.

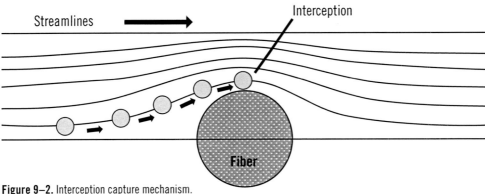

Figure 9–2. Interception capture mechanism.

SEDIMENTATION

Large particles 2 μm in size and larger are captured by sedimentation. This type of capture relies on gravity's effect on particles, and therefore airflow rate through the filter must be low (see Figure 9–3).

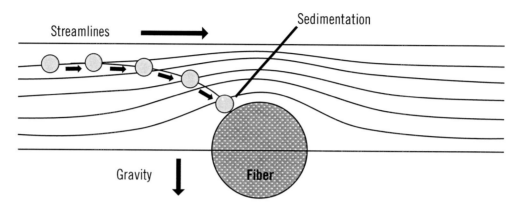

Figure 9–3. Sedimentation capture mechanism.

INERTIAL IMPACTION

As the airstream splits and changes direction to get around a filter fiber, particles with sufficient inertia (matter in motion will stay in motion unless a force is exerted on it) cannot change direction to avoid the fiber (Bollinger and Schutz 1987). The particles slam into the fiber and are captured by impaction (see Figure 9–4).

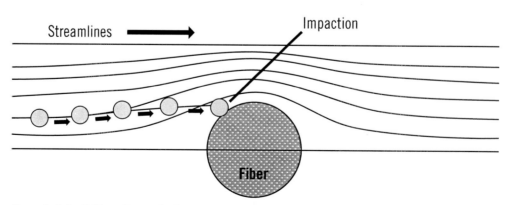

Figure 9–4. Inertial impaction mechanism.

DIFFUSION/BROWNIAN MOTION

Smaller particles, as they diffuse within the airstream, are affected by air molecules colliding with them and causing them to continually change their direction. The particles can randomly cross the airstream, encounter a fiber as they pass, and be captured. This random, Brownian motion is dependent on particle size and air temperature (see Figure 9–5).

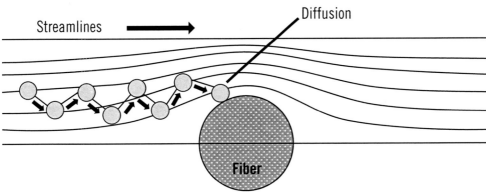

Figure 9–5. Diffusion capture mechanism.

ELECTROSTATIC PRECIPITATION

Particles can be electrically charged (positive or negative). The filter fibers have the opposite charge, which attracts particles that are charged or not charged (see Figure 9–6). The electrostatic filters use electret fibers. These fibers are plastic with a positive charge imbedded on one side of the fiber and a negative charge imbedded on the other. Charged particles are attracted to the opposite charge. Uncharged particles can also be attracted.

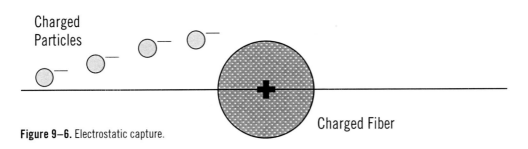

Figure 9–6. Electrostatic capture.

SUMMARY OF FILTRATION

Several factors determine how particles will be captured by a filter. Capture of particles depends on particle size, filter fiber size, particle velocity, charge of particle, and temperature and humidity. Filter media are designed to use a combination of the mechanisms described above to obtain maximum capture of particles. No filter media can be 100% effective in capture of particles. To achieve this design, the user would be unable to breathe through the media. Generally, large heavy particles are removed by impaction and interception; large light particles are removed by diffusion and interception. Very small particles are removed by diffusion. Sedimentation, interception, and inertial impaction combine to effectively remove particles 0.6 µm in size. Low flow rates through filters allow diffusion to remove particles 0.1 µm in size. Particles between 0.1 and 0.6 µm are most effectively captured through the electrostatic capture mechanism.

FILTERS

Filter classifications are determined for the most desirable compromise for the particulate to be filtered from the air. These classifications consider filter surface area, resistance to breathing, efficiency in filtering particles of specific size ranges, and time to clog the filter. Filters can be designed of random-laid, nonwoven fiber materials or fiberglass that can be loosely packed in a filter container or compressed into a flat sheet that is pleated (similar to furnace filters you see in your home). Most respirator filters are nonabsolute, which means they contain pores larger than the particles to be removed. Absolute filters use screening to remove particles that are larger than the filter pore size.

Filter efficiency is certified by the National Institute for Occupational Safety and Health (NIOSH) into nine classes of filters, which are divided into three series: N, P, and R.

The N-series filters are restricted to use in atmospheres free of oil aerosols. They may be used for any solid or liquid airborne particulate that does not contain oil. N-series filters include the following:

- N95 particulate filter—at least 95% filter efficient when tested with approximately 0.3-µm NaCl aerosol
- N99 (previously low efficiency) particulate filter—at least 99% filter efficient when tested with approximately 0.3-µm NaCl aerosol
- N100 (previously high efficiency) particulate filter—at least 99.97% filter efficient when tested with approximately 0.3-µm NaCl aerosol

The R-series filters are intended for removal of any solid or liquid particulate hazard including oil-based liquid aerosols. If the atmosphere does contain oil, the filters should be used for only a single shift (or 8 hours of continuous or intermittent use). R-series filters include the following:
- R95 particulate filter—at least 95% filter efficient when tested with approximately 0.3-μm dioctyl phthalate aerosol
- R99 (previously low efficiency) particulate filter—at least 99% filter efficient when tested with approximately 0.3-μm dioctyl phthalate aerosol
- R100 (previously high efficiency) particulate filter—at least 99.97% filter efficient when tested with approximately 0.3-μm dioctyl phthalate aerosol

The P-series filters are intended for removal of any solid or liquid particulate hazard including oil-based liquid aerosols. However, where oil-based aerosols are present, these filters can be used for longer periods than the R-series. NIOSH requires that the respirator manufacturer establish time-use limitation for all P-series filters. P-series filters include the following:
- P95 particulate filter—at least 95% filter efficient when tested with approximately 0.3-μm dioctyl phthalate aerosol
- P99 (previously low efficiency) particulate filter—at least 99% filter efficient when tested with approximately 0.3-μm dioctyl phthalate aerosol
- P100 (previously high efficiency) particulate filter—at least 99.97% filter efficient when tested with approximately 0.3-μm dioctyl phthalate aerosol

Filters become more efficient as material collects on them and plugs up the spaces between fibers.

VAPOR- AND GAS-REMOVING RESPIRATORS

In contrast to filters that are for the most part effective no matter what the particulate is in the air, cartridges and canisters used for vapor and gas removal are designed for protection against specific contaminants.

REMOVAL MECHANISMS
Vapor- and gas-removing respirators normally remove the contaminant by interaction of its molecules with a granular, porous material called the sorbent (Bollinger and Schutz 1987). The method by which the molecules are removed is called sorption. There are three removal mechanisms in vapor- and gas-removing respirators.

Adsorption

Adsorption retains the contaminant molecule on the surface of the sorbent granule by physical attraction over a large surface area. The intensity of this attraction varies with the type of sorbent and contaminant, but generally adsorption attraction holds molecules weakly (Bollinger and Schutz 1987). Activated charcoal is the most common adsorbent and is used primarily to remove organic vapors. However, impregnated activated carbon makes the bonds holding molecules stronger by chemisorption. Chemisorption is the formation of bonds between molecules of the impregnated granules and the contaminant. Activated charcoal impregnated with other substances can make cartridges or canisters more selective for specific gases and vapors.

Absorption

Absorbents are porous but do not have a large, specific surface area. Gas or vapor molecules penetrate deeply into the molecular spaces in the sorbent and are held there chemically. Absorption is slower than adsorption.

Catalyst

A catalyst is a substance that influences the rate of chemical reaction between other substances (Bollinger and Schutz 1987). These mechanisms are 100% efficient until the sorbent's capacity is exhausted. At this point, breakthrough occurs, and contaminant passes through the cartridge or canister and into the respirator. Change schedules need to be employed to change out the sorbent before breakthrough occurs. Humidity will affect the effectiveness of cartridges and canisters and the service life.

OSHA's respiratory protection program standards require change schedules. A change schedule is the part of the written respirator program that says how often cartridges should be replaced and what information was relied on to make this judgment. A cartridge's useful service life is how long it provides adequate protection from harmful chemicals in the air. The service life of a cartridge depends on many factors, including environmental conditions, breathing rate, cartridge filtering capacity, and the amount of contaminants in the air. It is suggested that employers apply a safety factor to the service life estimate to assure that the change schedule is a conservative estimate. As discussed by Nelson and Janssen (1999a) in the cartridge change guideline document, as well as the OSHA respiratory e-tool, there are three valid ways for an employer to estimate a cartridge's or canister's service life:
- Conduct experimental tests. Tests can help accurately determine the exposures and service life in a particular work environment. These can be reliable and save costs and validate change schedules.

- Use the manufacturer's recommendation. This can result in a more accurate estimate for the brand of respirator being used because it relies on the expertise and knowledge of the manufacturer. A downside, however, is that the recommendation might not accurately reflect a specific work environment and might raise the cost of changing cartridges or canisters.
- Use a math model. Mathematical equations have been used to predict the service lives of organic vapor respirator cartridges when used for protection against single contaminants. OSHA's website has a tool that allows predicting change-out schedules. This is an inexpensive way to develop these schedules. However, the tool is not as accurate as experimental testing, which makes it a conservative tool, and using it may result in service life estimates that are much shorter than necessary.

Some cartridges and canisters have an end-of-service-life indicator built in that allows the user to see when change out needs to be done. This usually occurs with a color change on the device.

TYPES OF RESPIRATORS

NIOSH and OSHA identify two types of respirators: **atmosphere-supplying respirators** and **air-purifying respirators**. Because there are several kinds of respirators associated with each type, it may be easier to identify the two as classes of respirators. The specific kinds of respirators in each class are discussed below.

AIR-PURIFYING RESPIRATORS

Air-purifying respirators are very common. They filter out aerosolized contaminants in the work environment or ambient air. They cannot be used in an oxygen-deficient atmosphere or atmospheres that are immediately dangerous to life or health (IDLH). These respirators offer protection from aerosols including dusts, particulate, fibers, mists, fumes, vapors, and gases. The useful life of these is limited to the concentration of air contaminants, breathing rate, temperature, humidity, filter loading, and cartridge/canister loading. The types of air-purifying respirators are as follows:

Filtering Facepiece
Filtering facepieces are disposable, one-time-use devices. (See Figure 9–7.)

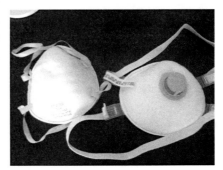

Figure 9–7. Filtering facepieces.

Figure 9–8. Advantage half-face, negative pressure, air-purifying respirator.

(Source: Used with permission from MSA.)

Figure 9–9. A full-face, tight-fitting respirator.

(Source: Used with permission from MSA.)

Disposable Air-Purifying Respirators

Filters or cartridges are on a half-face respirator that cannot be changed out.

Reusable Quarter-Face, Half-Face, or Full-Face Respirators

Filters or cartridges can be changed out, with the facepiece being reusable. A half-face respirator is shown in Figure 9–8. A full-face respirator is shown in Figure 9–9.

All the facepieces or respirators described above are negative-pressure air-purifying respirators. The user's lungs provide the power to draw air through the filter, cartridge, or canister. These types of respirators are tight fitting, providing a seal between the face and the device. This creates a negative pressure inside the facepiece or respirator when the wearer breathes in.

Mouthpiece Respirator

Mouthpiece respirators are for emergency use. Differing from the respirators mentioned previously, these do not form a seal around the face and are not practical for routine work.

Powered Air-Purifying Respirator (PAPR)

The powered air-purifying respirator has a battery-powered motor that pulls air through the filters, cartridges, or canisters. The wearer does not use lung power to draw air through the respirator. Therefore, pressure inside the respirator or hood is positive. A PAPR can be a tight-fitting or loose-fitting, positive-pressure air-purifying respirator. (See Figures 9–10, 9–11, 9–12.)

Figures 9–10, 9–11, 9–12. Positive-pressure PAPR.

(Source: Used with permission from MSA.)

ATMOSPHERE-SUPPLYING RESPIRATORS

Atmosphere-supplying respirators do not use ambient air or work environment air. Air is supplied in self-contained bottles carried by the user or by air compressors that remotely provide manufactured Grade D air. The types of respirators in this class contain clean air and provide a positive pressure in the breathing zone of the respirator. There are two types.

Self-Contained Breathing Apparatus

With a self-contained breathing apparatus (SCBA), the bottles are pressurized containers of breathing air provided directly to the user. These respirators are used in atmospheres that have unknown contaminants, that are IDLH, or that are oxygen deficient (< 19.5%). The air bottle is carried by the user, and air pressure is positive inside the respirator. Use is limited by the amount of air in the bottle. These bottles must last a minimum of 30 minutes per OSHA regulations and can last up to 1 hour depending on the consumption rate of the user and other factors.

Air-Line Respirators

Air-line respirators deliver compressed grade D (at a minimum) breathing air. Air quality for compressed air follows American National Standards Institute (ANSI)/Compressed Gas Association standards. Air is supplied remotely through an air line from the compressors or cylinders to the user's respirator.

Air pressure is positive. These respirators can be used for longer periods of time in the work environment because air lines can be left in the work environment and connected. An example of an air-line respirator system is shown in Figure 9–13.

Air-line respirator systems can be used in an IDLH atmosphere only if they are also equipped with an auxiliary self-contained air supply (see Figure 9–15). The SCBA is an air supply that is independent of the air-line air supply, allowing a worker to evacuate a contaminated area. The air line allows longer use time than the SCBA, which is used to evacuate areas such as confined spaces.

Figure 9–13. Constant-flow air-line respirator (supplied-air respirator).

(Source: Used with permission from MSA.)

Figure 9–14. PremAire Cadet Escape system on an air-line respirator (supplied-air respirator).

(Source: Used with permission from MSA.)

COMBINATION RESPIRATORS

NIOSH and the Mine Safety and Health Administration have approved combination air-line-type atmosphere-supplying and air-purifying respirators. This type of respirator would be approved under the provisions of the air-purifying class of respirator because these provide the least protection to the user. This combination provides protection when entering or leaving a hazardous area without being connected to a source of compressed air (i.e., walking to the actual work area and connecting to the air lines in the work area).

RESPIRATOR SELECTION

The OSHA Respiratory Protection Standards require that respirators be selected on the basis of the respiratory hazard in the workplace as well as work procedures (e.g., spraying versus painting). The hazards need to be identified and evaluated to determine the proper type of respirator, filters, cartridges, or canisters. If the hazard cannot be identified or the worker exposure determined, the work environment must be considered IDLH. Specific OSHA standards also need to be considered (e.g., asbestos) for selection (Colton 2012).

In OSHA's Respiratory Protection Standard, respirators selected must have an assigned protection factor (APF) that is adequate for the particular workplace exposure. To obtain a hazard ratio, divide the air contaminant concentration in the work area by the occupational exposure limit (OEL) or permissible exposure limit (PEL). From Table I: Assigned Protection Factors in 29 CFR 1910.134(d)(3)(i)[A], choose a respirator with an APF greater than or equal to the hazard ratio (Table 9–1).

Table 9–1. 29 CFR 1910.134(d)(3)(i)(A) Assigned Protection Factors[5]

Type of respirator[1,2]	Quarter mask	Half mask	Full facepiece	Helmet/ hood	Loose-fitting facepiece
1. Air-Purifying Respirator	5	[3]10	50	....	....
2. Powered Air-Purifying Respirator (PAPR)	....	50	1,000	[4]25/1,000	25
3. Supplied-Air Respirator (SAR) or Airline Respirator					
• Demand mode	....	10	50	....	....
• Continuous flow mode	....	50	1,000	[4]25/1,000	25
• Pressure-demand or other positive-pressure mode	....	50	1,000	....	....
4. Self-Contained Breathing Apparatus (SCBA)					
• Demand mode	....	10	50	50	....
• Pressure-demand or other positive-pressure mode (e.g., open/closed circuit)	....	....	10,000	10,000	....

Notes:

[1]Employers may select respirators assigned for use in higher workplace concentrations of a hazardous substance for use at lower concentrations of that substance, or when required respirator use is independent of concentration.

[2]The assigned protection factors in Table 1 are only effective when the employer implements a continuing, effective respirator program as required by this section (29 CFR 1910.134), including training, fit testing, maintenance, and use requirements.

[3]This APF category includes filtering facepieces, and half masks with elastomeric facepieces.

[4]The employer must have evidence provided by the respirator manufacturer that testing of these respirators demonstrates performance at a level of protection of 1,000 or greater to receive an APF of 1,000. This level of performance can best be demonstrated by performing a WPF or SWPF study or equivalent testing. Absent such testing, all other PAPRs and SARs with helmets/hoods are to be treated as loose-fitting facepiece respirators, and receive an APF of 25.

[5]These APFs do not apply to respirators used solely for escape. For escape respirators used in association with specific substances covered by 29 CFR 1910 subpart Z, employers must refer to the appropriate substance-specific standards in that subpart. Escape respirators for other IDLH atmospheres are specified by 29 CFR 1910.134 (d)(2)(ii).

(Source: OSHA, 29 CFR 1910.134 (D)(3)(i)(A))

In addition, the employer must select a respirator for employee use that maintains the employee's exposure to the hazardous substance at or below the maximum use concentration (MUC) in 29 CFR 1910.134(d)(3)(i)[B][1]. The MUC is determined mathematically by multiplying the APF specified for a respirator by the PEL, short-term exposure limit, ceiling limit, or OEL. The result is the upper limit at which the type of respirator is expected to provide protection. When exposures approach the MUC, the employer should select the next highest type of respirator.

Example

Calculating the MUC

The OSHA PEL for a contaminant is 0.1 ppm, 8-hour time-weighted average (TWA). Can workers use a full-face, air-purifying negative-pressure respirator if the contaminant concentration (cc) in the work area will be 10 ppm?

Solution:

The APF for this respirator is 50 (see Table 9–1).

50 (APF) × 0.1 ppm = 5 ppm

The answer is no because the work environment exceeds the MUC for this respirator. A worker would need to use at least a full-face PAPR.

By determining the hazard ratio, a respirator with an APF of at least 100 is needed for this work.

10 ppm/0.1 ppm = 100

MEDICAL EVALUATION

A respirator should not be donned, fit-tested, or used until the user has been determined to be physically fit and able to wear the respirator. OSHA's Respiratory Protection Program states that respirators may place a physiological burden on employees that varies with the type of respirator worn, the job and workplace conditions, and the medical status of the employee. The employer is responsible to have employees medically evaluated by a physician or licensed health care professional; this evaluation determines the work the employee will do with the respirator and the types of respirators intended to be used. The OSHA standards require that the physician or licensed health care professional utilize the medical questionnaire provided in the appendices of the standards. The questionnaire is confidential and covers the worker's history of respiratory disease, work history, and other pertinent medical evaluations. Follow-up medical testing is determined by the physician (e.g., pulmonary function test), who then provides written recommendations on the following for the worker:

- limitations of respirator use by the worker
- additional or follow-up medical evaluations
- that the worker can use the type of respirator being provided by the employer
- that the employer must provide an additional type of respirator
- whether the employee is medically fit to use a respirator

Medical evaluations must be done initially before fit-testing, annually thereafter as dictated by a specific standard (e.g., asbestos), and when any medical situation dictates.

FIT-TESTING PROCEDURES

A respirator fit-test must be performed by the employer on each employee prior to initially using a respirator. Subsequent fit-tests should be conducted annually thereafter or sooner if there is a change in respirator selection or the worker has had a medical condition/change, obvious weight gain/loss, or the employee notices that the fit is unacceptable. The OSHA standards require respirator fit-testing of all tight-fitting air-supplying respirators (positive and negative pressure), including the filtering facepiece, and all tight-fitting atmosphere-supplying respirators (Revoir and Bien 1997; 3M 2012). Continuous-flow atmosphere-supplied respirators or PAPRs equipped with hoods, helmets, or some other type of loose-fitting facepiece do not require fit-testing.

When respirators with elastomeric (natural or synthetic elastic material such as natural rubber, silicone) facepieces are being used by the employer, the employer needs to provide an assortment of types to fit-test employees. A good practice is to have at least three sizes for each type of respirator from at least two different respirator manufacturers (Revoir and Bien 1997). The employer has the choice of using a qualitative or quantitative fit-test with one exception: Qualitative fit-tests on negative-pressure air-purifying respirators can be used only in atmospheres up to 10 times the OEL or PEL. Positive-pressure respirators can be tested qualitatively or quantitatively in the negative-pressure mode only.

QUALITATIVE FIT-TESTING

In simple terms, a qualitative fit-test lets the employer know the answer to this question: Does the respirator fit or not? Qualitative fit-testing relies on the user being fit-tested to respond to a test agent; the user smells or tastes the agent, or the user is irritated, indicating that the respirator does not fit correctly. The advantage of this type of test is that it is very cost efficient. A disadvantage is that the user does not know exactly how well the respirator fits. A qualitative fit-test can be used to fit-test only negative-pressure air-purifying respirators that must achieve a fit factor of 100 or less. The OSHA Respiratory Protection Standards allow the following four different agents to be used:

- isoamyl acetate, an agent that has a sweet smell similar to bananas (also known as banana oil)
- saccharin, which has a sweet taste by mouth
- Bitrex®, which has a bitter taste by mouth
- irritant smoke, which causes the user to cough because of irritation

Qualitative fit-test protocols consist of three steps: threshold screening, respirator selection, and fit-testing (Colton 2012). The threshold screening step is performed without wearing a respirator to determine whether the subject can detect low levels of the test agent. This level would be similar to the amount inside the respirator if the facepiece-to-face seal had a small leak. During this test, the test subject also learns what to expect if the respirator is leaking.

The purpose of the respirator selection step is to find one that provides the most comfortable fit (Colton 2012). The fit-test consists of the test subject's wearing the respirator while exposed to the test agent and performing facial movements (exercises) to test the facepiece-to-face seal, which are described later. The test procedure is conducted with the user wearing the assigned respirator containing the test agent in the breathing zone by use of a fit-test hood (see Figures 9–15 and 9–16).

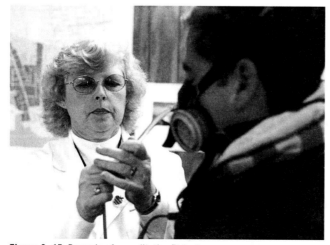

Figure 9–15. Preparing for qualitative fit-test.

(Source: Used with permission from MSA.)

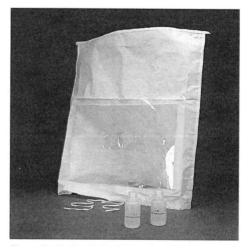

Figure 9–16. Qualitative fit-test hood/kit.

(Source: Used with permission from MSA.)

QUANTITATIVE FIT-TESTING

A quantitative fit-test lets the employer know the answer to this question: How well does the respirator fit? It quantifies the test by giving the test subject a number (a numerical fit factor) of how well it actually fits. The fit factor is a ratio of the outside concentration (ambient or generated aerosols) to the inside concentration in the respirator facepiece. The fit factor is different from the APF.

The OSHA Respiratory Protection Standards allow three different methods in quantitative fit-testing. They are as follows:

- Generated aerosol—use of equipment that uses sodium chloride or some other mist-generating system while users wear their assigned respirator inside an enclosure to keep the test agent in the breathing zone.
- Ambient aerosol—uses the existing ambient air with a device that measures particles outside the user's respirator, compared with particles that have entered the respirator facepiece. No enclosure is required for this test since it measures ambient air. The common device used for this is the TSI Porta-Count (Figure 9–17).
- Negative-pressure method—determines leakage (Colton 2012) by creating a negative pressure inside the facepiece and measuring the leakage rate of air. The respirator does not need a probe, but test adapter manifolds are placed on the respirator in place of filters or cartridges.

Figure 9–17. Quantitative fit-testing using a Porta-Count.

(Source: Used with permission from MSA.)

FIT-TEST PROCEDURES

The requirements for fit-testing are the same for both qualitative and quantitative fit-testing. The employer must first allow the user to pick an appropriate respirator and demonstrate how to put the respirator on and position it on the face. If the respirator selected is not acceptable, another size or brand should be selected. Once the respirator is acceptable, the user must wear the respirator for at least five minutes to assess comfort. The user will then conduct a seal (fit) check, and then have the exercises explained. A respirator cannot be worn or used if the user has a beard affecting the face seal of the respirator. The respirator must be worn for five minutes before the test exercises are conducted. Each test exercise is performed for one minute except for the grimace exercise, which lasts for 15 seconds. The specific tests are as follows:

- normal breathing
- deep breathing
- turning head side to side

- moving head up and down
- talking (reciting the rainbow passage)
- grimace (making faces)
- bending over or jogging in place
- normal breathing

If each test exercise is passed, the respirator is acceptable for the user. If any test exercise needs an adjustment to prevent leakage or a new size or brand of respirator is needed, all test exercises need to be repeated.

SEAL OR FIT CHECK

A seal check is different from a fit-test in that it is required every time a respirator will be donned. A positive- and negative-pressure seal check is performed once the respirator is on. A positive-pressure check is done by covering the exhalation valve and then exhaling. If the face-to-respirator seal is good, the respirator will balloon out as it contains air. If there is a leak, it will be noticed along the facepiece, and the respirator will not balloon out. The user would then adjust respirators, straps, and so forth to close the leak and perform the check again. A negative-pressure seal check (Figure 9–18) involves covering the cartridges or filters and breathing in. If there is a good face-to-respirator seal, the respirator will collapse, indicating a good seal. If there is a leak, the respirator will not collapse. The user would then adjust straps and such until there is a good seal and perform the check again.

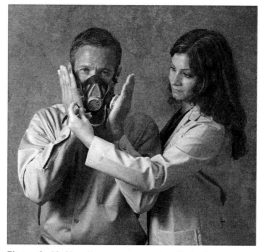

Figure 9–18. Negative-pressure seal or fit check.

(Source: Used with permission from MSA.)

SUMMARY

Inhalation is the most common route for worker exposures to toxic agents. The airborne pathway exposes workers to hazardous aerosols that are breathed in and absorbed by various portions of the respiratory tract. Respiratory protection is thus an important means to filter the hazardous agents out of the breathing air before they enter the body. It is important for industrial hygienists to have a solid understanding of the mechanisms by which these aerosols are removed from the breathing air and the details about the operations and capabilities of different designs of devices. It is also important to understand when supplied air is required to be provided directly to workers in environments where there is no oxygen and extremely hazardous chemicals are present.

REVIEW QUESTIONS

1. What are the nine requirements of an OSHA-approved respiratory protection program?
2. What are the required steps to be performed in a fit-testing procedure?
3. What are the five main methods through which aerosols are collected onto filter media?
4. What part of the lungs are the alveoli in and what do they do?
5. What types of respirator filters are used for removal of solid or liquid particulate aerosols?
6. What is the difference between gas or vapor adsorption and absorption in respirator filters?
7. When are SCBA respirators required to be used?

REFERENCES

Balmes, J. "The Lungs." In *Fundamentals of Industrial Hygiene*, edited by B. Plog. 38–45. Itasca, IL: National Safety Council, 2012.

Bollinger, N. J., and R. H. Schutz. *NIOSH Guide to Industrial Respiratory Protection. Publication No. 87-116*. Cincinnati: National Institute for Occupational Health and Safety, U.S. Department of Health and Human Services, 1987.

Brauer, R. L. *Safety and Health For Engineers*. Hoboken, NJ: John Wiley & Sons, 2006.

Colton, C. E. "Respiratory Protection." In *Fundamentals of Industrial Hygiene*, edited by B. Plog. 657–870. Itasca, IL: National Safety Council, 2012.

Goetsch, D. L. *Occupational Safety and Health for Technologists, Engineers, and Managers*. Upper Saddle River, NJ: Pearson Prentice Hall, 2005.

Nelson, T. J., and L. L. Janssen. *Developing Cartridge Change Schedules: What Are the Options? 3M Job Health Highlights*. Minneapolis, MN: 3M, 1999.

———. *Fit-Testing Requirements Under the Revised Respiratory Protection Standard 29 CFR 1910.134. 3M Job Health Highlights.* Minneapolis, MN: 3M, 1999.

Pennsylvania Department of Labor and Industry. *Respiratory Protection.* Presentation. 2001.

Revoir, W. H., and C.-T. Bien. *Respiratory Protection Handbook.* Boca Raton, FL: Lewis Publishers, 1997.

Scott, R. *Basic Concepts of Industrial Hygiene.* Boca Raton, FL: Lewis Publishers, 1997.

3M Personal Safety Division. *3M Respirator Selection Guide.* St. Paul, MN: 3M, 2012.

University of Kentucky Occupational Safety and Health Administration. *Respiratory Protection Training, OSHA 29 CFR 1910.134.* Lexington, KY: UKOSHA.

U.S. Department of Labor Occupational Safety and Health Administration. *Appendix C to Sec. 1910.134: OSHA Respirator Medical Evaluation Questionnaire (Mandatory).* Washington DC: USDL OSHA, 2012. www.osha.gov/pls/oshaweb/owadisp.show_document?p_table=STANDARDS&p_id=9783.

———. *Assigned Protection Factors for the Revised Respiratory Protection Standard.* OSHA 3352-02. Washington DC: USDL OSHA, 2009.

———. *OSHA Respiratory Protection Standards, 29 CFR 1910.134. Respiratory Protection eTool. Assigned Protection Factors.* Washington DC: USDL OSHA, 2011. www.osha.gov/SLTC/etools/respiratory/respirator_selection_apf.html.

———. *OSHA Respiratory Protection Standards, 29 CFR 1910.134. Respiratory Protection eTool. Respirator Change Schedules.* Washington DC: USDL OSHA, 2011. www.osha.gov/SLTC/etools/respiratory/change_schedule.html.

10
Dermal Hazards

LEARNING OBJECTIVES

After completing this chapter, readers should be able to do the following:
- Identify key anatomy and physiology of the skin.
- Understand how toxic agents are absorbed into the body through the skin.
- Describe several of the most significant adverse health effects of skin exposure to hazardous agents.
- Explain some of the latest methods of dermal exposure assessment.
- Develop controls necessary to minimize skin exposure to hazardous agents.

Photo credit: Judith Glick Ehrenthal/iStock

INTRODUCTION

Approximately 13 million workers in the United States have the potential for skin exposure to hazardous chemicals (CDC 2012). Workplace skin diseases account for 15–20% of all reported occupational diseases (CDC 2009), with the greatest number of occupational skin disease cases occurring in the agricultural and manufacturing industries. The economic burden in total annual costs associated with occupational skin disease is over $1 billion per year (DHSS 2011).

Although industrial hygiene traditionally focuses on inhalation exposures for workers, there has been increasing interest in dermal routes in the past 20 years. In 1980, skin exposure was first identified as a significant route of exposure. Throughout the 1980s, the first industrial hygiene field investigations for dermal exposure pathways and routes were conducted. In recent years, with the realization that most occupational exposures result from inhalation, there has been an emphasis on substituting highly volatile chemicals with those that do not evaporate easily. As a result, examining dermal exposure has grown in importance because more hazardous chemicals are now available to make contact with the skin. In addition, chemicals with lower volatility tend to be absorbed into the skin more readily.

ANATOMY OF THE SKIN

The dermal surface area of the average human body is between 1.5 and 2.0 m^2. It is the largest organ of the body, accounting for 10% of total body mass, and is made up of several layers of epithelial cells. Average skin surface areas are shown in Table 10–1.

The skin is made up of three major components: (1) epidermis (outer layer), (2) dermis (underlying layer), and (3) hypodermis. See Figure 10–1.

The epidermis is made up of a thin, waterproof layer of collagen fibers. This layer also contains melanin, which gives skin its color. The dermis layer, just beneath the epidermis, contains strong connective tissues, hair follicles, and sweat glands. The deeper, subcutaneous layer of tissue, called the hypodermis, is comprised of fat and connective tissue. The hypodermis acts as a cushion for the body and provides insulation from temperature extremes.

Table 10–1. Sample Skin Surface Areas for Adult Males

Body Region	Surface Area (cm^2)
Head	1,300
Face	650
Chest/stomach	3,550
Back	3,550
Upper arms	2,910
Forearms	1,210
Hands	820
Thighs	3,820
Lower legs	2,380
Feet	1,310

(Source: Franklin and Worgan 2005)

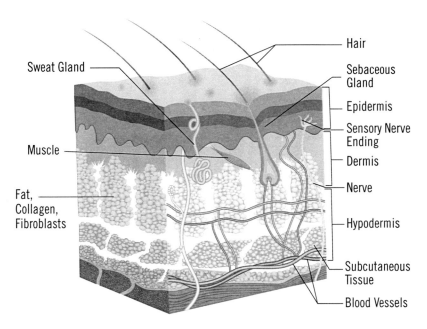

Figure 10–1. Simple skin anatomy.

(Source: snapgalleria/iStock)

One square centimeter of skin contains the following (approximately):
- 3 million cells
- 1 m of blood vessels
- 25 pressure sensors
- 10 hairs
- 3,000 nerve endings
- 200 pain sensors
- 100 sweat glands
- 15 oil glands

FUNCTIONS OF THE SKIN

The skin's primary function is to provide a barrier against the environment and protect the muscles and tissues that lie beneath. It keeps harmful materials, such as chemicals and biological agents, outside the body and protects inner organs and cells while letting necessary materials pass in and out. Movement is facilitated by diffusion processes and through metabolic activities within the skin.

The skin also provides mechanical support and shock absorption, and holds subcutaneous layers of the hypodermis in place. Additionally, it enables the neurosensory reception of messages from the outside environment needed for the body to respond appropriately to external stimuli such as heat, cold, pressure, and tissue damage.

The skin participates directly in the regulation of a broad range of physiological functions. For example, by releasing perspiration, the skin can cool the temperature of the body via evaporation. The blood vessels carrying blood near the surface of the skin, then, also become cooled, and the blood within the vessels carries the cooled blood to distant areas of the body. This is one of the skin's most important functions. Skin is also important for completing various metabolic functions such as synthesizing vitamin D, transporting electrolytes and hormones, and assisting immune functions of the body.

THE SKIN ABSORPTION PROCESS

Percutaneous absorption is the primary means by which hazardous chemicals enter the skin. The factors that are most associated with how much of a chemical gets into the body are the type and condition of the skin, the particular chemical, and specific exposure factors such as the size of the affected area of the body, the concentration of the chemical, the length of the contact time, and the temperature of the chemical. Damaged skin or skin with sunburn will generally absorb chemicals at a higher rate than that of intact skin.

Skin absorption occurs primarily by diffusion, which is when molecules spread from high- to low-concentration areas of the chemical. The three main mechanisms by which chemicals diffuse through the skin are (1) travel on the intercellular lipid pathway, (2) transcellular permeation, and (3) through appendages of the skin (CDC 2012).

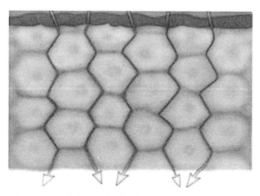

Figure 10–2. Intercellular lipid pathway.

(Source: Centers for Disease Control and Prevention)

Intercellular lipid travel occurs when the spaces between the corneocyte cells of the stratum corneum are filled with lipids, such as fats, oils, or waxes. Some chemicals penetrate the skin by diffusing through these lipids. (See Figure 10–2.)

Chemicals can also move through the skin by directly permeating the corneocyte cells—that is, by passing directly from one cell membrane to the next. This cell-to-cell travel, or transcellular permeation, is also based on diffusion and spreads out from areas of higher concentration to areas of lower concentration. (See Figure 10–3.)

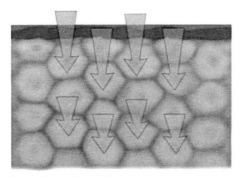

Figure 10–3. Transcellular permeation.

(Source: Centers for Disease Control and Prevention)

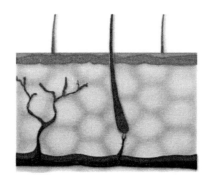

Figure 10–4. Absorption through the skin appendages.

(Source: Centers for Disease Control and Prevention)

Anatomical appendages contained within the skin, such as hair follicles and glands, are other pathways for absorption, but they tend to be less efficient due to their small surface area compared to the skin's total surface area (see Figure 10–4). They can, however, be substantial pathways for certain substances that are applied topically or move slowly.

SKIN FACTORS AFFECTING ABSORPTION

Several biological factors of the skin influence the rate and volume of chemicals that can be absorbed. For instance, the age of skin can affect absorption rates. Typically, older skin will absorb chemicals more slowly or less efficiently than newer skin.

In general, the skin's condition plays a major role in absorption rates. Wet skin has been shown to be more absorbent of some chemicals than dry skin. Skin that is broken or not completely intact can have higher absorption rates, too. If skin is already occluded (blocked) by a medical condition or externally applied agent, absorption may be reduced.

There is no hard-and-fast rule on when wet work is likely to be risky. However, contact of more than about 2 hours or more than 20–40 hand washes a day will likely lead to dermatitis, or inflammation of the skin (HSE 2009). Dermatitis from wet work is common in trades such as hairdressing, metal machining, food preparation, cleaning, and health care. (See the Contact Dermatitis section later in this chapter.)

The anatomical location of skin—and thus the type of skin at that location—can also influence absorption rates. For example, soft skin, which is located at the inner elbows, on the forearms, and behind the knees, tends to be very permeable. Differences in skin permeability relate to the thickness of the skin at different anatomical locations, as indicated in Table 10–2. As can be seen in the table, the skin on the sole of the foot is extremely thick, making it resistant

Table 10–2. Human Stratum Corneum Thickness by Anatomical Site

Anatomical Area	Thickness (microns)
Abdomen	15
Forearm	16
Back	10.5
Back of hand	49
Palm	400
Sole of foot	600

(Source: Scheuplein 1971)

to chemicals, whereas the skin on the forearm is only 16 µm thick, making it very permeable to chemicals.

Skin metabolism, which differs among individuals, also affects absorption rates. Workers with high metabolic rates may absorb more of a chemical than workers with low metabolic rates, even when they are all exposed for the same amount of time. In addition, circulatory effects may vary among workers and account for differences in absorption rates. In general, increasing circulation increases the absorption and distribution of toxic chemicals that come in contact with the skin throughout the body. Skin metabolism and circulatory rates may also be affected by certain health conditions and by taking certain medications or pharmaceuticals. Antihistamines such as Benadryl and Allegra, for example, tend to narrow blood vessels and reduce circulation near the skin, thus slowing the absorption of chemicals into the bloodstream.

EXPOSURE FACTORS AFFECTING PERCUTANEOUS ABSORPTION

There are numerous ways for the skin to be exposed to hazardous chemicals. A worker may come in direct contact by:
- touching contaminated surfaces
- being immersed in containers with liquids or semisolids
- sedimentation, impact, electrostatic attractions, or chemical splashes depositing aerosols on the skin

The duration of exposure is important in determining the amount of chemical absorbed because, in most cases, the longer the chemical is in contact with the skin, the greater the amount of chemical that will pass through the skin and into the body. Absorption may begin after a few seconds, but immediately washing or wiping off a chemical can considerably reduce the amount absorbed.

The surface area that a chemical touches affects the amount of chemical absorbed. In many cases, the relationship between absorption and surface area is nearly linear, with twice the surface area leading to twice the absorbed dose, for any given duration.

For many but not all agents, the relationship between concentration of the chemical and rate of absorption is linear. It is possible for some concentrations of chemicals to be so high that the skin reaches its absorption limit, and no additional chemical can be absorbed even at higher concentrations.

Other, less straightforward factors that affect percutaneous absorption in a work environment include the rate of work and the ambient temperature of the workplace. Each of these factors may alter the metabolic rate and capabilities of the skin, and they may alter the conditions of the skin by increasing or decreasing the moisture on the skin's surface, thus affecting the skin's absorption capability. The higher metabolic rate of a worker working quickly and strenuously would likely increase the flow and circulation of blood in the skin, which would hasten chemical absorption and circulation throughout the body. Working quickly in a hot environment might increase sweating and skin moisture content, which would most likely increase the absorption of chemicals through the skin. Conversely, skin exposure to chemicals in a dry, cold work environment would most likely reduce the absorption of chemicals through the skin.

Protective clothing is often used to reduce workers' skin exposure to hazardous chemicals. Gloves and chemical-resistant garments and face shields can reduce skin exposures to liquid, semisolid, and aerosol chemicals. When using protective clothing, it is important to understand permeation processes and know the permeation rate of the given chemical to ensure protective capabilities. For example, gloves used to protect a worker's hands from a chemical when they are submersed in a liquid eventually become saturated with the chemical and expose the worker's hands to the chemical. For certain glove and chemical combinations, permeation and saturation occur in a matter of minutes. Protective materials can also become degraded by the chemical's effects on the material, thus leading to increased permeation.

Workers are also commonly exposed to toxic chemicals when they accidentally cross-contaminate themselves by removing their contaminated personal protective equipment incorrectly. In addition, they may transfer chemicals unknowingly after touching a contaminated tool, or they might transfer chemicals to other parts of their body with their hands or gloves.

CHEMICAL FACTORS OF ABSORPTION

Depending on a chemical's characteristics, it may evaporate before it is absorbed into the skin. Or perhaps its molecules can penetrate quickly into the stratum corneum and be carried away by the blood to distant organs and parts of the body. Chemicals can also be metabolized near or at the location of exposure.

Molecular weight can influence absorption because of how it makes the chemical behave. Chemicals with high molecular weights tend to evaporate more slowly than chemicals with low molecular weights and also tend to stay together and bead rather than flow away. Chemicals that evaporate quickly leave less of the chemical on the skin to be absorbed. Less absorption can also be related to volatility, as discussed earlier, with volatile chemicals not remaining on the skin long and often not imparting a high dose (Frasch et al. 2014).

Chemicals that are hydrophilic, or highly soluble in water, can be absorbed relatively easily into the skin through normal permeation processes and metabolic functions (Tibaldi, ten Berge, and Drolet 2014). Some chemicals, such as benzene or carbon tetrachloride, can easily be carried by water molecules and be absorbed undetected by the body.

Chemicals that are lipophilic, or highly soluble in oils or fats, may not be absorbed deeply into the skin or readily transported away from the skin's surface, but their lipophilic nature allows them to bind easily with the lipid nature of the skin cells and be absorbed and "stuck" in place in the layers of the skin. These chemicals can be very difficult to remove by washing, and as they are metabolized, they can be extremely toxic to other organs and parts of the body when they are systemically transported by the blood. Some chemicals such as fat-soluble liquids, or vapors such as organic solvents like toluene can be easily absorbed by the skin.

The actions of certain chemicals may be exacerbated or reduced in the presence of other chemicals. Two chemicals may act synergistically to improve the permeability or metabolism of one or both of them. Conversely, some chemicals layered on the skin provide a protective barrier by blocking access to the skin or by reacting with the other chemical directly to reduce its permeation ability or metabolic characteristics (Semple 2004).

SKIN INJURY AND ILLNESS

When a worker is exposed to hazardous materials, skin injuries or disorders can result. In 2013, more than 33,000 new cases of skin diseases and disorders were diagnosed, making skin injury the most common occupational illness (NSC Injury Facts 2015). The effects of skin exposure to hazardous chemicals can range from contact dermatitis to diseases and physical damage. The most common health effect of skin exposure to chemicals is dermatitis, with estimated annual costs exceeding $1 billion (CDC 2014). Some chemical and physical agents cause cancer of the skin or enter the body through the skin and cause cancer in other body parts. Physical hazards such as radiation and sharp objects can cause direct damage to the skin.

CONTACT DERMATITIS

Contact dermatitis, also known as eczema, is an inflammation of the superficial regions of the skin that appears in large localized areas. Dermatitis is often the result of exposure to foreign substances, and chemical exposure is the most common cause of occupational dermatitis, which is typically characterized by inflammation and erythema (reddening) or the formation of scales. Other symptoms of dermatitis include itching, swelling, pain, burning, rashes, redness, blisters, and dry or flaky skin. Inflammation of the affected tissue is present in the epidermis and the outer dermis. Pustules (small, superficial pus-containing lesions) occur rarely and only with secondary infection.

Occupational contact dermatitis is frequently categorized as one of two types. **Irritant contact dermatitis** is a nonimmunologic reaction that results from direct damage following exposure to a hazardous agent and manifests as inflammation of the skin. The reaction is typically localized to the site of exposure. Available data indicate that irritant contact dermatitis represents approximately 80% of all cases of occupational contact dermatitis. Irritant contact dermatitis is associated with perfumes and coal tar, acute exposures to highly irritating substances (e.g., acids, bases, oxidizing/reducing agents), and chronic cumulative exposures to mild irritants (e.g., water, detergents, weak cleaning agents). In some cases, irritant contact dermatitis is caused by phototoxic responses. **Phototoxic contact dermatitis** develops only after exposure to UV light, such as sunlight.

Allergic contact dermatitis is an inflammation of the skin caused by an immunologic reaction triggered by dermal contact with a skin allergen. For allergic contact dermatitis to occur, a worker must first be sensitized to an allergen. Additional exposures to this allergen can lead to an immunologic reaction in the skin or systemic reactions in other parts of the body. Common sensitizers include plants (gardening), antibiotics (pharmaceutical industry), dyes (paint and cosmetics industry), metals, chromates (cement industry), adhesives, fragrances, rubbers, and resins. Agrochemicals (i.e., pesticides and fertilizers) and cutting oils used in machining are also known skin allergens (see Figures 10–5, 10–6, 10–7, and 10–8).

Figure 10–5. Exposure to cutting fluids used in metalworking operations.

(Source: Centers for Disease Control and Prevention)

Figure 10–6. Exposure to organic solvents for cleansing purposes.

(Source: Centers for Disease Control and Prevention)

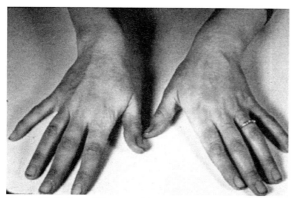

Figure 10–7. Dermatitis caused by the use of organic solvents for cleansing purposes.

(Source: Centers for Disease Control and Prevention)

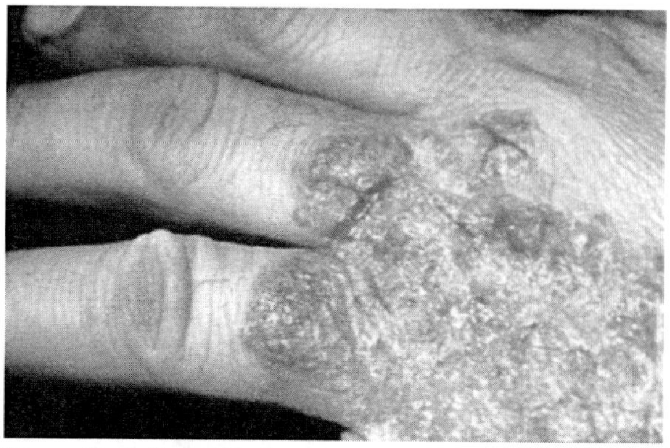

Figure 10–8. Dermatitis from exposure to cutting fluids used in machine shops and metalworking operations.

(Source: Centers for Disease Control and Prevention)

Exposure pathways that have been shown to lead to dermatitis or skin injury include the following:
- immersion—the skin is submerged into a liquid or powder
- splashes—from decanting or mixing liquids and powders
- deposition—when droplets, dusts, fumes, or aerosols contact the skin, either as part of a work activity or incidental to it (e.g., emissions from a nearby process)
- contact with contaminated surfaces via any of the following:
  - directly handling a contaminated workpiece
  - touching contaminated work surfaces
  - transferring residues on the hands to the eyes, nose, or mouth
  - transferring residues on the hands to tools, paperwork, or food
  - removing contaminated PPE incorrectly (HSE 2009)

URTICARIA

Urticaria is a transient skin condition that often appears as a wheal (swelling) and raised patches and is accompanied by intense itching. It can be caused by skin irritants or allergens and is different from dermatitis because it typically occurs rapidly following exposure and often goes away soon after exposure ceases. Urticaria often can be caused by detergents, pesticides, rubbers, and minerals. Because animals such as caterpillars and jellyfish and certain plants can cause urticaria, workers in forestry and fishing are often affected (EASHW 2008).

ACNE

Acne, an inflammatory disease of the sebaceous glands and hair follicles in the skin, leads to pimples and pustules (white-centered bumps) on the surface of the skin. Mechanics often develop acne because of their exposures to oil and grease, and roofers often develop acne because of their continuous exposures to pitch and tar.

SYSTEMIC DISEASES THAT RESULT FROM SKIN EXPOSURE

Numerous chemicals can pass through the skin and travel systemically throughout the body to cause disease. Organophosphate pesticides, for example, enter the body through the skin and damage the nervous system. Sometimes symptoms materialize soon after exposure, such as shortness of breath or respiratory irritation right after exposure to disinfectants or solvents. Systemic diseases such as bladder cancer, scrotal cancer, or diseases of the kidneys or heart may take years to appear in workers who were exposed to certain chemicals.

PHYSICAL DAMAGE TO SKIN

Direct contact with physical substances can damage the skin. Contact with sharp or rough objects can result in cuts, chapped skin, or abrasions, and burns can result from touching hot equipment or being exposed to hot sources such as furnaces or steam. Exposure to hot and wet materials such as in dishwashing can cause rashes. Working with oils, such as in machining processes, can cause pores to become blocked and infected, which is a physical cause rather than a chemical one. Chloracne and cysts can result from exposure to some agents that are difficult to wash from the skin and subsequently block sebaceous glands. Exposure to excessively cold or wet environments, as in the fishing industry, can cause frostbite. These physical damages represent a significant portion of hazardous workplace exposures and should be included in workplace risk assessments.

CANCER

Skin cancer is the most common form of cancer (Stern 2010). The cost of treating skin cancer in the United States is approximately $8.1 billion per year (Guy et al. 2014). Melanoma is the second most common form of cancer in 15- to 29-year-olds, and the most common form of cancer in adults ages 25 to 29 (Bleyer et al. 2006). Approximately 10,000 people die from melanoma each year, and occupational exposure is a significant risk factor associated with melanoma and with cancer in general (ACS 2015).

Both benign and malignant cancers have been linked to a variety of chemical and physical agents. Many chemicals commonly found in the workplace, such as polycyclic aromatic hydrogen compounds, can cause skin cancer. Other workplace chemicals, such as creosote and mineral oils, have been associated with benign skin cancer.

Ultraviolet and ionizing radiations have been linked to skin cancer and are closely associated with malignant melanoma. Outdoor workers are especially exposed to UV radiation over the duration of their careers, leading to an increased incidence in skin cancer among these workers. In addition, the combination of coal tar and UV radiation, which highway construction workers experience, has been shown to have a synergistic effect on skin cancer development. The current use of ultraviolet germicidal radiation in industry and health care has increased worker exposures to these hazardous rays.

DERMAL WORKPLACE EXPOSURES

Thousands of workers each year are exposed to hazardous chemicals and materials. Some of the principal industries, jobs, and chemicals implicated in dermal exposures are shown in Table 10–3.

Table 10–3. Common Industries with Hazardous Dermal Exposures

Industries	Jobs	Chemicals
Construction	Materials applications	Asbestos, fiberglass, solvents, wet cement, cement dust
Food service	Cleaner	Cleaning products, detergents, wet work
Printing/lithography	Printer	Dyes, solvents, bleaches
Agriculture	Pesticide application, animal handling	Pesticides, animal dander and excretions
Plastic manufacturing	Operator	Acrylates, petroleum products
Florists	Florist	Plant materials and allergens, wet work
Pharmaceuticals	Production, pharmacists	Bacampicillin, benzocaine,
Cosmetology	Hair stylists, nail technician	Hair dyes, solvents, preservatives, fragrances
Funeral homes	Embalmers	Formaldehyde, preservatives
Cleaning	Environmental service technicians	Solvents, detergents, disinfectants, wet work
Health care	Cleaning technicians, nurses, pathology	Solvents, detergents, disinfectants, latex, infectious agents, formaldehyde
Painting	Painters	Solvents, paints
Machine operations/repair	Mechanics, machinists	Cutting fluids, solvents, oils, nickel

A British study showed that nearly 8 percent of dermatitis cases are caused by the use of personal protective equipment (PPE), such as individuals' sensitivity to glove materials like latex or rubber over extended periods of exposure. In addition, approximately 35 percent of all dermatitis reports are related to wet work or working with soaps and cleaners (HSE 2013).

One of the most common misconceptions about dermal exposure, understandably, is that only the skin is affected. In many cases, however, the health effects are systemic or travel to organs that are distant from the skin.

DERMAL EXPOSURE ASSESSMENT

Despite the number of workers who are exposed to chemicals and other hazards via the dermal route and the number of health consequences of this exposure, occupational dermal pathway exposure assessment remains inadequate. Although methods to assess worker exposures are being developed, they are yet to be used proactively or extensively by industrial hygienists.

The value of dermal pathway exposure assessment includes being able to more effectively evaluate the hazards of various worker activities and work processes. Assessment can also identify where better controls are needed and determine which workplace controls are most effective at reducing worker exposures.

The knowledge and awareness resulting from better exposure assessment methods can in turn lead to more effective adherence to exposure limits. In addition, better quantitative assessment of dermal exposures can improve the epidemiological analysis of the health effects of a given chemical.

By examining the toxicity of a material or chemical in terms of frequency or intensity of dermal exposure, a more effective estimate of risk can be obtained for workplaces, jobs, and industry processes. For example, a job with a low level of risk might involve using a low concentration of a toxic chemical once a month in a fume hood while wearing gloves. In contrast, a high-risk job might involve spraying a high concentration of the same chemical on a daily basis and over a wide area. The relationship between dermal exposure and level of toxicity is depicted in Figure 10–9.

Dermal exposure assessment can be used in epidemiological studies to increase the understanding of the importance and contribution of the dermal route to subsequent health effects. Additional information about chemicals whose toxicity is currently underestimated will lead to increased awareness of the link between dermal exposure and the chemical's toxicity.

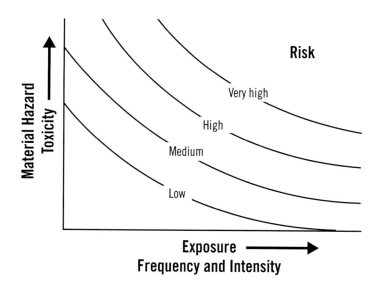

Figure 10–9. Relationship between dermal exposure and toxicity for a given toxic chemical.

Exposure assessment can begin with qualitative analysis of the exposure. This would include documenting the presence of workers mixing with their bare hands, leaking equipment, visible contamination of clothing, use of gloves, written procedures, and processes that generate droplets or splashes. Jobs or activities at a workplace could be categorized according to these or related aspects. The health data of workers in different categories of jobs could be compared to identify relationships between potential exposures and health outcomes.

Semiquantitative methods of exposure assessment can be used to rank worker exposures by frequency or intensity. Categories might include workers who have no contact with a hazardous agent, workers who have moderate contact with the agent, and workers who have frequent contact with the agent. These assessments should be made by trained and qualified experts to ensure consistency and accuracy.

Currently, quantitative dermal exposure studies are infrequently used. The goal of these studies is to identify and quantify not only how much of a chemical contacted the skin, but also how much of the chemical was absorbed into the skin. Parameters such as the mass of the chemical, the exposed surface area, the duration of exposure, and other factors affecting absorption need to be assimilated in a significant way to provide consistent and useful data that can be used in subsequent epidemiological analyses.

Quantitative sampling and exposure assessment are currently being expanded into three main areas. Surrogate skin techniques are being developed in which a patch covering a small area of skin is placed on a worker to measure the amount of chemical absorbed into the patch after the worker completes a particular number of tasks. Other types of surrogate skin techniques involve the use of gloves, shirts, or face shields that collect the chemical of interest and are then analyzed to find out how much chemical was absorbed and how much chemical touched the skin.

Removal techniques, a second method of quantitative sampling, involve using a device to collect the chemical from the worker's skin. For example, tape strips are used to pull the chemical off the skin. Wipes and physically washing the chemical or agent from the skin are also methods that have been developed to collect samples. Suction techniques that pull or vacuum dusts or powders off the skin's surface are also being developed.

Lastly, direct assessment methods of quantitative chemical dermal exposure have been developed. These methods involve embedding fluorescent tracers within the contaminant, or substituting them for the contaminant, to indicate the amount of chemical that contacts the skin. After a given area of the skin is exposed to the chemical, the chemical droplets can be counted or weighed. Video imaging can be used to track the contamination in progress and observe the interaction between the chemical and the skin or nearby surfaces.

Quantitative sampling does have some challenges. Measurements of exposure gathered from small body areas are difficult to extrapolate to whole-body exposure. Also, in any individual study, there could be considerable variability in the spatial distribution of splashes or other exposure pathways.

In addition, the evaporation that occurs after the splash or skin contact cannot be accounted for in the subsequent measurements. The variability in the numerous other parameters of absorption also makes the amount measured in a collection device less accurate than scientifically necessary, which makes it difficult to compare the assessments of different individuals and toxic chemicals.

COMPLIANCE AND INDUSTRY STANDARD GUIDANCE

The American Conference of Governmental Industrial Hygienists® (ACGIH) and the National Institute for Occupational Safety and Health (NIOSH) have each published lists of toxic chemicals that enter through the dermal route of exposure. These lists are designated as *skin notations* in the Threshold Limit Values® and as recommended exposure limits (RELs), respectively.

In 2009, NIOSH published a guide on skin exposure that provides additional assistance to assess skin hazards. The guide, *Current Intelligence Bulletin (CIB) 61: A Strategy for Assigning New NIOSH Skin Notations*, can be used to evaluate hazardous conditions and identify the best ways to protect workers. The NIOSH Pocket Guide to Chemical Hazards includes updated information on hundreds of chemicals and is very useful for assessing dermal hazards.

The ACGIH also publishes a set of Biological Exposure Indices (BEI)® for about 90 different chemicals. These indices correlate amount of chemical or metabolite in a biological sample to dose a worker would receive by inhaling the chemical at its airborne permissible exposure limit (PEL) for a 40-hour workweek. In some instances, these PELs can also indicate a worker's exposure through the dermal route. Despite the skin notations and RELs, no regulatory limits to workers' exposures to chemicals via dermal pathways have been established in the United States.

OSHA also provides guidance on and regulations for occupational dermal exposures. These regulations require skin exposure assessments in general terms and provide monitoring and measurement information. OSHA additionally provides direction for using a variety of Biological Exposure Indices (BEI) tools to analyze worker exposures through the dermal route. The *OSHA Technical Manual* provides best practices for skin and contaminated surface monitoring. For substances such as formaldehyde and cadmium, OSHA has promulgated regulations for workplace controls, PPE, workplace monitoring, and worker training.

CONTROLLING SKIN EXPOSURE

ELIMINATION AND SUBSTITUTION

As in other occupational exposure situations, controlling hazards from dermal exposure to toxic or hazardous chemicals can be achieved by eliminating the chemicals' use altogether. Elimination is most feasible at the **process design** stage. For existing systems, **elimination** usually means changing a process (e.g., removing paint by scraping instead of using solvents). However, if this change introduces other hazards, such as noise, dust, or musculoskeletal problems, the new hazards must be controlled, too.

When eliminating a process or product is not possible, substituting the hazardous material with one that is less toxic can reduce workplace risks. Examples are replacing solvent-based products with water-based ones and replacing surfactant degreasers with milder products. Substituting the physical form of a substance might also reduce skin contact. For example, using granulated or liquid forms rather than powders can reduce the spread of dust. Using prepackaged forms of a substance can eliminate scooping or weighing of the substance and reduce worker dermal exposures (WHO 2014).

Sometimes a change in the work process can reduce occupational exposures. Instead of pouring hazardous waste from an open container into a waste receptacle, it may be possible to use connecting hoses to transfer the liquids, thus eliminating the potential for splashes or spills.

> **Case: Alternative Chemicals Can Reduce Dermal Hazards**
>
> Health care pathology practice has historically consisted of collecting surgical specimens in aliquots of formaldehyde to preserve them for dissection and further study. This process exposes surgical workers, pathologists, forensic scientists, lab workers, and waste handlers to formaldehyde, which is a carcinogen and respiratory sensitizer.
>
> New preservatives, such as F-soly and FineFIX, were originally evaluated in hospital pathology departments, where they were found to provide the same amount of preservative and specimen support as formaldehyde but without formaldehyde's toxicity. These products are now successfully used in a variety of hospitals, where they provide high diagnostic quality in evaluating pathologies in specimens while reducing workers' dermal exposures.

ENGINEERING CONTROLS

Engineering controls commonly associated with respiratory protection can also be used to effectively reduce dermal exposures. Engineering controls are physical barriers that separate the worker from workplace hazards. In the case of dermal exposures, splash guards over tanks and processes can eliminate or reduce the likelihood of the chemical or agent getting on the skin of the worker. A spray-painting isolation booth not only eliminates inhalation of the aerosols generated by the process, but also eliminates or reduces the possibility of droplets reaching the worker's exposed skin.

Glove boxes, closed reactors, and partial enclosures with local exhaust ventilation can be used to minimize dermal exposures. Engineering controls should also be developed for maintenance and waste-handling activities and for cleaning and repair operations.

In some cases, increasing the distance between the worker and the chemical can reduce dermal exposure. A tool with an extension handle distances the worker from the hazardous agent. Other engineering controls are lids on open containers and backup latches or pressure-relief valves on closed containers.

Barriers separating contaminated and clean work areas can prevent the spread of contamination, and spillage controls such as drip trays also prevent the spread of contamination and make cleaning up spills easier. When surface contamination is inevitable, easy-to-clean work surfaces reduce the likelihood of dermal exposures. Contaminated surfaces should be cleaned regularly; alternatively, contamination can be reduced by using and regularly replacing disposable absorbent paper linings.

ADMINISTRATIVE OR WORK PRACTICE CONTROLS

Workers should be instructed in the risks of skin exposure to hazardous materials. Because emphasis is often given to respiratory routes of exposure and control, many workers do not realize that the symptoms they are experiencing could be related to a dermal exposure, particularly when the symptoms are present somewhere other than the skin. Training is needed to ensure that workers are aware of dermal pathway hazards and the methods they should use to protect themselves from dermal exposure. Workers should also be trained on the appropriate emergency response procedures to follow when dermal exposure occurs. Good hygiene practices should be encouraged and included as a routine part of the industrial hygiene and safety program.

Administrative or work practice controls that restrict access to areas where dermal exposure might occur can reduce the number of workers who are exposed by the dermal route. Locked doors and warning lights and alarms are examples of these controls.

Written procedures regarding the safe handling of hazardous materials should be developed, as should the measures to be taken to prevent splashes and contamination. Procedures to clean up spills should be developed long before any chemicals are spilled, and the necessary cleanup materials should be purchased, stored near work locations, and inspected periodically.

PPE

PPE, such as footwear, aprons, face shields, overalls, and gloves, is an important and effective way to minimize workers' dermal exposure to hazardous materials. PPE can be used to protect against a wide range of chemicals, including radioactive materials, and gloves and PPE are often the final barrier between a worker and the hazard.

Besides protecting against chemical, radioactive, and biological agents, gloves protect the hands from hot and cold surfaces and environments and from abrasions and cuts. The ways in which PPE can protect the skin should be evaluated for each workplace hazard in all job exposure assessments, and the process used to select PPE to protect the skin should be documented.

Because most occupational skin exposures take place on the hands, gloves are one of the most important and common PPE items used. If forearm exposure is likely, gloves should extend well beyond the wrist.

The gloves appropriate for a job depend on the type of work to be done and on the substances that will be handled. Not all glove materials offer protection from all chemicals. Manufacturers' literature should be reviewed to learn which gloves protect against which chemicals. How long workers' hands will likely be in contact with the chemical should also be considered. Breakthrough times can range from seconds to hours, but all glove materials eventually allow chemicals to break through to the skin. Gloves and other chemical protective PPE need to be changed whenever they become saturated or physically degraded. For some chemicals and processes, it is prudent to wear two pairs of gloves in case the outer glove has a tear.

Workers need to be aware of the limitations of their gloves. Sometimes wearing a glove that is completely saturated with a chemical is no better than putting the bare hand into the chemical solution.

The worker's comfort and the dexterity needed to perform work tasks are also important considerations when selecting gloves. The suitability of the glove for different workplaces and jobs must be considered as well.

In some cases, the gloves themselves are a hazard. For example, latex gloves used in health care have been linked to hypersensitivity reactions in physicians, nurses, dentists, veterinarians, and other health care staff. Debilitating health effects often become apparent only after several years of use, so it can be difficult to persuade workers to switch to available alternatives that perform equally well but do not have toxic properties.

Gloves and PPE have numerous limitations. Gloves might have holes or tears that allow chemicals through, and workers must be knowledgeable about and have training in the PPE for it to be fully effective. This training includes the proper methods to don and remove the gloves and other PPE to prevent cross-contamination. It also includes proper maintenance of dermal protection PPE and knowing when the PPE needs to be disposed of.

Case: Preventing Dermatitis

Joe was a part-time worker at a fast food restaurant. His duties were to fry food and clean the equipment and stove at the end of the day. Joe found that handling the moist foods and stove cleaners stung his hands, and this, along with lots of hand-washing, made his skin dry. After seeing a doctor, Joe learned that he had irritant contact dermatitis from wet work. Joe told his boss, who then introduced some changes at work:

- An alternative cleaner that is still effective but less hazardous was substituted.
- Thick rubber gloves are used for cleaning.
- Vinyl food handler's gloves are used for handling moist foods.
- Workers use a moisturizer before breaks and at the end of shifts.
- Workers' skin is checked periodically.
- After a short time of using the less abrasive cleaner and wearing the appropriate gloves, Joe's hands improved and the dermatitis subsided.

Drawbacks of protective gear are that it is typically only 1 or 2 millimeters thick, it can be easily penetrated or torn, and it usually can be worn by only one worker for limited durations. The use of PPE on a regular basis can be expensive because new gear has to be continually purchased, replaced, and discarded (often as hazardous waste). Also, PPE must be used appropriately to be effective, and it often gives workers a false sense of security.

SKIN CARE PRODUCTS

Several skin protection products have been developed over the past few years. Some provide treatment of damaged skin, and some provide actual protection from hazardous environments or agents.

Some creams trap potentially hazardous contaminants before they are absorbed into the skin. These creams are typically lipid-based oils, lanolin, or emulsifiers that attract the hazardous agent. After the contaminant is trapped, the cream and contaminant are easily washed off the skin. Other skin-protective creams form a film over the skin that repels the hazardous chemical. Creams that contain silicone, stearates, or beeswax repel water-based chemicals. Creams that contain glycerin repel oils and organic solvents. Some types of skin creams interact with the hazardous agent to make it less toxic or less able to penetrate the skin. Lastly, UV-radiation-protective skin creams protect the skin from harmful rays from the sun or manmade sources.

Skin moisturizers can effectively treat or protect the skin from mild abrasions or rashes. Industrial hygienists cannot prescribe medications to treat skin illnesses or injuries, but they can suggest possible preventive measures. Information on a chemical's possible skin hazards and protective measures against those hazards may be available. Keeping in mind that the lotions or soaps a worker uses off the job might be the source of irritation or dermatitis, the industrial hygienist should inquire about a worker's personal hygiene and offer advice if appropriate.

Skin cleaners can remove contaminants, but washing too frequently and with abrasive cleaners can also be damaging and a source of dermatitis. Surfactants (soap) can be used to break apart dirt or contaminants on the skin's surface so that the contaminants can be rinsed off with water; abrasive cleaners can break up and release skin contaminants; and organic solvents can be used to chemically interact with surface contaminants and neutralize them. However, each of these cleaning methods also has its own particular risks. The industrial hygienist should thus select the appropriate cleaning method only after thoroughly analyzing the benefits and drawbacks of each and should make adjustments as needed. It is also important to communicate the cleaning methods and agents that workers should use and to make the cleaning agents available near the workstations (HSE 2014).

SUMMARY

Dermal exposure to toxic and hazardous chemical and physical agents is a significant occupational health concern. When the skin is exposed to hazardous materials, the important structures and functions of the skin lose their abilities to protect the body. Hazardous agents' processes of interaction, absorption, and distribution differ and lead to different health effects in the body. Methods assessing the dermal route of worker exposure had been inadequate, but are now being developed. Wherever possible, the best prevention of worker exposure is chemical elimination or substitution, followed by engineering controls to isolate hazardous processes or agents. The use of PPE is a final barrier to protect workers from hazardous materials and continues to be used pervasively as needed for worker protection.

REVIEW QUESTIONS

1. Why is the study of dermal exposure to hazardous agents important today?
2. Why have dermal routes of occupational exposure become more prevalent in the past several years?
3. How many square centimeters of skin do average hands and forearms have in total?
4. What are three of the primary functions of the skin?
5. What are the three main means of permeation of agents through the skin?
6. List at least three skin factors that may increase chemical absorption and permeation through the skin.
7. What are three exposure factors that may increase chemical absorption and permeation through the skin?
8. What are three chemical factors that may increase chemical absorption and permeation through the skin?
9. What are the two primary differences between irritant contact dermatitis and allergic contact dermatitis?
10. The best ways to minimize or control exposures through the dermal route are through (1) _____, (2) _____, and (3) _____.
11. Splash guards, lids, and glove boxes are examples of _____ controls for dermal exposures.
12. Why does chemical PPE need to be changed periodically?
13. What are some of the drawbacks of using PPE as a worker protection strategy?

REFERENCES

American Cancer Society. "Cancer Facts & Figures 2015." Accessed June 23, 2015. www.cancer.org/acs/groups/content/@editorial/documents/document/acspc-044552.pdf.

Bleyer, A., M. O'Leary, R. Barr, and L. A. G. Ries, eds. *Cancer Epidemiology in Older Adolescents and Young Adults 15 to 29 Years of Age, Including SEER Incidence and Survival: 1975–2000*. Bethesda, MD: National Cancer Institute, 2006.

Centers for Disease Control and Prevention. "Chemical Skin Hazard Strategy Revised by NIOSH to Provide More Useful, Detailed Notations." Accessed April 13, 2015. www.cdc.gov/niosh/updates/upd-07-17-09.html.

———. "Skin Exposures and Effects." www.cdc.gov/niosh/topics/skin/, 2012.

DHSS (NIOSH). *Effects of Skin Contact with Chemicals—Guidance on Occupational Health Professionals and Employers*. Publication No. 2011-200. August 2011.

European Agency for Safety and Health at Work (EASHW). *Occupational Skin Diseases and Dermal Exposure in the European Union (EU-25). Policy and Practice Overview*. European Observatory Report, 2008.

Franklin, C., and J. Worgan. *Occupational and Residential Exposure Assessment for Pesticides*. Hoboken, NJ: John Wiley & Sons, Ltd., 2005.

Frasch, H., G. Dotson, A. Bunge, C. Chen, J. Cherrie, G. Kasting, J. Kissel, J. Sahmel, S. Semple, and S. Wilkinson. "Analysis of Finite Dose Dermal Absorption Data: Implications for Dermal Exposure Assessment." *Journal of Exposure Science and Environmental Epidemiology* 24 (2014): 65–73.

Guy, G. P., S. R. Machlin, D. U. Ekwueme, and K. R. Yabroff. "Prevalence and Costs of Skin Cancer Treatment in the U.S., 2002–2006 and 2007–2011." *American Journal of Preventive Medicine* 104, no. 4 (2014): e69–e74.

Health and Safety Executive. *Managing Skin Exposure Risks at Work: Health and Safety Directive*. Merseyside, UK: HSE, 2009.

———. *Work-Related Skin Disease in Great Britain 2013*. Merseyside, UK: HSE, 2014. www.hse.gov.uk/statistics/causdis/dermatitis/skin.pdf.

Scheuplein, R., and I. Blank. "Permeability of Skin." *Physiological Reviews* 51, no. 4 (1971).

Semple, S. "Dermal Exposure to Chemicals in the Workplace: Just How Important Is Skin Absorption?" *Occupational and Environmental Medicine* 61 (2004): 376–82.

Stern, R. S. "Prevalence of a History of Skin Cancer in 2007: Results of an Incidence-Based Model." *Archives of Dermatology* 146, no. 3 (2010): 279–82.

Tibaldi, R., W. ten Berge, and D. Drolet. "Dermal Absorption of Chemicals: Estimation by IH SkinPerm." *Journal of Occupational Medicine* 11, no. 1 (2014): 19–31.

World Health Organization. Inter-Organization Programme for the Sound Management of Chemicals, International Programme on Chemical Safety. Environmental Health Criteria 242—Dermal Exposure. 2014.

11
Noise

LEARNING OBJECTIVES

After completing this chapter, readers should be able to do the following:

- Explain the importance of noise control to protect worker hearing.
- Understand the propagation of sound through space and perform associated calculations.
- Relate the concept of sound pressure levels to how noise is detected by the human ear and interpreted by the brain.
- Describe how excessive noise levels damage hearing.
- Identify safe levels of noise in the workplace and determine whether they are being exceeded.
- Use monitoring equipment to measure noise in the workplace and perform job site surveys.
- Perform mathematical calculations of noise levels.
- Implement controls to reduce or eliminate excessive noise in the workplace.

Photo credit: Leslie Achtymichuk/iStock

INTRODUCTION

Worker exposure to excessive levels of noise is pervasive in industrialized society. Excessive noise levels can be found in several areas of labor including manufacturing, mining, agriculture, construction, and transportation, to name a few. Even workers in hospitality and entertainment are routinely subjected to excessive levels of noise.

According to OSHA, approximately 30 million workers in the United States are exposed to excessive workplace noise on a regular basis. Hearing loss is the No. 1 area of worker compensation claims, with a total annual cost of $242 million.

Three factors make workers' overexposure to noise particularly significant. First, unlike many other occupational health hazards, hearing loss resulting from exposure to noise is irreversible. Once occupational noise limits have been exceeded and workers have been overexposed, the damage has already been done. Unlike many hazardous exposures that the body can detoxify, metabolize, or reduce the toxicity of, once noise has damaged the sensitive portions of the inner ear, there is no way to repair that damage. Continued exposure causes additional, accumulating damage.

The second factor is that damaging noise levels are generally easy to assess. It is simple to determine when workers are being exposed to hazardous levels of noise, even without expensive or complicated survey or monitoring equipment. As a general rule of thumb, if it is difficult to hear someone about a meter away speak, unless that person shouts, the noise levels are around 90 decibels. If a worker is exposed to this level of noise for more than a few hours, current standards and regulations would consider him or her to be overexposed. Recent research findings regarding health effects indicate that even this level may be excessive, and there is a building consensus for the allowable levels to be lowered further. As one can imagine, a noisy restaurant or bar, drilling or excavation, and many mechanical operations we encounter every day easily exceed the allowable and potentially hazardous noise levels.

The third fundamental reason that occupational noise exposure is an area that warrants more attention is that overexposure is almost always preventable. Design of systems and equipment can be used to eliminate the creation of noise at the source in many applications. In addition, engineering controls are often available to enclose the source or block excessive noise from reaching workers. Lastly, a variety of personal protective equipment in the form of earplugs, headphones, and other sophisticated noise-control technologies is available to reduce worker exposures to within acceptable, safe levels.

THE SCIENCE OF SOUND CREATION AND PROPAGATION

Sound is the propagation of energy through space and matter similar to waves moving across a pond after a stone hits the surface. Much as water waves move with peaks and valleys of energy, sound moves as a wave of pressure variances through the air. The peaks represent the air compressions, and the valleys represent the rarefactions of low pressure that follow. Because the amplitude of the waves is related to the intensity of the energy source, the harder the rock hits the water, the greater the amount of energy that is transferred and the bigger the waves. In the case of sound, the harder the hands clap together, the greater the pressure variance and amplitude of the sound wave.

If you slap your hand on a table, the abrupt physical intensity of pressure from your hand is transferred to the air around the table. This pressure differential travels as a wave, creating pressure differentials around the source.

The wave created travels with an intensity, frequency, and wavelength that are all related to the size of your hand, the hardness of the table, and the force with which you struck the table. The pressure differential can be represented by the diagram in Figure 11–1.

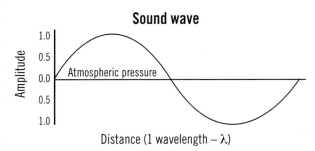

Figure 11–1. Properties of a sound wave.

In Figure 11–1, the height of the wave is called the amplitude and represents the force used to create the sound. The frequency is the number of times one complete wave cycles in a given duration of time. Figure 11–2 represents the frequency of a sound wave that cycles one complete wave in 1 second. This frequency, called a Hertz (Hz), is used to quantify sound waves.

Frequency represents the number of peaks and valleys per second and is related to the source, as sound is created by imparting energy to the transporting medium, or air.

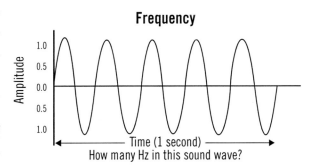

Figure 11–2. A sound wave with a frequency of 1 Hz.

Large sources of energy tend to impart low-frequency sounds. These large sources may create large waves with long wavelengths. Small sources, such as a small rock, create small and short wavelengths that tend to have a high frequency.

Through air, sound moves at a constant speed (c) of 344 m/s. Through solid materials and liquids, sound moves even faster. In old cowboy movies, characters would put their ears to the iron railroad tracks to listen for whether a train miles away was coming, which they could detect well before they could hear the train's sound through the air. Because frequency (f) and wavelength (l) are interdependent, they can each be expressed by this formula:

$$\lambda = c/f$$

As the frequency increases, the wavelength decreases.

Example

What is the wavelength in meters of a pure sound at 1,000 Hz?

Solution:

$\lambda = c/f$
$\lambda = (344 \text{ m/s})/1{,}000 \text{ Hz}$
$\lambda = 0.344 \text{ m}$

What about a sound at 250 Hz?
$\lambda = 1.4 \text{ m } (\sim 4.5 \text{ ft})$

What about 5,000 Hz?
$\lambda = 0.07 \text{ m } (\sim 2.7 \text{ in.})$

Because sound moving through a medium is really a set of discrete variations in pressure, the power of these pressure variations is measured in units of Pascal (Pa). The human ear's threshold for hearing these pressure differences is about 20 mPa (20×10^{-6} Pa) (atmospheric pressure) at the frequency of 1,000 Hz. The measureable threshold of pain, for average humans, is about 20 Pa.

Because the range of pressure variations in human hearing is so broad, logarithms are used to quantify them for comparison. The decibel (dB) is a dimensionless unit based on the log of the ratio of the noise pressure to a reference quantity. The root-mean-square of the sound pressure level (SPL) related to the atmospheric reference level can be converted to decibels using the following equation.

$$SPL_p = 10 \log(P/P_{ref})^2$$

where

SPL_p = Sound pressure level (dB)
P = Measured pressure (Pa)
P_{Ref} = Reference pressure (20 µPa)

Example

What is the sound pressure level in decibels of a noise with a pressure of 700 Pa?

Solution:

$SPL_p = 10 \log (P/P_{Ref})^2$

$L_p = 10 \log [700 \text{ (Pa)} / 20 \times 10^{-6} \text{Pa}]^2$

$L_p = 90 \text{ dB}$

Another method of measuring and analyzing sound is to measure the power as it radiates through free space in terms of intensity, which is measured in units of watts (W). The sound reference level for humans at 1,000 Hz is about 10^{-12} W. In order to convert the intensity of a sound wave to decibels, the following equation may be used:

$$SPL_p = 10 \log (I/I_{Ref})^2$$

where

SPL_p = Sound pressure level (dB)
I = Measured pressure (Pa)
I_{Ref} = Reference pressure (20 µPa)

Using the above equations, we can convert differences in sound pressure levels into relative sound intensities that we can analyze and compare. Table 11–1 provides a list of sound intensities in terms of decibels, along with typical sources.

Table 11–1. Typical Sound Pressure and Sound Pressure Level Comparisons

Sound Pressure (µPa)	Sound Pressure Level (dB)	Source
20,000,000,000	180	Rocket launch pad
200,000,000	140	Threshold of pain, jet
20,000,000	120	Ambulance siren
6,300,000	110	Rock concert
630,000	90	Subway, lawn mower
200,000	80	Noisy restaurant
63,000	70	Vacuum cleaner
20,000	60	Conversation
6,300	50	City residence
2,000	40	Audiometric test booth
630	30	Whisper
20	0	Threshold of hearing (1,000 hz)

When there are multiple noise sources, it is possible to calculate the total level from the contributions of several sources using the following equation:

$$L_{total} = 10 \log \Sigma^N (10^{L_i/10})$$

Example

What is the total sound pressure level of three different noise sources with individual sound pressure levels of 87 dB, 85 dB, and 82 dB?

Solution:

$L_{total} = 10 \log \Sigma^N (10^{L_i/10})$
$L_{total} = 10 \log \Sigma^N (10^{87/10} + 10^{85/10} + 10^{82/10})$
$L_{total} = 89 \text{ dB}$

WHAT HAPPENS WHEN SOUND HITS THE EAR?

As sound energy moves through the air, the ear detects the waves of pressure variation through its sophisticated and sensitive anatomical structure. First, the **outer ear**, made of folded cartilage called the **pinna**, is shaped in a way to collect the sound energy. This energy moves through the **auditory canal**, which is about 1.5 in. long, and converts the pressure differentials into vibratory motion in the eardrum, or tympanic membrane. This delicate membrane, which separates the outer ear from the inner ear, through slight vibrations sends information about the strength and frequency of the sound to the middle ear.

The **middle ear** is an air-filled cavity that contains three bones called ossicles. These bones are the **malleus (hammer)**, the **incus (anvil)**, and the **stapes (stirrup)**. Through several anatomical and mechanical processes, these bones convert the vibrations of the eardrum and transfer them to the oval window and the inner ear.

The **inner ear** is a fluid-filled organ with two main parts. The first part is a series of semicircular canals called the **cochlea**. The fluid in these canals absorbs the energy from the oval window and transmits it up through fluid-filled loops. Within these loops tiny hair-like cells vibrate to different sound frequencies. The energy in the cochlea is converted to electrical impulses that are transmitted via the second main part of the inner ear, the **organ of Corti**. The organ of Corti then transmits electrical signals through the auditory nerves to the brain, where they are registered as sound. The basic anatomy of the ear is shown in Figure 11–3.

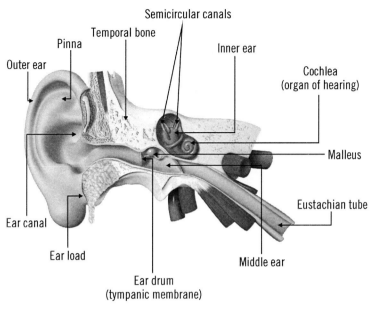

Figure 11–3. The anatomy of the ear.

(Source: pixologicstudio/iStock)

HOW NOISE DAMAGES HEARING

Noise is defined as unwanted sound. So when we consider sound that can potentially damage hearing, we speak of it in terms of noise.

When different sound pressure levels strike the ear, there are variations in intensity, harmonic content (or timber), frequency, and direction. These attributes combined determine whether the sound is useful, pleasant, or harmful.

Just like any other tool, the human ear is damaged by normal wear and tear over time. With age, nearly everyone shows the loss of hearing called **presbycusis**. It is believed that this loss develops over a lifetime of normal noise exposure and from various parts of the ear losing their abilities to function.

In addition, several risk factors that accelerate the loss of hearing have been identified. Some of these factors are regular exposure to excessive noise levels, the frequency of the noise, the amount of noise each day, and a duration of exposure over several years.

For the most part, occupational hearing loss is associated with a cumulative exposure to noise. People who are routinely exposed to excessive levels of noise

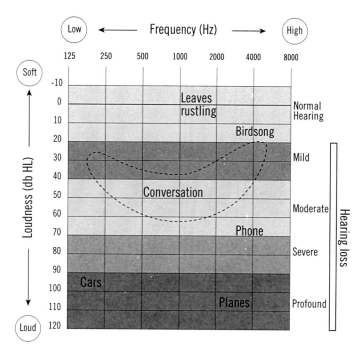

Figure 11–4. Human hearing loss at different loudness levels.

(Source: ©Thomas Haslwanter/Wikimedia Commons/CC-BY-SA-3.0/ GFDL)

over a long period of time tend to lose their hearing sooner and more noticeably than those who do not have this cumulative exposure. Again, this type of hearing loss is irreversible.

The actual loudness that people hear is closely related to sound pressure level, but it is also affected by frequency. The range of normal hearing is between 20 Hz and 20,000 Hz. Because the human ear is more sensitive to higher-frequency sounds, those are the frequencies that tend to cause hearing loss first. Figure 11–4 demonstrates that humans lose hearing more drastically in the higher loudness levels. Hearing loss is also typically more severe in the 2,000 to 6,000 Hz frequency ranges. For example, for the top line on the graph, a loudness level of 120 dB in the 4,000-Hz frequency range is heard at only 110 dB. At the same frequency, another person can hear only 50 dB, which indicates that he or she has significant hearing loss.

In addition to losing hearing in these midlevel frequency ranges more easily than at the upper and lower ranges, humans tend to hear better in these midlevel ranges, which also happen to be in the range of human voice frequencies, 500–4,000 Hz. For that reason, noise exposures in these ranges tend to be more significant, and noise measurements need to be weighted in these areas to better protect workers from exposures in these frequencies. The range of frequencies where most people hear is called the A range. Because this range is so important, exposure limits and most occupational measurements are conducted in the A ranges and are referred to as dBA.

PROTECTING WORKERS' HEARING

In an attempt to protect workers' hearing, recommended exposure limit criterion levels have been set by regulatory bodies and through consensus standards. The original sound pressure level limits set by the Occupational Safety and Health Administration (OSHA) were based on the level that OSHA determined would not produce damage in a worker exposed 40 hours per week over 40 years of work. The permissible exposure limit is a threshold level average of 90 dBA for an 8-hour workday.

Along with the limit of 90 dBA, OSHA allowed a dBA exchange rate of 5 dB for every 4 hours of exposure. That is, at 90 dBA, people could work 8 hours, but at 95 dBA they could work only half the time, or 4 hours. In addition, workers could be exposed to a sound pressure level of 100 dBA for only 2 hours. To calculate the daily noise dose in terms of percentage of allowed exposure, the following equation can be used:

$$DND = [C_1/T_1 + C_2/T_2 + L + C_n/T_n]$$

where

DND = Daily noise dose
C_i = Number of h exposed at a given SPL_i
T_i = Number of h allowable at a given SPL_i

After setting its original noise exposure limits, OSHA created a Hearing Conservation Amendment (HCA) to make the noise limits more restrictive. At that time, the criterion reference was changed to include the occupational exposures at levels down to 80 dBA in the worker's daily exposure average, with the same 5-dB doubling attribute. OSHA's two different noise exposure limits are shown in Tables 11–2A and 11–2B.

Tables 11–2A and B. OSHA's Noise Exposure Limits

11–2A. PEL to Determine the Need for Engineering or Administrative Controls		11–2B. PEL to Determine the Need for Hearing Conservation Program. HCA Limits	
Duration (hr/day)	Sound level (dBA)	A-weighted sound level duration	Reference hours
8	90	80	32
6	92	85	16
4	95	90	8
3	97	95	4
2	100	100	2
1½	102	105	1
1	105	110	.50
½	110	115	.250
¼	115	120	.125
		125	.063
		130	.031

> **Example**
>
> If workers are exposed to 90 dBA for 1 hour, 95 dBA for 1 hour, 100 dBA for 30 minutes, and 85 dBA for 5.5 hours, what percent of their daily noise dose would they receive?
>
> **Solution:**
>
> $$\begin{aligned} DND &= [C_1/T_1 + C_2/T_2 + L + C_n/T_n] \\ &= [1/8 + 1/4 + 0.5/2] \\ &= 62\% \text{ of the allowable daily dose (OSHA permissible exposure limit)} \end{aligned}$$
>
> What would be the daily noise dose percentage if the hearing conservation limits were applied?
>
> $$\begin{aligned} DND &= [C_1/T_1 + C_2/T_2 + L + C_n/T_n] \\ &= [1/8 + 1/4 + 0.5/2 + 5.5/16] \\ &= 96\% \text{ of the allowable dose} \end{aligned}$$

While the HCA limits are more restrictive and protective of workers, the American Conference of Governmental Industrial Hygienists (ACGIH®) promotes an even more conservative approach. This professional consensus body publishes recommended Threshold Limit Values (TLVs®) for worker noise exposure. The ACGIH TLVs begin with a limit of 85 dBA for an 8-hour noise exposure. It also promotes the use of a conservative, 3-dB doubling criterion, as shown in Table 11–3.

Table 11–3. ACGIH Limits on Noise Exposure

Duration (hr/day)	Sound Level (dBA)
24	80
16	82
8	85
4	88
2	91
1	94
0.5	97
0.25	100

Source: American Conference of Governmental Industrial Hygienists

> **Example**

If the ACGIH TLV® limits are applied to the same worker noise levels as in the previous example, by what percentage would the levels exceed recommendations?

Solution:

$$DND = [C_1/T_1 + C_2/T_2 + L + C_n/T_n]$$
$$= [1/1.75 + 1/1 + 0.5/0.25 + 5.5/8]$$
$$> 400\% \text{ of the allowable daily dose (ACGIH TLVs)}$$

It is obvious that the ACGIH consensus limits are more conservative and protective of worker hearing than either of the limits set by OSHA. Recent research indicates that the more conservative approach to protecting worker hearing is probably appropriate and will likely lead to fewer workers' compensation claims from hearing loss in older workers.

Because workplace noise levels often do not change much throughout the day, it is easy to determine the time allowed at a given noise level by using the following equation:

$$T \text{ (hours)} = \frac{8}{2^{[(SPL-90)/5]}}$$

> **Example**

How long could a worker stay in an area of 97 dBA without wearing hearing protection?

Solution:

$$t = \frac{8}{2^{(SPL-90)/5}}$$

$$t = \frac{8}{2^{(97-90)/5}}$$

$$t = 3.0 \text{ h}$$

SOUND LEVEL MEASUREMENT

To accurately determine the noise hazards associated with a given job or work space, it is necessary to conduct surveys and measurements of exposure levels. These may be done with sound level meters to measure the environmental levels, or personal noise dosimetry can be used to evaluate exactly what each worker is exposed to on the job.

Sound level meters are typically noise measurement instruments that consist of the following:
- microphone
- amplifier
- attenuator
- weighting network
- metering system

Figure 11–5. CEL 254 sound level meter.

They usually measure in the SPL range of about 40–140 dB. A sound level meter is shown in Figure 11–5.

METER WEIGHTING

In special cases, an octave-band analyzer type of sound level meter can be used to determine the various noise levels at individual frequencies. This can be useful for filtering out the noises in the lower and higher frequencies, to which the human ear is less sensitive.

In fact, even sound level meters that do not provide a readout of specific frequencies can typically be set to measure three different frequency ranges. Commonly, meters may be weighted to measure frequency ranges of interest. A flat meter measures all frequencies without any weighting factors, which is considered a linear response. A-scale devices weight the frequencies to simulate the response and sensitivities of the human ear. These devices attenuate the contributions from the lower frequencies to the total sound pressure level readings. Lastly, C-scale meters, which provide little weighting and a fairly flat response, are sometimes used for specific industrial purposes.

RESPONSE TIME

Sound meter devices may respond to sound pressure level variations at different speeds. Slow-response devices may take up to a second or more to register the total noise level in the area. This is a good feature when noise levels fluctuate, as it allows the technician to read the meter without the reading changing too much.

Fast-response meters can separate individual noises by registering each one as quickly as 125 ms. This feature is useful for distinguishing among the individual sound pressure contributions from different machines or processes.

Lastly, impact response time reading devices can perceive 90% of the peak sound pressure levels in as little as 50 ms. In some industrial conditions, these devices are the only way to measure the noise levels.

METER MICROPHONES

Perhaps the most important feature of a sound level meter is its microphone. There are two basic types of microphones. Type 1 microphones use electret condensers and provide excellent linearity and response to high-frequency sounds. These devices are extremely sensitive to moisture and vibration, however. The type 2 microphone is the more common and rugged type. However, this type of microphone is less able to record noise at extremely high frequencies.

SOUND LEVEL SURVEYS AND MEASUREMENTS

Sound level measurements are made for a variety of reasons. Ambient measurements are often made to gather initial assessments about compliance with occupational noise regulation limits. Sometimes noise measurements are made to determine compliance with environmental limits. In addition, these sound level measurements can provide information and guidance regarding opportunities and needs for noise reduction and controls to be put in place to reduce noise emissions or exposures. Sometimes sound surveys are conducted to collect information on the emissions from a single noise source or machine.

When collecting noise level data, a fairly standard survey format is used.

NOISE DOSIMETRY

When specific, individualized noise exposures are desired, noise dosimeters can be used to record the exact sound pressure levels that a worker is exposed to at any given time. A lightweight dosimeter is attached to the worker's torso, and a microphone is typically placed near the ear to pick up the noises workers are exposed to in the workplace. The worker can wear the dosimeter for all or part of the workshift (Figure 11–6). The dosimeter records data about the worker's daily noise dose. Some dosimeters can provide graphical representations of the time of noise exposure as well as information about frequencies.

Figure 11–6. Noise dosimeter used to measure personal exposures.

ADDING CONTRIBUTIONS FROM MULTIPLE NOISE SOURCES

In certain instances, noise may result from a combination of sources, such as different machines on a factory line. In these cases, if the contribution of each of the machines is known, it is possible to calculate the total SPL at a given location.

The total sound power level can be calculated via logarithmic mathematics using the following equation:

$$L_{total} = 10 \log \left(\sum_{i=1}^{N} 10^{L_i/10} \right)$$

where

L_{Total} = Total SPL from all sources, and
L_i = SPL from each separate source

Example

What is the total noise level from four different machines producing 72 dB, 87 dB, 85 dB, and 82 dB?

Solution:

$L_{Total} = 10 \log (10^{72/10} + 10^{87/10} + 10^{85/10} + 10^{82/10})$
$L_{Total} = 89.8$ dB

CONTROLLING NOISE LEVELS

Wherever possible, noise should be minimized through engineering controls. Selling points for reducing workplace noise include the following:
- conservation of worker hearing
- fewer workers' compensation claims
- better OSHA compliance
- fewer Environmental Protection Agency or public health department issues
- better employee morale and productivity
- improved public relations

In order to design effective controls, some basic concepts should be considered. In general, higher frequencies are more disturbing than lower ones; however, low frequencies are more difficult to control with barriers, and they tend to travel farther in space. Higher-frequency sounds have less power, are more directional, and are more reflective than lower frequencies. For example, when you hear a car with the music blaring, you tend to hear the lower bass frequencies rather than the higher notes and vocals. For these reasons, higher frequencies tend to be easier to control through engineering design.

As sound propagates through space, it interacts differently depending on the environment. In a **free field**, there is no reflection of sound waves, and the sound pressure decreases with increasing distance. In the **near field**, or close to the source, sound waves are slightly irregular. But in the **far field**, far from the source, the sound pressure decreases more consistently with an inverse squared relationship. In indoor, industrial environments, with walls and other surfaces, sound travels in a **reverberant field**, and the sound pressure remains relatively constant with distance from the source.

CONTROLLING THE SOURCE

Substitution is the first option to be considered for reducing the noise created by machines or systems. Typical motors in manufacturing range from 100 to 110 dBA. Yet quieter machines that emit only 80–85 dBA are now available. The types of machines used to perform various functions can also emit different noise levels. For example, belt drives are quieter than gears, and hydraulic presses are quieter than mechanical ones.

Controlling noise from processes is another way to reduce sound levels. One way to reduce noise is to reduce the driving force of motions, such as by using welding instead of riveting, using hot work instead of cold work, and reducing driving forces behind industrial movements. Adding perforations to solid surfaces such as equipment reduces the amount of noise emitted when they vibrate. Reducing the size and length of continuous materials can lessen the noise transmitted from the surfaces.

Noise levels can also be reduced by lowering rotational speeds, isolating vibrating surfaces through separation and damping, and reducing the surface area of vibrating pieces. Altering the mass of materials can be used to change the resonating frequencies.

Fan blades can be designed to be more aerodynamic and produce less noise. Small fans with many blades can be exchanged for larger fans with fewer blades to increase the frequency of noise created. The relationship between the number of fan blades, their speed, and frequency is shown in the following equation.

$$f_B = \frac{(rpm)(N)}{60}$$

where

f_B = Noise frequency created by blades
N = Number of blades

Example

What is the frequency of the predominant tone that would be emitted from an axial fan with four blades rotating at 6,000 rpm? What if the fan had 10 blades instead?

Solution:

$$f_B = \frac{(rpm)(N)}{60}$$

$$f_B = \frac{(6000)(4)}{60} = 400 \text{ Hz} \qquad f_B = \frac{(6000)(10)}{60} = 1{,}000 \text{ Hz}$$

Lastly, processes that involve the flow of air or water can be altered to reduce the noise emitted from the systems. Air or water at lower pressures and flows will create less noise. Minimizing flow restriction by eliminating T sections and 90-degree elbows can reduce turbulence and noise created by fluid systems. Reducing the responses of vibrating surfaces by damping and decreasing the radiating surfaces can reduce operational noise. If pipes are hung with dampened supports, it can make a drastic change in the level of noise throughout the plant.

DISTANCE

One way to reduce noise exposure is to use distance to separate workers from the source. As the distance from noise sources increases, the SPL tends to fall off at a rate of 6 dB each time the distance is doubled. This relationship can be represented mathematically by this equation:

$$SPL_2 = SPL_1 + 20 \log (d_1/d_2)$$

where

SPL_2 = Sound pressure level at d_2
SPL_1 = Sound pressure level at d_1
d_n = Distances from source

Example

If the SPL at 12 ft is 104 dBA and the machine acts like a point source, what is the SPL at 4 ft?

Solution:

$SPL_2 = SPL_1 + 20 \log (d_1/d_2)$
$SPL_2 = 104 + 20 \log 12'/4'$
$SPL_2 = 104 + (20 \times .48)$
$SPL_2 = 104 + 9.54$
$SPL_2 = 113.5$ dBA

BARRIERS

Noise levels near the worker can be reduced by adding barriers between the worker and the source. Barriers can take the form of enclosures, mufflers, or sound shields (which reflect the sound in other directions). The main factors influencing the effectiveness of the barrier are its height and its proximity to the source. For a given noise emanating from a source, a reduction in the noise level that the worker receives can be calculated:

$$\text{Noise reduction} = 10 \log [20H^2/\lambda(R)]$$

where

H = Height of barrier
λ = wavelength = c (344 m/s)/Hz (cy/s)
R = Distance from receiver to barrier

Example

What would the SPL be at a point 20 m from a 5,000-Hz source after you build a sound barrier wall 3 m high and 2 m away from the source, if the original SPL at 20 m is 85 dBA?

Solution:

NR = 10 log [20 (barrier height)^2/(wavelength × R)]
 = 10 log [20(9)/(0.0688 m)(18 m)]
 = 10 log (145.3) = 21.6 dB
New SPL = 85 − 21.6 = 63 dB

Because high-frequency sounds have less energy and are greatly reduced just by passing through air, changing the frequencies produced to higher levels reduces the power transmitted. However, lower frequencies are less disturbing to the human ear.

Enclosures are a special category of barrier. In order for enclosures to effectively contain noise, they must meet the following conditions:

- They must be airtight; otherwise, noise will escape, similar to air leaking out of a balloon.
- Acoustical absorption must be added to the interior. In addition, acoustic absorption materials must be the correct type and shape for the noise frequencies and sources.
- The correct wall materials must be used; heavier walls absorb sound better.
- Absorption must take place on the interior; absorption on the outside of an enclosure does not work.

- Enclosures should be dampened so that vibrations traveling from the source to the enclosure are minimized.

ADMINISTRATIVE CONTROLS

Various administrative controls can be used to reduce occupational noise exposures. It is often possible to rotate workers to make sure that any one individual's exposure to is within acceptable levels, thus spreading out the overall exposure. The problem with this administrative approach is that it does not lower the noise levels or solve the problem.

Another administrative approach is to schedule noisy machine operations for when fewer workers are in the vicinity so that overall noise exposures are reduced. It may even be possible to have workers leave an area before noisy operations or activities take place.

Whenever workers have an average daily exposure of 85 dBA, they must be included in a Hearing Conservation Program (HCP). The HCP includes requirements for monitoring and recording workplace exposures, performing annual audiometric testing of employee hearing, training workers, and using hearing protection equipment.

Hearing Protection Devices

The last resort in reducing the exposure of workers to noise is the use of personal protective equipment in the form of hearing protection devices. These devices include earplugs and earmuffs and are typically sold with an assigned Noise Reduction Rating (NRR), which indicates the approximate amount of dB that the device removes from the ambient noise levels. For example, if a set of earmuffs has an NRR of 15 and the ambient noise level is 98 dBA, then the earmuffs would reduce the level received by the worker wearing the earmuffs to approximately 98 − 15 = 83 dBA.

Drawbacks of hearing protection devices are that they must be properly fitted to ensure their effectiveness and that it is often difficult to ensure a proper fit.

SUMMARY

The health effects associated with excessive occupational noise exposure are well recognized. It is relatively easy to measure hazardous levels of noise and take actions to eliminate them from the workplace. Whenever possible, excessive levels of noise should be controlled at the source, but sometimes moving the source or the workers has to be done instead. Barriers may also be installed between the source and the workers to block the noise path. Lastly, personal protective equipment in the form of earmuffs and earplugs can be used to protect workers when all other options have been implemented and noise levels are still high.

REVIEW QUESTIONS

1. Find the wavelength in meters of a pure sound at the following frequencies:
   a. 2,000 Hz
   b. 850 Hz
   c. 25,000 Hz
2. How loud in decibels is an audio system with a sound power rating of 280 W per channel?
3. How loud in dB is a drill press with a sound power rating of 0.003 W?
4. What is the SPL in decibels coming from a source generating 2.3 Pa?
5. You conducted a survey and obtained the following results (see table below). Was the OSHA permissible exposure limit exceeded? Was the HCA exceeded?

Sound Pressure Level (dBA)	Duration in hours	PEL	HCA
90	½	0.06	0.06
80	2	0	0.06
95	2	0.5	0.5
100	½	0.25	0.25
85	3	0	0.187
		Total 0.81	1.0575

6. If you measured the 3,000-Hz noise coming from a compressor outdoors to be 110 dBA at a distance of 10 ft, what would you estimate the noise level to be at a distance of 200 ft?
7. Referring to Question 6, what would the noise level be at 200 ft from the source if you placed a wall 20 ft high, 15 ft away from the source?
8. At which frequency would a noise-induced hearing loss notch most likely occur?
   a. 1,000 Hz
   b. 2,000 Hz
   c. 3,000 Hz
   d. 4,000 Hz

9. What is the maximum noise level in decibels of a receiver that is rated at 200 W per channel?
10. The root-mean-square sound pressure level (SPL) of the receiver is rated at 80 N/m². What is the maximum noise level in decibels from this system?
11. The NRR for a pair of earplugs is 24 dBA, and the 8-h time-weighted average in a work area is 100 dBA. At how many decibels will the earplugs reduce the noise level using ACGIH's safety factor of 2?
12. If you want to install five machines with the individual noise levels listed below, what would the total noise level be?

Machine	SPL (sBA)
1	86.0
2	88.1
3	88.1
4	82.2
5	84.5

13. What is the frequency of the sound emitted from a fan with 35 blades operating at 4,000 rpm?
14. What percent of the OSHA criterion allowable dose does a worker receive if she works in an area with 100 dB for 0.5 hours, 70 dB for 4 hours, 90 dB for 3.0 hours, and 105 dB for 0.5 hours?
15. What is sensorineural hearing loss, and why is it typically irreversible?
16. The audible range of an average young person with unimpaired hearing is _____.
    a. 4,000–8,000 Hz
    b. 1,000–10,000 Hz
    c. 16,000–20,000 Hz
    d. 2,000–4,000 Hz
17. OSHA requires audiograms for all employees exposed to an 8-hour time-weighted average of _____ dBA.

REFERENCES

American Conference of Governmental Industrial Hygienists. 2013 TLVs® and BEIs®. Cincinnati, OH, 2013.

Bell, L., and D. Bell. *Industrial Noise Control: Fundamentals and Applications*. 2nd ed. New York: Marcel Dekker, 1994.

Berger, E., L. Royster, J. Royster, D. Driscoll, and M. Layne. *The Noise Manual*. 5th ed. Fairfax, VA: American Industrial Hygiene Association, 2003.

Bruce, R., C. Bommer, C. Moritz, K. Lefkowitz, and N. Hart. "Noise, Vibration and Ultrasound." In *The Occupational Environment—Its Evaluation, Control, and Management*, edited by Daniel Anna. 3rd ed. Fairfax, VA: American Industrial Hygiene Association, 2011.

Occupational Safety and Health Administration. *Occupational Noise Exposure*. Code of Federal Regulations, Title 29, Part 1910.95, subpart G.

Pelton, K. "Noise." In *Applications and Computational Elements of Industrial Hygiene*, edited by M. Stern and S. Mansdorf. Boca Raton, FL: CRC Press, 1999.

Standard, J. "Industrial Noise." In *Fundamentals of Industrial Hygiene*, edited by Barbara Plog. Itasca, IL: National Safety Council, 2012.

12
Radiation

LEARNING OBJECTIVES

After completing this chapter, readers should be able to do the following:

- Describe electromagnetic radiation.
- Perform calculations demonstrating the relationship among the speed of light, frequency, and wavelength.
- Use the inverse-square law to calculate radiation intensity at different distances.
- Understand the basic principles of radiation shielding and the concept of the Half Value Layer
- Describe radioactive decay, and perform relevant calculations.
- Discuss physical and biological half-lives and perform relevant calculations.
- Understand how radiation quality factors affect the hazard levels of different types of radiation.
- Describe how ionizing radiation causes biological damage and how occupational exposure limits are determined.
- Demonstrate the use of the ALARA principle.
- Identify different levels of laser hazards and devices.
- Calculate power levels for microwave transmission devices.
- Identify hazard levels, biological effects, and safety controls for various nonionizing sources of radiation including radiofrequency, microwaves, optical light, UV light, and lasers.

Photo credit: RobertKovacs/iStock

INTRODUCTION

Radiation is energy that moves through space. Radiation is present in the natural world in the form of sunlight, heat, cosmic rays, and emissions from various radioactive elements found in rocks and even in living creatures. Modern technology also uses and creates various kinds of radiation.

The risks of different types of radiation to human health depend on the nature of the radiation and how it interacts with matter. Ionizing radiation consists of electromagnetic waves or high-velocity subatomic particles that are capable of causing ionization—the ejection of electrons from atoms—in biological tissue. In general, it takes at least 33 electron volts (eV) of energy to cause ionization in tissue (EPA 2015). Nonionizing radiation consists of electromagnetic waves that are not capable of causing ionization.

BASICS OF ELECTROMAGNETIC RADIATION

Electromagnetic radiation consists of an electric field and a magnetic field that oscillate together and propagate through space as a wave traveling at the speed of light (see Figure 12–1). An electromagnetic wave is entirely characterized by its frequency or wavelength. Figure 12–2 illustrates the electromagnetic spectrum (see also Figure 6–9). Notice that as the frequency increases, the type of radiation changes from radio waves to infrared (IR) radiation (heat) to visible light to UV light to ionizing radiation. Frequency and wavelength are inversely related to each other, as shown in the following equation:

$$f\lambda = c \quad \text{or} \quad \lambda = \frac{c}{f} \quad \text{or} \quad f = \frac{c}{\lambda}$$

where f = Frequency measured in hertz (Hz) or cycles per second
l = Wavelength measured in meters
c = The speed of light (2.998×10^8 m/s in a vacuum)

Example

What is the wavelength of a 2.45-GHz microwave? (This is the frequency of radiation produced by a microwave oven.)

Solution:

2.45 GHz = 2.45×10^9 Hz = 2.45×10^9 cycles/s

$$\lambda = \frac{c}{f} = \frac{2.998 \times 10^8}{2.45 \times 10^9} = 0.118 \text{ m} = 11.8 \text{ cm}$$

Picture a sunburst with rays streaming out from the center. When a stream of radiation diffuses from a source, the law of conservation of energy and simple geometry tell us that the energy rays must become thinner the farther from the source they travel. For sources that radiate uniformly in all directions and out to distances much larger than the physical size of the source itself (as in isotropic point sources), the energy transferred through any defined area follows the inverse-square law. The inverse-square law, which is given in the equation below, says that the energy transfer per unit area is inversely proportional to the square of the distance from the source.

$$\frac{I_2}{I_1} = \left(\frac{d_1}{d_2}\right)^2$$

where I_1 = The energy transfer per unit area at location 1
I_2 = The energy transfer per unit area at location 2
d_1 = The distance from the source to location 1
d_2 = The distance from the source to location 2

Different measures of energy transfer are used for different regions of the electromagnetic spectrum. In this chapter, we will use the inverse-square law in several examples pertaining to different types of electromagnetic radiation.

Although electromagnetic rays move as waves and have no mass, they behave in some ways like particles—tiny bits of electromagnetic energy that are called photons. The energy of a photon is proportional to its frequency. Ionizing radiation in the form of x-rays or gamma rays, which are described in more detail later in this chapter, is the most energetic type of electromagnetic radiation. The boundary between ionizing and nonionizing electromagnetic radiation falls around a frequency of 3×10^{15} Hz, which is equivalent to a photon energy of 12.4 eV.

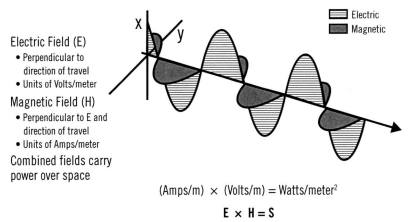

Electric Field (E)
• Perpendicular to direction of travel
• Units of Volts/meter

Magnetic Field (H)
• Perpendicular to E and direction of travel
• Units of Amps/meter

Combined fields carry power over space

(Amps/m) × (Volts/m) = Watts/meter²
E × H = S

Figure 12–1. An electromagnetic wave.

(Source: Plog 2012)

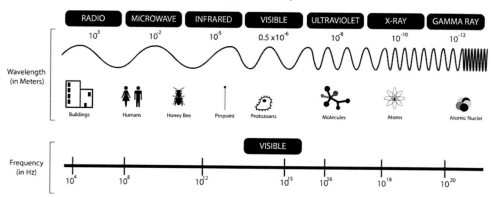

Figure 12–2. The electromagnetic spectrum.

(Source: Jonathan S. Urie / CC-BY-SA-3.0)

IONIZING RADIATION

ATOMIC STRUCTURE AND ISOTOPES

An atom is composed of three types of subatomic particles: the proton, which has an electrical charge of +1; the electron, which has an electrical charge of –1; and the neutron, which has no charge. Protons and neutrons are clustered together to form the nucleus, which is the core of an atom and contains nearly all its mass. Electrons, which are much smaller and lighter than protons and neutrons, are held in orbit around the nucleus through electrical attraction. Although the orbit of an electron is commonly depicted as a circular path, electrons actually exhibit wavelike behavior that forms them into cloudlike shapes called **orbitals** (de Broglie 1925).

The number of protons in the nucleus of an atom is called the **atomic number** and is denoted by the letter Z. The atomic number identifies the chemical element. For example, any atom that possesses three protons (Z = 3) is an atom of lithium (chemical symbol Li).

The term **nuclide** refers to a species of nucleus that is defined by its number of protons, its number of neutrons, and its energy state. Every element except hydrogen (Z = 1) must have neutrons in its nucleus in order to hold together. Atoms with the same number of protons but different numbers of neutrons are called **isotopes** of the element. Different isotopes of an element are identified by their mass number (also called the **atomic mass**), which is the number of protons plus the number of neutrons in the nucleus. For example, lithium has two naturally occurring isotopes: lithium-6, which has three protons and three neutrons in its nucleus, and lithium-7, which has three protons and four neutrons in its nucleus (Figure 12–3). Isotopes are commonly denoted by their chemical symbol and their mass number, for example, ⁶Li or Li-6. All isotopes of a given element exhibit the same chemical behavior.

MODES OF RADIOACTIVE DECAY

If a nucleus contains too many or too few neutrons relative to the number of protons, it is unstable and will eventually undergo a spontaneous change—called radioactive decay, disintegration, or transformation—in which ionizing radiation will be emitted from the atom. For example, although lithium-6 and lithium-7 are stable, lithium-8 is highly unstable and rapidly disintegrates, releasing high-velocity particles. The elements beyond lead or bismuth (Z > 83) have no stable isotopes; they are all radioactive.

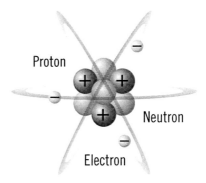

Figure 12–3. A simplified model of the structure of a lithium-7 atom.

(Source: Oliver Hoffmann/iStock)

Types of ionizing radiation resulting from nuclear transformation include alpha-particles, beta-particles, neutrons, and gamma rays. The kinetic energy of ionizing particles and the energy of electromagnetic rays are usually measured in kiloelectron volts (keV) or megaelectron volts (MeV).

An alpha-particle consists of two protons and two neutrons that are ejected from the nucleus during some radioactive decay. An alpha-particle is exactly the same as the nucleus of the helium ion (⁺⁺He-4) except that an alpha-particle moves very quickly and therefore has a very high kinetic energy, typically in the range of 4–7 MeV. Many isotopes of heavy elements, such as uranium, thorium, and radium, are alpha-emitters. Some smoke detectors sense alpha-particles emitted by tiny amounts of americium-241. Alpha-particles are easily absorbed (stopped) by matter; for example, they can be stopped by a sheet of paper or several inches of air and prevented from penetrating the dead outer layer of human skin. On the other hand, alpha-particles emitted by radionuclides inside the body are very harmful.

A beta-particle is an electron or positron that is ejected from the nucleus during a nuclear transformation. Positrons (positive beta-particles) are the antimatter versions of electrons; they have a charge of +1 and a mass equal to an electron's mass. Beta-particle energies range from about 10 keV for some radioisotopes to several megaelectron volts for other radioisotopes. The penetrating ability of beta-particles increases with their energy; beta-particles of energy greater than about 70 keV can penetrate far enough through the skin to reach living cells. The term **beta-emitter** is often reserved for the many radioisotopes that eject electrons (negative beta-particles) from the nucleus. Many commonly used radionuclides are beta-emitters. Positron-emitters have more limited uses, such as in positron emission tomography scans.

Neutrons are emitted during fission of heavy nuclei in nuclear reactors. Another source of neutrons is portable devices used in moisture gauges or well-logging equipment. Portable neutron sources are made by combining an alpha-emitter such as americium-241 with certain light, stable nuclides such as beryllium-9, which absorb alpha-particles and then emit neutrons. Neutrons are capable of passing through matter for an indefinite distance, but they also tend to lose energy in random collisions with hydrogen and other atoms and then get absorbed into a nucleus. Absorption of a neutron by a nucleus can give rise to the emission of secondary radiation, such as other subatomic particles or gamma-radiation, or may produce a radioactive nucleus, a process called *neutron activation*. Many useful radionuclides are produced intentionally through neutron activation in nuclear reactors.

Gamma rays are electromagnetic waves emitted from the nucleus during a nuclear transformation. Gamma-ray emission is often associated with alpha-particle or beta-particle emission. Nuclides in a metastable nuclear energy state, such as technecium-99m used in diagnostic medical procedures, emit gamma rays when the nucleus transforms to its ground state. Gamma rays are also produced in matter-antimatter reactions, as when a positron and an electron meet and annihilate each other. Gamma photon energies can range up to several megaelectron volts. Like neutrons, gamma photons are capable of passing through matter for an indefinite distance before being absorbed. On average, high-energy gamma rays tend to travel farther through matter than low-energy gamma rays.

Certain radionuclides, such as cadmium-109, undergo nuclear transformation by capturing an orbital electron. This results in the emission of electromagnetic radiation as the remaining electrons rearrange themselves among the orbitals. Ionizing electromagnetic rays originating from orbital electrons are called x-rays. Cadmium-109 and other radionuclides that undergo electron capture are commonly used as portable x-ray sources, for example, in x-ray fluorescence instruments used for detecting lead in paint. Because they are photons of pure electromagnetic energy, x-rays are indistinguishable from gamma rays; they differ only in where they originate.

Depending upon the radionuclide, the daughter nucleus may be stable or unstable after radioactive decay. For example, carbon-14 undergoes beta-decay to nitrogen-14, which is stable. On the other hand, strontium-90 undergoes beta-decay to yttrium-90, which is also radioactive, and decays through beta-particle emission to the stable isotope zirconium-90. Radon-222, an important source of natural background radiation, is the alpha-decay product of radium-226, which itself is the product of a series of decays starting from uranium-238. Decay products of radon-222, the so-called radon progeny or radon daughters, include alpha-emitters and beta-emitters. Radon progeny attach themselves to airborne dust and are inhaled into the lungs, where they irradiate lung tissue with alpha-particles and beta-particles.

ACTIVITY AND RADIOACTIVE DECAY CALCULATIONS

The amount of radioactive material is usually expressed not in terms of mass but in terms of the decay rate of the material, called its activity. In the International System of units (**Système Internationale** [SI]), the unit of activity is the becquerel (Bq), defined as 1 disintegration per second. Another commonly used unit of activity is the curie (Ci), which is equal to 3.7×10^{10} disintegrations per second.

Decay of an unstable nucleus is a random event that is not influenced by the chemical's environment or any other condition outside the nucleus. Each radionuclide has a characteristic half-life that can be used to predict how the activity of a sample of a radionuclide will decrease over time. For example, lithium-8 has a half-life of just 0.84 seconds. In contrast, potassium-40 is a naturally occurring radioisotope of potassium with a half-life of 1.25 billion years. If the activity of a radioactive source is known at one point in time, its activity at any later time can be calculated using the following equation:

$$A(t) = A_0 e^{-0.693 \frac{t}{T_{1/2}}}$$

where $A(t)$ = The activity remaining at time t
A_0 = The original activity
t = The time elapsed
$T_{1/2}$ = The half-life

Note that the time elapsed and the half-life must be expressed in the same units.

A plot of the decay of 1 curie of cesium-137, which has a half-life of 30 years, is given in Figure 12–4.

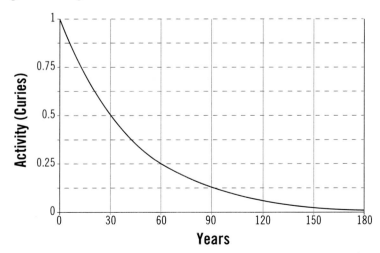

Figure 12–4. Radioactive decay of Cs-137, which has a half-life of 30 years. For each activity point along the plot, 30 years later, that activity has decreased by half.

> ### Example
>
> Tritium (H-3), a radioisotope of hydrogen, has a half-life of 12.3 years. Tritium is used in self-luminous products such as watch dials and exit signs. Consider an exit sign that contained 25 Ci of tritium when it was purchased. Calculate the tritium activity remaining after 20 years.
>
> **Solution:**
>
> $A_0 = 25$ Ci
>
> $t = 20$ years
>
> $T_{\frac{1}{2}} = 12.3$ years
>
> $A(t) = 25\, e^{-0.693 \frac{20}{12.3}} = 25 e^{-1.12} = 25 \times 0.32 = 8.1 \text{Ci}$

Specific information about modes of decay, radiological half-life, and radiation energies for each radionuclide can be found in a chart of nuclides, which is available in print or online.

If a radionuclide is taken into the body, it will irradiate tissues on the inside as it decays. How long the radionuclide will persist in the body depends not only on its radiological (physical) half-life but also on its biological half-life. Biological half-life is a measure of how rapidly the body is able to metabolize and remove the radionuclide-containing chemical. Remember that all isotopes of an element—even a radioisotope—exhibit the same chemical behavior. So, for example, radioactive cesium-137 is metabolized in the body the same way that stable isotopes of cesium are metabolized. Cesium usually exists chemically in salts, with a biological half-life of about 110 days, that are excreted from the body. The combined effects of radioactive decay and biological removal can be expressed as an effective half-life, $T_{\frac{1}{2}\text{eff}}$, which is shown in the following equation:

$$T_{\frac{1}{2}\text{eff}} = \frac{T_{\frac{1}{2}\text{phys}} \times T_{\frac{1}{2}\text{bio}}}{T_{\frac{1}{2}\text{phys}} + T_{\frac{1}{2}\text{bio}}}$$

where $T_{1/2\text{eff}}$ = The effective half-life
$T_{1/2\text{phys}}$ = The physical (radiological) half-life
$T_{1/2\text{bio}}$ = The biological half-life

Note that the half-lives in this equation must all be expressed in the same units. Also note that the effective half-life is always shorter than either the physical half-life or the biological half-life.

> **Example**
>
> Find the effective half-life of cesium-137. The physical half-life of Cs-137 is 30 years, and the biological half-life of cesium is 110 days.
>
> **Solution:**
>
> Convert $T_{1/2phys}$ to days:
>
> 30 y × 365 d/y = 10,950 d
>
> $$T_{\frac{1}{2}eff} = \frac{T_{\frac{1}{2}phys} \times T_{\frac{1}{2}bio}}{T_{\frac{1}{2}phys} + T_{\frac{1}{2}bio}} = \frac{10{,}950 \times 110}{10{,}950 + 110} = 109 \text{ days}$$

X-RAY-GENERATING MACHINES

As mentioned before, x-rays are emitted from some radionuclides as a result of nuclear transformation by electron capture. However, x-rays are also commonly generated by machines containing x-ray tubes. In an x-ray tube, a high voltage accelerates a stream of electrons through a vacuum and toward a positively charged metal target (the anode). When the fast-moving electrons strike the anode, they kick electrons in the anode out of their orbitals, which causes the remaining orbitals to rearrange themselves and in so doing emit x-rays with energies that are characteristic of the element forming the anode. The fast-moving electrons themselves also emit x-rays (called **bremsstrahlung**) as they are deflected by heavy nuclei in the anode. Bremsstrahlung x-rays can have energies up to the peak energy level of the electrons in the stream. For example, if an electron is accelerated across a voltage of 250 kilovolts (kV), the kinetic energy of the electron will be 250 keV when it reaches the anode, and it can generate x-rays with energies ranging up to 250 keV.

X-ray machines are widely used not only in health care and security but also as quality control devices in many manufacturing processes. Other devices that generate streams of high-energy electrons, such as electron microscopes and cathode-ray tubes (CRTs), also produce x-rays.

DETECTING AND MEASURING IONIZING RADIATION

Ionizing radiation cannot be perceived with the human senses. To detect ionizing radiation, we must instead detect either the ionization caused by the radiation or the physical or chemical changes in a material brought about by the deposition of radiation energy.

A common way to measure exposure to x-rays or gamma rays is to measure the amount of charge created by the radiation when it ionizes the gas in a small chamber called an ionization chamber or ion chamber. The amount of charge created by x-rays or gamma rays in a specified amount of air or gas is often measured in a unit called the roentgen (R). The roentgen was traditionally defined as the amount of x-rays or gamma rays that generates 1 statcoulomb of charge (1 statcoulomb = 3.33×10^{-10} coulomb) per 1 cm^3 of air. This might not seem like much of a charge, but it is actually equivalent to 2 billion ionizations. X-rays or gamma rays with photon energies greater than 3 MeV are likely to pass through an ionization chamber without causing enough ionizations to create a measurable amount of charge. This is not really a problem, however, because the roentgen is defined in terms of ionization by x-rays or gamma rays less than 3 MeV. The SI unit of exposure to x-rays and gamma rays is the coulomb per kilogram (C/kg). A roentgen is equivalent to 2.58×10^{-4} C/kg.

When an x-ray or a gamma-ray source is continuously emitting radiation, it is useful to characterize the strength of the ionizing radiation field around the source in terms of the rate of ionization, or **exposure rate**, measured in roentgens per hour (R/h). Radiation survey meters are often scaled in R/h.

Geiger-Mueller counters are gas-filled detectors that are highly sensitive to even very small amounts of ionization. Geiger-Mueller probes typically have a window made of an extremely thin plastic membrane that many alpha-particles and beta-particles are able to penetrate. Any alpha-particle, beta-particle, or gamma ray that enters and causes ionization within the probe is likely to be counted by the detector. Geiger-Mueller counters are used to survey surfaces for the presence of radioactive contamination, which is measured in units of counts per minute (cpm).

Solid scintillation probes are also used in radiation survey instruments. In a scintillation detector, the radiation excites orbital electrons in the scintillating medium, and the electrons then emit flashes of light that are detected with a photomultiplier tube. Different types of scintillators are suitable for detecting alpha-particles, beta-particles, or gamma rays.

"Absorbed dose" is defined as the amount of ionizing radiation energy deposited in an absorbing material per unit mass of absorber. The SI unit of absorbed dose is the gray (Gy), which is defined as 1 J/kg. Another commonly used unit of absorbed dose is the rad, defined in the centimeter-gram-second (cgs) system of units as 100 ergs per g. Because there are 10^7 ergs to a joule and 10^3 g to a kilogram, the conversion factor between the units of absorbed dose is 100 rad = 1 Gy. A useful rule of thumb in radiation dose assessment is that 1 R of exposure to x-rays or gamma rays produces approximately 1 rad of absorbed dose in tissue.

The absorbed dose includes the energy deposited by any type of ionizing radiation: alpha-particles, beta-particles, and neutrons as well as x-rays and gamma rays. An absorbed dose can come from external sources of ionizing radiation that are energetic enough to penetrate the outer layer of human skin. An absorbed dose can also come from radiation emitted by radionuclides inside the body.

Dosimetry badges are often used to estimate the absorbed dose from external sources of gamma rays, x-rays, and strong beta-rays. Some dosimeters, such as film badges, undergo measurable chemical changes caused by deposition of radiation energy. Other dosimeters, such as thermoluminescent dosimeters, trap radiation energy in a form that can later be released and measured in the laboratory. Specialized survey meters and dosimeters are needed to detect neutron radiation.

BIOLOGICAL EFFECTS OF IONIZING RADIATION

Ionizing radiation interacts with matter by causing ionizations or by imparting enough energy to excite orbital electrons. Either way, absorption of ionizing radiation disrupts chemical bonds, damaging molecules. Important biomolecules such as DNA can be damaged directly by ionization or excitation or indirectly by free radicals and other highly reactive chemical species formed from the ionization of intracellular water. Damage to a molecule of DNA can lead to a genetic mutation. If the cell does not repair the damaged DNA, the mutation might ultimately result in cancer or other genetic disorders and diseases. If a genetic mutation occurs in the precursor to an egg cell or sperm cell, the mutation might be inherited by the offspring as a genetic defect. If a cell receives a high dose of ionizing radiation in a short time, DNA damage may be so extensive that the cell dies.

A good indication of the ability of radiation to cause biological damage is linear energy transfer (LET). LET is a measure of the energy imparted to a material, including living tissue, per unit path length of the ionizing particle. Heavy charged particles such as alpha-particles have a high LET because they lose their kinetic energy over a short path. When alpha-particles move through matter, they cause ionizations spaced about a nanometer or less apart. The ribbon-shaped DNA molecule is usually wadded up so that it measures about 30 nm across, and an alpha-particle can rip through a DNA molecule and disrupt one chemical bond after another, possibly even breaking both strands of the double helix. In contrast, the ionizations caused by beta-particles may be spaced hundreds of nanometers apart. X-rays and gamma rays interact with biological tissue by imparting energy to orbital electrons and kicking these electrons out of their orbitals with high enough velocity that they travel some distance through the tissue, causing ionizations along their path. The spacing of these ionizations is similar to the spacing in beta-particle tracks.

Beta-particles, gamma rays, and x-rays are considered low-LET radiation, whereas the damage caused by an alpha-particle is potentially much greater than the damage caused by a beta-particle or photon with the same energy. Therefore, a given absorbed dose of alpha-radiation is likely to be much more damaging than the same absorbed dose of beta-radiation, gamma-radiation, or x-rays.

A measure called the **dose equivalent** is used to adjust for the different biological effectiveness of different types of ionizing radiation. The SI (International System) unit of dose equivalent is the sievert (Sv). The cgs (centimeter-gram-second) unit of dose equivalent is the rem (the acronym for **roentgen-equivalent man**). The conversion factor between the units of dose equivalent is 100 rem = 1 Sv. The dose equivalent is obtained by multiplying the absorbed dose by a quality factor, Q, that represents the relative biological effectiveness of the radiation.

Example

What is the dose equivalent corresponding to 1 rad of absorbed dose from alpha-particles?

Solution:

According to the U.S. Nuclear Regulatory Commission, Q = 20 for alpha-particles (10 CFR 20.1004). The dose equivalent from 1 rad of absorbed dose from alpha-particles is therefore as follows:

1 rad × 20 = 20 rem

As discussed previously, 1 Gy = 100 rad, so in this example the absorbed dose in SI units is 0.01 Gy and the dose equivalent calculated in SI units is as follows:

0.01 Gy × 20 = 0.02 Sv

For beta-radiation, gamma-radiation, and x-rays, Q = 1, so 1 rad of absorbed dose from these radiations would give 1 rem of dose equivalent. The quality factor for neutrons ranges from 2 to 11, where the higher quality factors are typically associated with neutrons of higher energy. When neutrons collide with nuclei in tissue, imparting kinetic energy to them, these energetic nuclei behave like heavy charged particles, with high LET.

The use of the dose equivalent allows us to add up doses from different types of radiation on an equal basis. For example, 1 rem from gamma-radiation, 1 rem from beta-radiation, or 1 rem from neutrons is considered to cause the same risk of harm as 1 rem from alpha-radiation.

Now that we have defined units of absorbed dose and dose equivalent, we can discuss the biological responses of humans to various dose levels of ionizing radiation. First, it should be noted that cells differ in their sensitivities to ionizing radiation. Cells that divide rapidly are more sensitive to radiation than cells that divide slowly. The hematopoietic (blood-forming) tissues and the developing embryo or fetus are highly radiosensitive. Nerve cells and muscle cells, which divide very slowly, are the most resistant to the acute effects of radiation.

Acute radiation syndrome is observed in people who receive a large, whole-body dose of ionizing radiation in a short time. Such doses have resulted from nuclear weapon use or radiation accidents. The speed of onset (days or weeks) and severity of symptoms increase with dose. The acute whole-body effects of gamma-radiation doses have been well characterized:

- At about 0.25 Gy, reduced blood counts may be detected due to damage to the hematopoietic system, but few people will experience symptoms.
- Above about 1 Gy, damage to the hematopoietic system may cause mild flulike symptoms such as nausea and malaise, as well as increased susceptibility to infection.
- Around 3–6 Gy, the flulike symptoms and risk of infection are more severe, and hemorrhage, hair loss, and diarrhea may also occur. If not treated, about 50% of people who receive doses in this range will die, typically within 1–2 months.
- Above about 6 Gy, the intestinal epithelium is damaged; death is likely within 1–2 weeks.
- Above about 10 Gy, the central nervous system is damaged, resulting in disorientation, convulsions, and loss of consciousness, followed by death within 1–2 days.
- The lethal dose to 50 percent of an exposed population is 5 Gy to the whole body. (ENS 2015)

Acute radiation doses to the skin from beta-radiation, gamma rays, or x-rays may cause burns and other damage. Transient erythema (reddening) typically occurs within hours of an acute dose of 2 Gy to the skin. Higher doses to the skin cause more prolonged reddening and other severe effects that may appear days or weeks after the dose was received. These effects and their threshold doses include the following:

- temporary hair loss (3 Gy)
- dry desquamation (peeling), invasive fibrosis (10 Gy)
- spider veins (12 Gy)
- moist desquamation (thinning and weeping of the skin) (15 Gy)
- necrosis (18 Gy)
- secondary ulceration (20 Gy)

Men are more susceptible than women to temporary impairment of fertility from radiation because males are continually producing sperm by the division of precursor cells (spermatogonia), whereas egg cells are already formed and stored in the ovaries before a female is even born.

- Temporary infertility in the male may be caused by an acute dose of 0.15 Gy to the testes.
- Temporary infertility in the female may be caused by an acute dose of 0.65–1.5 Gy to the ovaries.
- Permanent sterility in the male can be caused by doses of 3–5 Gy to the testes received either acutely or over a short time.
- Permanent sterility in the female can be caused by an acute dose of 2.5–6 Gy to the ovaries.

Radiation doses received in utero may be teratogenic and cause defects in the developing embryo. The risk is greatest during the first three months of gestation.

Other effects of radiation, notably cancer and inheritable genetic defects, may manifest themselves years after the radiation dose was received. In any population, cancers and inheritable mutations will occur spontaneously. An ionizing radiation dose increases the rate of the occurrence of these events, but not the severity. For example, on the basis of public health statistics, if we choose 100 people at random, we would expect that roughly 20–30 of them will eventually die of cancer. If each of these 100 people received a dose equivalent to 1 Sv over his or her working lifetime, it is estimated that four more of these people will die of cancer than would otherwise be expected. There is no way of knowing whether any particular case of cancer was caused by the radiation or would have happened anyway. It would even be difficult to discern an elevated cancer rate in this hypothetical group of 100 people, given the normal statistical variability in cancer rates among small groups of people. The estimates of cancer risk were derived from studies of large cohorts of people, including atomic bomb survivors, radium dial painters, uranium miners, and medical patients in the 1930s–1950s who received radiation treatment for nonmalignant diseases. People in these cohorts received higher doses over a shorter period of time than workers would typically receive today, which contributes to the uncertainty of the risk estimates (Mantowski et al. 2001).

LIMITING IONIZING RADIATION DOSE

Annual dose limits have been established by the U.S. Nuclear Regulatory Commission to protect workers and the public from the risk of cancer and genetic defects. These limits are summarized in Table 12–1. The term **effective**

dose equivalent used in the dose limits refers to a whole-body dose in which the dose equivalents to various organs and tissues of the body, such as the gonads, bone marrow, and so on, are weighted by the risk of cancer or genetic defects from irradiation of the organ. These dose limits do not include doses from natural background radiation (averaging about 310 mrem per year) or radiation doses received by a patient during diagnostic or therapeutic procedures.

Table 12–1. Dose Limits for Workers and the Public Resulting from Operations Licensed by the U.S. Nuclear Regulatory Commission

Description	Dose Limit (rem)
Annual occupational dose limits for adults • total effective dose equivalent • dose equivalent to any organ • dose equivalent to the lens of an eye • shallow dose equivalent to the skin	 5 50 15 50
Annual occupational dose limits for minors	10% of the occupational dose limits for adults
Dose limit to embryo/fetus of a declared pregnant woman	0.5
Annual dose limit for individual members of the public • total effective dose equivalent	0.1

(Source: U.S. Nuclear Regulatory Commission)

A worker who is exposed to the annual occupational dose limit of 5 rem for 20 years would accumulate a whole-body dose of 100 rem (1 Sv). As discussed earlier in this chapter, among a group of 100 workers receiving a cumulative dose of 100 rem, it is estimated that the radiation would cause four more deaths from cancer than would otherwise occur. It is evident that the annual radiation dose limit is less protective than the exposure limits established by the Occupational Safety and Health Administration (OSHA) for many carcinogenic chemicals; these limits are intended to reduce the number of excess cancer deaths to no more than 1–10 deaths in 1,000 workers exposed over their working lifetimes. Does this mean that the standard for protection from ionizing radiation is less stringent than the standards for protection from carcinogenic chemicals?

Actually, protection from ionizing radiation is greatly strengthened by the additional rule that radiation doses must be kept as low as is reasonably achievable (ALARA). Under the ALARA principle, it isn't enough to keep doses within the applicable limits. Every reasonable effort must also be made to keep doses as far below the dose limits as is practical. In deciding what is reasonable, both the potential economic costs of incremental reductions in dose and the societal benefits of using ionizing radiation can be taken into account.

With this in mind, let's look at how radiation doses can be controlled. The control of the doses from external radiation sources is often summed up in terms of **distance, time,** and **shielding.**

Distance as a radiation protection concept refers to the inverse-square law presented earlier. Let's look at the inverse-square law as it applies to gamma-radiation or x-rays:

$$\frac{I_2}{I_1} = \left(\frac{d_1}{d_2}\right)^2$$

where I_1 = The dose rate or exposure rate at location 1
I_2 = The dose rate or exposure rate at location 2
d_1 = The distance from the source to location 1
d_2 = The distance from the source to location 2

According to the inverse-square law, doubling a person's distance from a source reduces the dose rate by a factor of four. Under the ALARA principle, people should be no closer to a radiation source than is necessary.

Example

The exposure rate 10 ft from a certain gamma-ray source is 120 mR/h. At what distance from the source does the exposure rate fall to 2 mR/h? (Tip: Carry the units throughout the equation to ensure the problem balances at the end with the correct units.)

d_1 = 10 ft
I_1 = 120 mR/h
I_2 = 2 mR/h

Solution:

$$\frac{I_2}{I_1} = \left(\frac{d_1}{d_2}\right)^2 \rightarrow d_2^2 = d_1^2\left(\frac{I_1}{I_2}\right) \rightarrow d_2 = d_1\sqrt{\frac{I_1}{I_2}}$$

$$d_2 = 10\sqrt{\frac{120}{2}} = 10\sqrt{60} = 77.5 \text{ ft}$$

$$\frac{I_2}{I_1} = \left(\frac{d_1}{d_2}\right)^2$$

$$d_2^2 \frac{I_2}{I_1} = d_1^2$$

$$d_2^2 = d_1^2 \frac{I_1}{I_2}$$

$$\sqrt{d_2^2} = \sqrt{d_1^2 \frac{I_1}{I_2}}$$

$$d_2 = d_1\sqrt{\frac{I_1}{I_2}} = 10 \text{ ft}\sqrt{\frac{120 \text{ mR/hr}}{2 \text{ mR/hr}}} = 10 \text{ft}\sqrt{60} = 77.5 \text{ ft}$$

Time as a radiation protection concept refers to the fact that radiation dose is accumulated over time. If there is a reduction in how long a person is exposed to radiation from a source, the person's dose is also reduced. Under the ALARA principle, people should spend no more time in a given radiation field than is absolutely necessary.

Shielding refers to putting an absorptive medium between people and a source. Alpha-particle sources are easily shielded by a sheet of paper or a few inches of air. Beta-particles have a limited range in any material; even the most energetic beta-particles can be stopped by 1 in. of plastic, ¼ in. of aluminum, or 1/16 in. of lead. However, lead or other elements with high atomic numbers are not ideal shielding materials against beta-particles because beta-particles produce bremsstrahlung x-rays as they move through high-Z materials. Plastics such as polyacrylate that contain only low-Z elements—carbon, hydrogen, and oxygen—are the most suitable shielding materials against beta-particles. In the case of high-energy gamma rays and x-rays, any shielding material, even solid lead, acts like a screen or filter rather than an impenetrable barrier. The thicker and denser the shield, the larger the number of gamma-ray or x-ray photons filtered out. Gamma rays and x-rays are effectively filtered out by dense, high-Z materials such as lead. Concrete and steel are also commonly used as gamma-ray or x-ray shields. Under the ALARA principle, shields for gamma-radiation and x-rays should be as dense as is reasonable and practical.

The ability of a given material to shield a radiation beam is characterized by the Half Value Layer (HVL). The HVL is the thickness of a material that can reduce the intensity of the radiation beam to one-half of its original value. The HVL depends on the energy of the incident beam and on the density and atomic mass of the material itself. If a 50 keV x-ray beam with an intensity of 10.0 mSv hits an aluminum shield that is 2.0 cm thick and emerges from the other side with only 5.0 mSv, then it could be stated that for a 50-keV x-ray beam, the HVL of that type and density of aluminum is 2.0 cm. For an x-ray beam of 75 keV, it would be assumed that the HVL would need to be higher—that is, for a more energetic x-ray, thicker aluminum would be needed to reduce the intensity to one-half. The HVL description of radiation interaction with matter is the most fundamental aspect of radiation-shielding science. Advanced radiation-shielding science considers a wide range of energies and physical interactions between radiation and materials that lead to extremely complex analyses.

To prevent the intake of radioactive material, the materials must be isolated or contained. Radioactive materials used in industrial devices such as gauges or irradiation sources are typically sold as sealed sources. In a sealed source, the radionuclide is usually incorporated into a solid matrix such as ceramic and sealed inside a metal canister. Gaseous radionuclides such as krypton-85, which is used in static eliminators, and tritium gas, which is used in self-luminous signs, are sealed in metal or glass tubes. For beta-radiation or alpha-radiation sources, the radionuclide may be integrated into a foil. Sealed sources or foils should be tested periodically for leakage of radioactive material as specified by the manufacturer. A qualified person should perform this testing by wiping the source with a swab to see if there is removable radioactive contamination.

The use of unsealed radioactive materials, such as radiopharmaceuticals and radiolabeled tracers as well as chemicals containing natural uranium and thorium, requires strict contamination control programs. The ALARA principle applied to internal doses means that companies should do what is reasonably practical to prevent or reduce radioactive contamination of air, water, and surfaces accessible to people.

REGULATIONS AND RESPONSIBILITIES FOR IONIZING RADIATION PROTECTION

The possession and use of radioactive materials in the United States are subject to the regulations of the U.S. Nuclear Regulatory Commission (NRC) or Agreement States. Agreement States have signed formal agreements with the NRC that give those states regulatory responsibilities for some classes of radioactive materials. Possession of some types and quantities of radioactive materials requires a specific license issued by the NRC or Agreement State. The prospective owner must file an application to obtain a specific license before acquiring the radioactive material. On the other hand, under a general license that is published in the NRC or Agreement State regulations, many manufactured devices containing radioactive sources may be owned without prior application to the regulatory authority. The possession and use of x-ray-generating machines and particle accelerators are subject to regulation by the states, which may require permitting and registration of such sources.

Within a facility that uses ionizing radiation sources, administrative responsibility for ensuring compliance with regulations usually resides with an individual who has been designated the Radiation Safety Officer (RSO). NRC and Agreement State regulations and guidelines describe the necessary qualifications for an individual to serve as RSO in the various categories of licensees. At facilities with diverse or complicated radiological issues such as hospitals and nuclear reactors, the role of RSO is often filled by health physicists who have undergraduate or graduate degrees in this specialty. Facilities that have limited use of ionizing radiation sources, for example, in industrial gauges, might designate an industrial hygienist or other environmental health and safety professional to serve as RSO. Widely available RSO training courses cover the fundamentals of ionizing radiation protection in 40 hours of instruction.

NONIONIZING RADIATION

Although nonionizing radiation by definition does not cause ionization in biological tissues, some portions of the nonionizing electromagnetic spectrum are known to cause harmful biological effects. These effects include skin cancer and burning from UV radiation; eye damage from UV, visible, and IR radiation; and overheating of tissue from IR and radiofrequency (RF) radiation.

Table 12–2 lists the regions of the nonionizing electromagnetic spectrum, including the names designating major bands in the spectrum. The names of electromagnetic wave bands and the division points between bands are not always consistent in different references. In this chapter, the bands are divided according to common usage in the context of human health. Optical radiation includes UV, visible, and IR radiation. Particular points on the optical radiation spectrum are usually specified in terms of their wavelength rather than their photon energy, as we did for x-rays and gamma rays. At wavelengths longer than 1 mm, we enter the RF region, where points on the spectrum are usually specified by their frequency rather than by their photon energy or wavelength. (Refer back to the first equation in this chapter to see the relationship between frequency and wavelength.) In the RF region below about 100 MHz and throughout the subradiofrequency region, it is necessary to think of the oscillating electric and magnetic fields separately rather than as energy that radiates from the source.

Table 12–2. Nonionizing Electromagnetic Radiation Bands

Region or Band	Spectral Range
Optical radiation	100 nm–1 mm
UV-C (germicidal)	100–280 nm*
UV-B (erythemal)	280–315 nm
UV-A (black light)	315–400 nm
Visible (light)	400–780 nm
IR-A	780–1400 nm
IR-B	1.4–3 µm
IR-C	3–1000 µm
RF radiation	30 kHz–300 GHz
Microwaves	300 MHz–300 GHz
Extremely high frequency (EHF)	30–300 GHz
Superhigh frequency (SHF)	3–30 GHz
Ultrahigh frequency (UHF)	0.3–3 GHz
Very high frequency (VHF)	30–300 MHz
High frequency (HF)	3–30 MHz
Medium frequency (MF)	0.3–3 MHz
Low frequency (LF)	30–300 kHz
Subradiofrequency fields	< 30 kHz
Very low frequency (VLF)	3–30 kHz
Ultralow frequency (ULF)	0.3–3 kHz
Extremely low frequency (ELF)	< 300 Hz

*UV wavelengths shorter than 190 nm are strongly absorbed by air and are therefore rarely encountered on earth.
Source: Seuss 1989

UV RADIATION SOURCES AND HAZARDS

UV radiation is the most energetic nonionizing radiation. UV photons have enough energy to spur orbital electrons into an excited state, leading to photochemical reactions. DNA is directly damaged by absorption of UV-B and UV-C radiation and indirectly damaged by reactive oxygen species generated in photochemical reactions induced by UV-A, UV-B, and UV-C radiation. Common types of UV sources and the bands that they typically may emit are listed in Table 12–3.

UV is energetically absorbed by human tissue. The skin and the cornea, as the outermost tissues of the body, absorb nearly all the UV-B and UV-C energy irradiating the body and thus bear the brunt of UV damage while shielding the underlying tissues. UV-A radiation can reach the lens of the eye and, to a lesser extent, the retina.

UV dose, which accumulates over time, is also called the **radiant exposure** and is measured in terms of energy per unit area of irradiated surface and expressed as millijoules per square centimeter (mJ/cm^2) or as joules per square meter (J/m^2). The UV dose rate, called the **irradiance**, is measured in terms of power per unit area of irradiated surface and is expressed as milliwatts per square centimeter (mW/cm^2) or as watts per square meter (W/m^2). If the irradiance is constant over the exposure period, the UV dose can be calculated using the following equation:

$$H = I \times t$$

where H = Dose in mJ/cm^2
I = Irradiance in mW/cm^2 (remember that 1 W = 1 J/s)
t = Exposure time in seconds

Table 12–3. UV Sources and Their Main Emission Bands

Source	UV Bands
Sunlight at earth's surface	UV-A, UV-B
Welding arcs	UV-A, UV-B, UV-C
Xenon lamps	UV-A, UV-B, UV-C
UV photo-curing lamps	UV-A, UV-B, UV-C
Germicidal lamps	UV-C
Tanning lamps	UV-A, some UV-B
Mercury vapor lights	UV-A, UV-B, UV-C
UV light-emitting diodes (ULEDs)	UV-A
Black lights	UV-A

Example

A worker is exposed to a UV-A photo-curing source for 15 min at an irradiance of 0.5 mW/cm^2. What is the worker's dose?

Solution:

I = 0.5 mW/cm^2

t = 15 min = 900 s

H = I × t

H = 0.5 mW/cm^2 × 900 s = 450 mJ/cm^2

The primary short-term adverse effects of overexposure to UV are inflammation of the cornea (photokeratitis, sometimes called welder's flash or snowblindness) and sunburn (erythema). These conditions tend to develop hours after the exposure. Photokeratitis, like sunburn, is temporary but can be very painful. The cornea is most sensitive to radiation that has a wavelength of 270 nm, and a dose of about 4 mJ/cm² accumulated over a day can be enough to cause photokeratitis. At 254 nm, the wavelength emitted by germicidal lamps, which is twice that dose, would be needed to cause the same effect. At 315 nm, a dose of about 1,000 mJ/cm² would be needed to cause photokeratitis. Because many UV sources emit a range of wavelengths, the dose or irradiance at each wavelength must be weighted by its relative effectiveness in order to assess the overall hazard.

The sensitivity of the skin to UV burning varies greatly among individuals and depends on their pigmentation and genetic susceptibility. For many fair-skinned people, the threshold dose for erythema is slightly higher than the threshold dose for photokeratitis. The relative capabilities of different wavelengths to cause erythema is similar to their relative capabilities to cause photokeratitis.

Long-term effects of UV exposure include skin cancer, premature aging of the skin, cataracts, and suppression of the immune system. Malignant melanoma, a very deadly type of skin cancer, is associated with a history of severe sunburn in childhood. Two other types of skin cancer, basal cell carcinoma and squamous cell carcinoma, are associated with long-term, repeated exposure to UV.

UV sources that emit wavelengths shorter than 242 nm can produce ozone by their photochemical reactions with oxygen in the air. Ozone is a powerful irritant gas that can damage the lungs and even cause death at high-enough concentrations.

VISIBLE RADIATION HAZARDS FROM NONLASER SOURCES

Visible radiation is simply light that the human eye can see. Intense light sources can damage the retina, permanently impairing vision. There are two mechanisms by which light can damage the retina: photochemical and thermal.

Photochemical damage occurs when pigments in the retina absorb excessive amounts of light, leading to bleaching or possibly cell death. The effect is greatest with wavelengths of 400–500 nm. Even though this is called the **blue-light hazard,** the light doesn't have to appear blue to cause damage; in fact, intense white-light sources such as projection lamps and floodlights emit large amounts of blue light. Violet light and blue-green light are less dangerous than blue light, but at high-enough intensities, they can also cause photochemical damage. Chronic exposure to blue light might lead to macular degeneration, which is a breakdown of the most sensitive part of the retina.

Thermal damage occurs when light energy, from any wavelength of light, that is focused on the retina is converted to heat, causing a retinal burn. This happens with exposure to intense light sources over a time frame of seconds or less. The blink reflex helps to protect the eye by limiting to about 0.25 s the length of viewing intense light. However, people can overcome this protective reflex when they want to view a light source, for example, when staring at the sun.

IR RADIATION SOURCES AND HAZARDS

IR radiation is emitted by hot sources such as welding arcs, furnaces, heat lamps, molten metal or glass, and, of course, the sun. IR radiation is robustly absorbed by tissues and contributes to heat stress as radiant heat. (See Chapter 13 on Thermal Stressors.) IR-A can reach and be absorbed by the lens of the eye, causing cataracts, and by the retina, causing burns. IR-A and IR-B can cause burns or redness of the skin. Unlike UV burns, however, burns from IR radiation are not based on a dose accumulated over hours.

ASSESSMENT AND CONTROL OF BROADBAND OPTICAL RADIATION HAZARDS

Limits to broadband (nonlaser) optical radiation exposure have been developed by the International Commission on Non-Ionizing Radiation Protection (ICNIRP). Because optical radiation hazards are strongly related to wavelength, some of these exposure limits involve complicated spectral weighting of the radiation dose or irradiance.

Measuring optical radiation (UV, visible, and IR) is called radiometry. Radiometers use a semiconductor device such as a photodiode that converts photon energy into electrical current. Detection probes in radiometers may be equipped with filters that approximately match the instrument's spectral response to relevant spectral hazard weighting functions published by ICNIRP.

Where possible, UV sources and intense light sources should be enclosed or oriented so that they cannot be viewed. When viewing these light sources cannot be prevented, or when the skin is irradiated by UV or IR, the concepts of distance, time, and shielding are useful for determining appropriate exposure controls.

Regarding distance, the inverse-square law is appropriate when we seek to reduce corneal and skin hazards from UV and IR point sources. In this case, we state the inverse-square law in terms of the irradiance:

$$\frac{I_2}{I_1} = \left(\frac{d_1}{d_2}\right)^2$$

where I_1 = Irradiance at location 1
I_2 = Irradiance at location 2
d_1 = Distance from the source to location 1
d_2 = Distance from the source to location 2

Example

The effective irradiance 8 ft from a small UV source is 0.025 mW/cm². What is the effective irradiance at 5 ft?

Solution:

$$\frac{I_2}{I_1} = \left(\frac{d_1}{d_2}\right)^2$$

$$I_2 = \left(\frac{d_1}{d_2}\right)^2 I_1 = \left(\frac{8}{5}\right)^2 0.025 = 0.064 \text{ mW/cm}^2$$

Although retinal hazards can be reduced by increasing the distance to the light source, the hazard reduction is not described entirely by the inverse-square law. Rather, increasing the distance to the source reduces the size of the retinal area that is irradiated by the light source, and the risk of thermal injury to the retina is reduced because heat more efficiently dissipates from a small irradiated area. Increasing the distance to a blue-light source also reduces the potential size of the photochemical; however, the severity of the photochemical injury is not reduced.

Limiting exposure time during a day is useful for controlling UV and blue-light hazards because photochemical damage has a cumulative-dose threshold. On the other hand, thermal damage occurs when energy is deposited in a tissue faster than it can dissipate. The time frame of thermal injury to the eye from IR is on the order of minutes, and thermal injury to the retina from intense light can occur literally within a blink of an eye.

Appropriate shielding depends very much upon the spectral components of the radiation. In general, materials that are opaque to light also block UV radiation. Ordinary clear glass and polycarbonate plastic are effective at filtering out UV-B and UV-C radiation but transmit a significant amount of UV-A radiation. Amber or orange-tinted eyewear absorbs UV-A and blue light but may still transmit a significant amount of these harmful wavelengths. A standardized system of filter shades has been established for eye and face protection from welding, plasma, and carbon arcs. Tightly woven clothing, including long-sleeved shirts, pants, and neck drapes, should be worn to protect the skin from UV radiation. Reflective clothing can provide limited whole-body protection from intense IR sources.

LASER HAZARDS AND CONTROL

Laser is an acronym for "light amplification by stimulated emission of radiation." To understand how a laser differs from other light sources, it helps to think of light as a stream of photons and each photon as a tiny wave like the one illustrated in Figure 12–1. In a beam of light from an ordinary light source such as a flashlight, the photons don't all move in exactly the same directions, so the beam continues to spread out the farther it travels. When you look into the beam of a flashlight, only a fraction of the photons emitted by the flashlight are directed at the pupil of your eye, where they enter the lens and are focused on your retina. Also, the oscillations of different photons in the flashlight beam aren't all in phase with each other—that is, the crests of the waves don't all line up. In contrast, lasers emit streams of photons that move in almost perfect alignment and in phase, which is referred to as coherent radiation. If a laser beam happens to line up with your pupil, all those photons will pass through the lens and deposit the full power of the laser on a tiny spot on your retina in perfectly synchronized oscillations. Depending on laser wavelength and power, the beam can cause thermal burns or inflammation of the retina or, in the case of some pulsed lasers, even vaporize tissue (photodisruption), which may result in a permanent blind spot. UV lasers and IR lasers may damage the cornea and the lens as well as the retina. High-powered IR lasers may also cause skin burns.

UV and IR lasers are especially dangerous to work with because they emit radiation at a single wavelength (monochromatic radiation) that is completely invisible. In contrast, most nonlaser sources of UV and IR also emit light. For example, the mercury vapor in germicidal lamps emits blue light along with UV-C radiation. You can tell that a germicidal lamp is on just by looking it. Similarly, many IR sources are so hot that they glow. UV and IR lasers don't give these kinds of visual clues that the laser beam is present.

Lasers are classified according to their potential for causing damage. An international classification system for lasers has been established by the International Electrotechnical Commission (IEC). In the United States, manufacturing and importing lasers are regulated by the Food and Drug Administration (FDA), which uses a slightly different classification system from that of the IEC. The FDA requires every laser to be labeled with its classification. Laser classifications are summarized in Table 12–4.

Table 12–4. Laser Classifications Established by the U.S. FDA and the IEC

FDA Laser Class	Description of Hazard Level	IEC Laser Class	Description of Hazard Level
I	Not capable of causing eye damage unless viewed with focusing optics.	1	Eye safe under all conditions, or a more hazardous laser is enclosed to prevent human access to the beam.
		1M	Expanded laser beam that is safe for unaided viewing but can exceed limit if viewed with focusing optics.
II	Visible beam; may cause retinal damage if direct beam is viewed for extended period or with focusing optics.	2	Visible beam; safe for brief, accidental viewing under all conditions.
		2M	Expanded visible beam that is safe for brief, accidental viewing unless viewed with focusing optics.
IIIa	May be hazardous if beam is viewed intentionally or with focusing optics.	3R	May be hazardous if beam is viewed intentionally or with focusing optics.
IIIb	Risk of injury from direct beam or reflection off mirrorlike surface.	3B	Risk of injury from direct beam or reflection off mirrorlike surface.
IV	Risk of injury from direct beam. Diffuse reflection* may injure retina. May ignite combustible materials.	4	Risk of injury from direct beam. Diffuse reflection may injure retina. May ignite combustible materials.

*A diffuse reflection occurs when the laser beam strikes a surface and rays of light bounce off in all directions. If a laser beam strikes a surface and you can see spots of light on the surface, whatever your viewing angle, you are seeing a diffuse reflection.

Source: U.S. Food and Drug Administration; International Electrotechnical Commission

Other hazards associated with high-powered lasers come from the high-voltage sources used to power the lasers. Using high-powered lasers to cut materials can produce a plume of toxic byproducts.

The control measures necessary to prevent injury depend upon the laser class. Laser safety guidelines can be found in the American National Standards Institute (ANSI) Z136 series of standards. Injury-prevention measures for Class 3B and Class 4 lasers may include administrative controls to keep personnel outside the beam path and engineering controls to enclose the beam if possible and prevent unexpected or unintentional activation of the laser.

RF SOURCES AND HAZARDS

RF radiation is widely used in modern society for wireless communications including broadcast television and radio, mobile phones, cordless phones, wireless networks, and Bluetooth. RF radiation is also used in radar systems, tracking devices, magnetic resonance imagers, and heating devices such as industrial and household microwave ovens, induction heaters, dielectric heaters, and medical diathermy units. RF radiation for communications, tracking, and radar is intentionally emitted via an antenna. Many antennas are placed in plain sight, but others are concealed for aesthetic or practical reasons. RF heating devices may emit RF fields through unintentional leakage.

Similar to an antenna, electromagnetic fields have a very complicated configuration. The spatial region close to the antenna is called the near field. Electromagnetic fields settle into the simple plane wave configuration illustrated in Figure 12–1 some distance from the antenna, in a spatial region called the far field. The boundary between the near field and the far field is given in the equation below for antennas that are longer than one-half the wavelength emitted:

$$d \approx \frac{2L^2}{\lambda}$$

where d = Distance from the antenna
L = Longest dimension of the antenna
l = Wavelength of the radiation

For antennas that are shorter than one-half the wavelength, the far field begins at a distance greater than two wavelengths from the antenna.

In the near field, the energy isn't radiated in accordance with the inverse-square law. Rather, part of the energy emitted by the antenna is stored in the near-field electric field and magnetic field. In the far field, the energy moves outward in an orderly way as predicted by the inverse-square law. The energy transfer rate in the far field is referred to as the power density and is measured in units of watts per square meter (W/m²) or milliwatts per square centimeter (mW/cm²). Notice that power density is essentially the same as irradiance.

Example

Estimate the distance to the far field for a WiFi antenna that is 0.9 m long and operates at 2.4 GHz.

Solution:

First, it is necessary to estimate the wavelength:

$$\lambda = \frac{c}{f} = \frac{2.998 \times 10^8}{2.4 \times 10^9} = 0.125 \text{ m}$$

Because the antenna is longer than ½λ,

$$d \approx \frac{2L^2}{\lambda} \approx \frac{2 \times (0.9)^2}{0.125} = 13 \text{ m}$$

In both the near field and the far field, absorption of RF energy into biological tissues causes heating. The energy transferred to tissue is expressed in terms of the specific absorption rate (SAR), measured in watts per kilogram (W/kg) of tissue. The SAR depends on the frequency of the radiation, the dimensions of the body or body part being irradiated, and the orientation of the electromagnetic wave. Whole-body SAR is highest when the electric field is parallel to the long axis of the body and when the body length is about 40% of the radiation wavelength if the body is ungrounded or 20% of the wavelength if the body is grounded. The frequency corresponding to this wavelength is called the resonant RF frequency of the body. Peak absorption of RF energy by humans occurs in the VHF band, 30–300 MHz.

Example

Estimate the resonant RF frequency of an ungrounded person who stands 1.7 m (5 ft 7 in.) tall.

Solution:

Resonant wavelength $\approx 1.7/0.4 = 4.25$ m

$$f = \frac{c}{\lambda} = \frac{2.998 \times 10^8}{4.25} 7 \times 10^7 \text{ cycles/s} = 70 \text{ MHz}$$

Excessive whole-body heating can contribute to heat stress. Heating of the testes reduces fertility; temporary sterility can occur at an SAR of 5.6 W/kg. Excessive heating of the eyes can cause cataracts. Localized RF heating can cause deep burns. The threshold exposure for thermal effects comes at RF power densities greater than 10 mW/cm^2.

At RF frequencies below 100 MHz, strong RF fields can induce currents in the human body or cause a shock or burn from contact current when a person touches a metal object.

The ubiquitous use of mobile phones and other wireless devices has given rise to concerns that the RF radiation emitted by these devices might cause cancer. So far, the evidence is inconclusive. Some studies have found that cancers of the brain and salivary glands tend to occur more often on the same side of the head as the mobile phone is held. In a major international study, mobile phone use was associated with a slightly elevated risk of cancer among people with the heaviest cumulative lifetime uses. Much of this use involved older mobile phone technology (INTERPHONE 2010). Because wireless technology has changed so rapidly over the past two decades, it is unclear whether the results of these studies are relevant to the RF exposures from the present generation of mobile phone technology.

SUBRADIOFREQUENCY FIELDS

The subradiofrequency region of the electromagnetic spectrum encompasses frequencies less than 30 kHz. These frequencies occur in power generation and transmission (60 Hz in North America) and are also associated with electrically powered mechanical equipment that operates on a cycle, such as motors and compressors. Subradiofrequency magnetic signals are also used for communicating in mines and caves.

Subradiofrequency wavelengths are so large that far-field plane waves are not observed. The electric field and magnetic field must be assessed separately. As with RF fields below 100 MHz, strong subradiofrequency fields can induce currents in tissue and can cause shocks from contact current if a person touches an ungrounded conductor within the field. Induced currents above about 1 milliampere per square meter (mA/m^2) in the body can potentially interfere with nerve function. Strong, extremely low-frequency fields can also potentially interfere with cardiac pacemakers.

Some large, well-designed studies have been conducted to determine whether leukemia or brain cancer is related to exposure to 50- or 60-Hz magnetic fields from electric power transmission. Weak but consistent associations have been found between some surrogate measures of exposure and certain types of leukemia. These findings leave open the possibility that extremely low-frequency magnetic fields are weakly carcinogenic (NIH 2015).

ASSESSING AND CONTROLLING RF AND SUBRADIOFREQUENCY FIELDS

RF-emitting devices in the United States are regulated by the Federal Communications Commission, which enforces limits on the power and frequencies emitted. The Federal Communications Commission limits on the SAR from use of portable communications devices are as follows:
- 0.4 W/kg for whole-body exposure in occupational or controlled settings
- 8 W/kg for partial-body exposure in occupational or controlled settings
- 0.08 W/kg for whole-body exposure in general or uncontrolled settings
- 1.6 W/kg for partial-body exposure in general or uncontrolled settings (FCC 2001).

Consumer devices must meet the SAR limits as well as power density or field limits for general or uncontrolled settings. Controlled settings include large base station and broadcast antennas. In these settings, administrative controls and security systems are important to keep unauthorized and untrained people outside the hazard zone, where power densities or fields could exceed health and safety limits.

RF power density in the far field theoretically follows the inverse-square law. However, the reflection and focusing of RF radiation from conductors may increase the power density above the level predicted by the inverse-square law. Possible conductors include the ground, sheet metal, and vehicles. Even metal screens and fences can reflect RF radiation if the openings between the wires are smaller than one-half the wavelength. In the near field, the change in field strength with distance from the source is often difficult to predict and must be measured.

RF radiation in the microwave region (300 MHz–300 GHz) is usually measured with an electric field meter. The meter may convert the electric field reading to a power density reading using this formula:

$$S \approx \frac{E^2}{377}$$

where S = Power density in W/m²
E = Root-mean-square electric field in volts/meter (V/m)

Below 300 MHz or in near-field conditions, separate measurements of the electric field and the magnetic field should be made. Meters are available that include both an electric field probe and a magnetic field probe. It is important to select a meter that responds in the frequency band of interest. Some meters allow spectral analysis of the frequencies present. Measurement of electric fields is complicated by the fact that the human body, being a conductor, can distort electric fields. For this reason, electric field measurements should be taken with the meter held away from the person performing the measurement. A nonconductive pole or tripod can be used for this purpose.

Health and safety guidelines allow an averaging time of 6 minutes for RF measurements in the 30 kHz–3 GHz band so that transient readings above the applicable exposure limit can be averaged out. At higher frequencies, the averaging time is gradually decreased to 30 s at 30 GHz and 10 s at 300 GHz. Nonuniform fields or power densities should be spatially averaged via measurements taken at multiple points spanning the height of the human body.

SUMMARY

Radiation includes electromagnetic waves and high-energy particles. Different types of radiation interact in different ways with biological tissue, causing distinct types of injuries. Potential control measures, including containment, distance, time, and shielding, must be selected in accordance with the specific nature and properties of the radiation. Therefore, an industrial hygienist must understand the basic concepts of ionizing and nonionizing radiation in order to make safe and accurate choices.

REVIEW QUESTIONS

1. What types of human cells tend to be the most sensitive to biological damage from ionizing radiation?
   a. Very large cells
   b. Nerve and brain cells
   c. Muscle and bone cells
   d. Those that divide and replicate often and rapidly
2. If a patient is to be treated with 2.3 mBq of radioactive iodine-131, how much will be in his or her system 80 hours later if the physical half-life is 8.0197 days and the biological half-life is 120 days?
3. If the dose rate from a stationary source is 2.0 Sv at 45 cm, what is the dose rate at 10 cm?
4. Why are children more sensitive to the harmful effects of ionizing radiation?
5. Why is an alpha-particle capable of doing such a large amount of biological damage to living tissue?
6. What part of the body is commonly exposed to and especially sensitive to alpha-particles?
7. In radiation protection, what is the becquerel (Bq) used to quantify?
   a. Dose
   b. Exposure
   c. Energy
   d. Radioactivity
8. Explain the concept of ionization and describe why it is biologically significant.
9. What does the physical half-life of a radioactive material refer to?
10. If you went into a radioactive materials laboratory with your Geiger counter and noticed high readings all over a benchtop near where the scientists had previously been pouring a liquid containing radioactive phosphorous, this would be an example of _____.
    a. electromagnetic radiation
    b. chronic affects
    c. cumulative exposure
    d. radioactive material contamination

11. What is the effective half-life of a radionuclide with a physical half-life of 28 hours and a biological half-life of 50 days?
12. If you start with 300 μCi of iodine-131, how much will you have after 25 days? (I-131 has a half-life of 8.04 days.)
13. Radioactivity may be defined as what?
14. On the basis of material-penetrating ability, indicate in the spaces provided the type of ionizing radiation (beta, gamma, alpha, or neutron) that would be stopped by the materials shown.

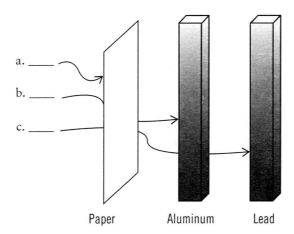

    Paper   Aluminum   Lead

15. Who is affected by a teratogenic health effect?
16. What is biological half-life?
17. What is the lethal dose of whole-body exposure to ionizing radiation?
    a. 5 Gy
    b. 2 Gy
    c. 0.5 Gy
    d. 10 Gy
18. What does the acronym ALARA stand for?
19. What part of the body is most sensitive to the effects of ionizing radiation?
    a. Bones
    b. Blood
    c. Brain
    d. Muscles
20. What is the photon energy level of ionizing radiation?
    a. > 100 ergs/g
    b. > 250 BTU
    c. > 100 joules
    d. > 13.0 eV

21. What distinguishes broadband optical sources from lasers?
    a. All broadband sources emit less light power than lasers.
    b. All broadband sources operate at very high temperatures.
    c. All broadband sources emit noncoherent light.
    d. Radiation from broadband sources cannot burn the skin.
22. Which of the following is *not* a way to control the blue-light hazard?
    a. Orient the light source so that it can't be viewed.
    b. Use white light instead of blue light.
    c. Limit viewing duration.
    d. Use orange-tinted shielding.
23. The daily dose limit for germicidal UV radiation at 254 nm is 6.0 mJ/cm². Suppose a worker is exposed to an (unweighted) irradiance of 0.01 mW/cm² at this wavelength. How long can the worker be exposed without exceeding the dose limit?
24. What is the maximum irradiance at 254 nm that a worker can be exposed to for 8 hours without exceeding the dose limit (6.0 mJ/cm²)?
25. A worker is exposed to an irradiance of 0.01 mW/cm² at a distance of 6 ft from a source. What will the irradiance be if the worker moves another 3 ft away?
26. A 100-MHz RF electric field is measured as 40 V/m. What is the equivalent plane wave power density in watts per square meter?
27. What is the frequency and band designation of RF radiation with a 2-m wavelength?
28. What is the wavelength of 1 terahertz (THz) of radiation? (Tera = 10^{12})
29. Estimate the distance to the far field for a laptop WiFi antenna that is 10 cm long and radiates at a frequency of 2.4 GHz.

REFERENCES

American Conference of Governmental Industrial Hygienists. *2013 TLVs® and BEIs®: Threshold Limit Values for Chemical Substances and Physical Agents and Biological Exposure Indices.* Cincinnati, OH: ACGIH, 2013.

De Broglie, L. "Recherches sur la théorie des quanta" [Researches on the quantum theory]. *Annales de Physique* 10, no. 3 (1925): 22.

Environmental Protection Agency. "Radiation Protection." Accessed June 8, 2015. www.epa.gov/radiation/understand/index.html#ionizing.

European Nuclear Society. Lethal Dose. 2015. Accessed July 5, 2015. www.euronuclear.org/info/encyclopedia/l/lethal-dose.htm.

Federal Communications Commission. *Evaluating Compliance with FCC Guidelines for Human Exposure to Radiofrequency Electromagnetic Fields, Additional Information for Evaluating Compliance of Mobile and Portable Devices with FCC Limits for Human Exposure to Radiofrequency Emissions.* Supplement C (Edition 01-01) to OET Bulletin 65 (Edition 97-01), 2001.

———. *Questions and Answers about Biological Effects and Potential Hazards of Radiofrequency Electromagnetic Fields.* OET Bulletin 56. Washington DC: FCC, 1999.

Fuller, T. P. "Nonionizing Radiation." In *The Occupational Environment: Its Evaluation, Control, and Management.* 3rd ed. Fairfax, VA: American Industrial Hygiene Association, 2011.

Institute of Electrical and Electronics Engineers International Committee on Electromagnetic Safety (SCC39). *IEEE Standard for Safety Levels with Respect to Human Exposure to Radio Frequency Electromagnetic Fields, 3 kHz to 300 GHz,* C95.1-2005. New York, NY: IEEE, 2005.

International Commission for Non-Ionizing Radiation Protection. "Guidelines on Limits of Exposure to Broad-Band Incoherent Optical Radiation (0.38 to 3 mm)." *Health Physics* 73 (1997): 539–54.

International Electrotechnical Commission, Safety of Laser Products – Part 1: Equipment classification and requirements. IEC 60825-1:2014.

INTERPHONE Study Group. "Brain Tumor Risk in Relation to Mobile Telephone Use: Results of the INTERPHONE International Case-Control Study." *International Journal of Epidemiology* 39 (2010): 675–94.

Laser Institute of America. *American National Standard for Safe Use of Lasers,* Z136.1-2000. Orlando, FL: Laser Institute of America, 2000.

Mantowski, G., J. Boice, S. Brown, E. Gilbert, J. Puskin, and T. O'Toole. "Radiation Exposure and Cancer: Case Study." Supplement, *American Journal of Epidemiology* 154, no. 12 (2001).

Martin, J. E., and C. Lee. *Principles of Radiological Health and Safety.* Hoboken, NJ: Wiley-Interscience, 2003.

Miller, G., and M. Yost. "Nonionizing Radiation." In *Fundamentals of Industrial Hygiene*, edited by B. A. Plog and P. J. Quinlan, 5th ed. Itasca, IL: National Safety Council, 2001.

National Institutes of Health. "Magnetic Field Exposure and Cancer." Accessed June 8, 2015. www.cancer.gov/about-cancer/causes-prevention/risk/radiation/magnetic-fields-fact-sheet.

Phillips, M. L., and A. Butler. "Nonionizing Radiation: Broadband Optical." In *Patty's Industrial Hygiene*, edited by V. E. Rose and B. Cohrssen, 6th ed. Hoboken, NJ: John Wiley & Sons, 2010.

Seuss, M. J., and D. A. Benwell-Morison. *Nonionizing Radiation Protection.* Copenhagen: WHO Regional Publications, 1989.

Shapiro, J. *Radiation Protection: A Guide for Scientists, Regulators, and Physicians.* Cambridge, MA: Harvard University Press, 2002.

Sliney, D., and M. Wohlbarsht. *Safety with Lasers and Other Optical Sources: A Comprehensive Handbook.* New York: Plenum Press, 1980.

Turner, J. E. *Atoms, Radiation, and Radiation Protection.* Weinheim, Germany: Wiley-VCH Verlag GmbH & Co, 2007.

U.S. Food and Drug Administration, "Laser Products", Code of Federal Regulations, Title 21, Section 1040.10.

U.S. Nuclear Regulatory Commission. Title 10, Part 20 of the Code of Federal Regulations (10 CFR 20), *Standards for Protection Against Radiation.* Accessed November 6, 2015. www.nrc.gov/reading-rm/doc-collections/cfr/part020/.

13
Thermal Stressors

LEARNING OBJECTIVES

After completing this chapter, readers should be able to do the following:
- Describe the health effects and hazards associated with extremely hot or cold work environments.
- Identify the thermal balance and the factors of heat exchange and metabolism in the human body.
- Explain the various methods and instruments used to measure and analyze the thermal work environment.
- Identify available engineering, administrative, and personal protective equipment controls that can minimize or reduce the hazards associated with thermal work environments.

Photo credit: sdlgzps/iStock

INTRODUCTION

Workers who are exposed to extremely hot or cold work environments—indoors or outdoors—are at risk of thermal stress, which can lead to occupational illnesses, injuries, and fatalities.

Hot conditions can be found in many indoor industrial operations such as glass-manufacturing facilities, electrical utilities (particularly boiler rooms), iron and steel foundries, smelters, nonferrous foundries, brick-firing and ceramic plants, rubber-manufacturing plants, chemical plants, underground mines, and steam tunnels, as well as bakeries, confectioneries, commercial kitchens, laundries, and food-processing facilities. These operations are usually sources of high humidity, elevated ambient air temperature (t_a) or dry-bulb temperature, intense radiant heat, inappropriate air movement, and burns when the body touches hot objects. In addition, these operations often require workers to engage in laborious physical activities. Extreme thermal (hot and cold) conditions are also present in outdoor occupations performed in hot climates and direct sunlight. These occupations include construction, farming, hazardous-waste-site operations, landscaping activities, emergency response operations including fire fighting, and oil and gas well activities. Major areas of cold stress include freezer plants, meatpacking houses, cold storage facilities, cattle ranching, and lumbering.

Several important factors affect a worker's physiological reactions to extreme temperatures, including the worker's:
- weight
- age
- amount of acclimatization
- metabolic rate (activity)
- state of physical fitness
- alcohol or drug consumption (prescription or not)
- individual susceptibility
- type of clothing worn
- medical circumstances (e.g., breathing problems, hypertension, and prior thermal injury)

Occupational heat stress can cause serious illnesses including heat rashes, heat cramps, heat exhaustion, and heatstroke. In hot work environments, workers might experience dizziness, sweaty palms, fogged-up glasses, or burns from contacting hot surfaces, hot chemicals, or steam; and the possibility of injuries also increases.

However, thermal stress is preventable. It is a company's responsibility, usually with the help of the industrial hygienist or safety professional, to provide endangered workers with the appropriate training and protective environments, tools, and clothing. Workers should be educated in the nature of heat stress, in the harmful effects of heat stress on health and safety, and how they can assist in preventing heat stress.

HEAT EXCHANGE

TERMS AND DEFINITIONS
Heat: the measure of energy

Temperature: the measure of the intensity of heat

Specific heat: the quantity of heat that is necessary to raise 1 g of material from 16.5°C to 17.5°C

Heat capacity: the amount of heat necessary to raise one unit mass of a substance 1°C

Heat of vaporization: the quantity of heat required to vaporize one unit mass of a liquid without changing its temperature

Heat of fusion: the quantity of heat necessary to melt one unit mass of a solid without changing its temperature

Exothermic process: a process in which heat is given up

Endothermic process: a process in which heat is absorbed

THERMAL BALANCE
Normal body functioning depends on the body's ability to maintain a fairly constant core temperature of 37 ± 1°C. A simplified version of net heat exchange between the human body and the environment is presented in the following equation:

$$dH = M \pm R \pm C - E$$

where dH = Body heat storage load
M = Metabolic heat gain
R = Radiant heat load
C = Convection heat load
E = Evaporative heat loss

Each element in the equation represents the rate of energy transfer. Therefore, the unit of each element is in energy per unit of time. The most appropriate unit is watts, but other commonly used units are kcal/h (1.162 W) and British Thermal Units (BTU)/h (0.2931 W). The rates of energy transfer are usually normalized to be per body surface area, such as watts per meter squared (W/m^2).

BODY HEAT STORAGE LOAD
Ideally, a person's body heat storage load (dH) is zero. The body adjusts its

functions to maintain a balance between (1) the heat gained from metabolic heat (basic metabolism plus physical activity) and the environmental heat imposed on the body and (2) the heat lost by sweating (evaporation) and other means. If the thermal balance is not maintained in a hot environment, heat builds up in the body, causing the body's temperature to rise. In cold environments, heat is lost from the body, causing the body's temperature to fall.

A set of comprehensive equations can be used to calculate heat exchange and show conditions of (1) the seminude individual, (2) the worker wearing conventional long-sleeved work clothing and trousers, and (3) the standard worker, who is defined as a person who weighs 70 kg (154 lb) and has a body surface area of 1.8 m² (19.4 ft²).

To solve the thermal equation for dH, at a minimum the following parameters should be determined:

- metabolic heat production (physical activity; W, kcal/h)
- air temperature (°C, °F)
- water vapor pressure in air (Pa, mmHg)
- wind speed (m/s, ft/min)
- mean radiant temperature (tr; °C, °F)
- type of clothing (Clo)

Wearing appropriate clothing protects a worker from hazardous agents and plays a major role in the rate and amount of heat exchange between the skin and the ambient air. Therefore, to obtain accurate results, it is necessary to apply correction factors to the thermal balance that reflect the type, amount, and characteristics of the clothing.

METABOLIC HEAT GAIN

Metabolic heat gain (M) is composed of (1) **basal heat**, which is generated as cells perform their functions; and (2) *work heat*, which is generated as a byproduct of muscular activity, which in turn can be a major contributor to metabolic heat gain. Muscular activity also increases muscle temperature, leading to an increase in the body's core temperature. As discussed in a later section of this chapter, a worker's metabolic heat gain can be determined either directly or indirectly. Methods for estimating metabolic rate are shown in Table 13–1.

Table 13–1. Values Needed to Estimate the Metabolic Rate for a Task (in watts)

Task	Intensity	Average (W)	Range (W)
Basal metabolism (B)		70	
Posture metabolism (P)			
Sitting		20	
Standing		40	
Walking		170	140–210
Activity (A)	**Intensity**		
Hand	Light	30	15–85
	Heavy	65	
One arm	Light	70	50–175
	Heavy	120	
Both arms	Light	105	70–245
	Heavy	175	
Whole body	Light	245	175–1,050
	Moderate	350	
	Heavy	490	
	Very heavy	630	

Climbing (V)
V vert = Rate of vertical ascent (m/min)
V = 56 V vert

Adapted from Plog, B. A., J. Niland, and P. J. Quinlan, eds. *Fundamentals of Industrial Hygiene*. 4th ed. Itasca, IL: National Safety Council, 1996.

> **Example**
>
> While walking, a worker uses both arms to perform a light task. Using Table 13–1, estimate the worker's total metabolic rate.
>
> **Solution:**
>
> Basal metabolism = 70 W/m²
> Standing = 170 W/m²
> Using both arms to perform a light task = 245 W/m²
> Total Metabolic Rate = Basal + Standing + Arm Work
> = 70 + 170 + 245 = 485 W/m²

RADIANT HEAT EXCHANGE

The rate of radiant heat exchange (R), which is usually measured in W/m², between the person and the surrounding solid objects may be calculated using the following equations:

- Clothed worker heat gain: $R = 4.4(t_r - t_{sk})$
- Unclothed worker heat gain: $R = 7.3(t_r - t_{sk})$

where

t_r is the mean radiant temperature of the surrounding solid objects in degrees Celsius. The value of t_r is determined using an equation that will be presented in a later section.

Also, t_{sk} is the mean weighted skin temperature in degrees Celsius; t_{sk} is usually assumed to be 35°C.

- Clothed worker heat gain: $R = 4.4(t_r - 35)$
- Unclothed worker heat gain: $R = 7.3(t_r - 35)$

When t_r is > 35°C, there will be a gain in body heat by radiation from the ambient air.

When t_r is < 35°C, there will be a loss (negative gain) in body heat by radiation from the ambient air.

> **Example**
>
> A clothed worker is standing in front of a furnace. The t_r of the surrounding environment is 39°C. What is the radiation heat exchange (R) between the worker and the environment?
>
> **Solution:**
>
> R = 4.4(t_r − 35)
> R = 4.4(39 − 35)
> = 17.6 W/m²; this is a heat gain

CONVECTION HEAT EXCHANGE

The rate of convection heat exchange (C; W/m²) is a function of the (1) difference in temperature between the ambient air (ta; °C) and the mean weighted skin temperature (tsk; °C) (usually assumed to be 35°C) and (2) air speed over the skin (V; m/s).

- Clothed worker heat gain: $C = 4.6V^{0.6}(ta - tsk)$
- Unclothed worker heat gain: $C = 7.6V^{0.6}(ta - tsk)$

If tsk is assumed to be 35°C, the equations become as follows:
- Clothed worker heat gain: $C = 4.6V^{0.6}(ta - 35)$
- Unclothed worker heat gain: $C = 7.6V^{0.6}(ta - 35)$

When ta is > 35°C, there will be a gain in body heat by convection from the ambient air.

When ta is < 35°C, there will be a loss (negative gain) in body heat by convection from the ambient air.

> **Example**
>
> A clothed worker is performing a task in a warm environment, where the ambient temperature is 30°C and the air speed is 0.7 m/s. What is the convection heat exchange (C) between the worker and the environment?
>
> **Solution:**
>
> $C = 4.6V^{0.6}(ta - 35)$
> $C = 4.6(0.7)^{0.6}(30 - 35)$
> $= 0.81 \times 5$
> $= -18.6$ W/m²; this is a heat loss

EVAPORATIVE HEAT LOSS (E)

Sweat can be an important source of cooling when it evaporates. The rate of evaporation of sweat is controlled by the amount of humidity in the air and the air velocity (V). For workers performing physical work in hot environments, the maximum sweat rate (Emax) of 0.625 L/h (650 g/h) is recommended for an unacclimatized worker, and the Emax of 1.00 L/h (1,040 g/h) is recommended for an acclimatized worker.

Emax (W/m²) can be described with the following equations:
- Clothed worker heat loss: $Emax = 7.0V^{0.6}(Psk - Pa)$
- Unclothed worker heat loss: $Emax = 11.7V^{0.6}(Psk - Pa)$

where

V is in m/s, Pa is water vapor pressure of ambient air (kPa), and Psk is water vapor pressure on the skin. Note that kPa = 7.5 mmHg. When skin temperature is 35°C, Psk is 5.6 kPa (42 mmHG).

> **Example**
>
> What is the Emax for a clothed worker when the air speed is 1.5 m/s and the ambient temperature is 40°C?
>
> **Solution:**
>
> Since water vapor pressure at 40°C is 7.4 kPa, the evaporative heat loss for a clothed worker will equal
>
> $7.0V^{0.6}(5.6 - Pa) = 7.0(1.5)^{0.6}(5.6 - 7.4) = -16.1$ W/m^2

> **Example**
>
> A clothed worker is involved in an activity that has a total metabolism of 210 W/m^2. The worker's heat exchange is −56 W/m^2 by radiation and +91 W/m^2 by convection. If the maximum evaporative heat loss is 190 W/m^2, calculate the allowable exposure time (AET) for the worker.
>
> **Solution:**
>
> Ereq = M ± R ± C
> Ereq = 210 − 56 + 91 = 245 W/m^2
> Emax = 190 W/m^2
> AET = (40.7 × 60)/(Ereq − Emax), in min
>   = (40.7 × 60)/(245 − 190) = 44.4 min

HARMFUL EFFECTS

Exposure to extreme temperatures can lead to harmful physiological responses. Workers in high-temperature environments can experience dangerous heat disorders. At the same time, extremely cold workplaces can put workers at risk for injuries and illnesses.

HEAT DISORDERS

Exposure to extreme heat can result in a variety of heat-related illnesses. Physical symptoms can include rashes, cramps, exhaustion, and heatstroke. Excessive heat exposure may also have adverse psychological and behavioral effects, which include increased irritability and fatigue, lack of attention, a reduced work rate, and a higher frequency of accidents.

Heat-related illnesses can be divided into four categories: heat rash, heat cramps, heat exhaustion and heatstroke. Heat disorders are interrelated and seldom occur separately.

- *Heat rash:* The sweat glands are plugged with retention of sweat, and a rash is the inflammatory response.
- *Heat cramps:* This illness is characterized by spastic contractions of the voluntary muscles (mainly the hands, arms, feet, and legs). It is generally associated with insufficient salt intake and profuse sweating with no significant body dehydration.
- *Heat exhaustion:* This illness is characterized by muscular weakness; distress; nausea; vomiting; dizziness; pale, clammy skin; and fainting. It is generally associated with an inadequate water intake, lack of heat acclimatization, and poor physical fitness. Although the oral temperature may be normal or low, the core temperature often is elevated to 38.5°C (101.3°F).
- *Heatstroke:* This is a serious illness characterized by an excessive rise in body temperature and a failure of the body's temperature-regulating mechanism. Symptoms of this acute medical emergency include a sudden and sustained loss of consciousness preceded by vertigo, nausea, headache, cerebral dysfunction, bizarre behavior, and core body temperature in excess of 40.5°C (104°F).

COLD DISORDERS

In a cold environment, an individual attempts to remain physically active, shivering is activated, and blood vessels contract. The main factors contributing to cold injuries are exposure to very low temperatures, high winds, high humidity, inadequate clothing, and wet and cold objects such as metals. Advanced age, poor physical health, and consumption of certain medications can also contribute. Cold injuries are classified as follows:

- *Hypothermia:* This condition occurs when the core temperature drops below the normal range. The first symptoms are uncomfortable shivering and sensations of cold. The heartbeat slows and sometimes becomes irregular, the pulse weakens, and the blood pressure drops. Other symptoms are vague or slow slurred speech; memory lapses; incoherence; and drowsiness. The worker can become restless, can become confused, and may make little or no effort to keep warm. When the core temperature drops down to about 29°C (85°F), significant drops in blood pressure and respiration occur, and serious health problems develop that can kill the exposed individual.
- *Blood vessel abnormalities:* These include the following:
  - *Raynaud's phenomenon:* This involves the blanching of the distal portion of the digits. Numbness, itching, tingling, or a burning sensation may occur during intermittent attacks. This reaction is triggered

by the cooling of the skin. Of great importance in industry is the association between this phenomenon and the use of vibrating hand tools, a condition sometimes called White Finger Disease.
- *Acrocynosis:* This condition is caused by exposure to cold and is characterized by a reduced level of hemoglobin in the blood and by the hands and feet acquiring a slightly blue, purple, or gray coloring.
- *Frostbite:* Frostbite, the freezing of the fluids around the cells of body tissues, can occur because of inadequate circulation and insulation. The freezing point of the skin is −1°C (30°F). As wind velocity increases, heat loss also increases and frostbite develops more rapidly. If the skin comes in contact with objects that have temperatures colder than freezing, frostbite may develop at the point of contact, even in a warm environment. The symptoms of frostbite are uncomfortable sensations of coldness, numbness, tingling, stinging, or aching. Damage from frostbite can be serious and can lead to scarring, tissue death, and amputation. A photo of toes 12 days after frostbite is shown in Figure 13–1.

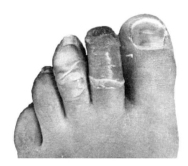

Figure 13–1. Toes 12 days after frostbite.

(Source: Dr. S. Falz / CC-BY-SA-3.0)

- *Frostnip:* This condition occurs when the extremities are exposed to a cold wind, which results in the skin turning white.
* *Trench foot:* This occurs when a foot is exposed to persistent dampness and cold (but without freezing). Symptoms are swelling (edema), tingling, itching, severe pain, blistering, death of skin tissue, and ulceration. *Chilblain* is the name of this condition when other parts of the body are affected.

THERMAL MEASUREMENT

The following factors are commonly evaluated to measure and analyze thermal hazards:
* *factors related to environment*
  - air temperature
  - V (air velocity)
  - radiant heat
  - relative humidity
* factors related to individuals
  - activity
  - clothing

- physiological factors such as body temperature, heart rate, and sweat rate
- personal characteristics such as age, weight, fitness, and work habits

AMBIENT AIR TEMPERATURE (DRY-BULB TEMPERATURE)
- This temperature is measured using a thermometer: mercury- (or alcohol-) in-glass, thermoelectric thermometer, or thermistor (semiconductor).
- Each thermometer should be calibrated over its range against a known standard.
- A dry-bulb thermometer should be shielded from any source of radiant heat.
- Temperature is usually measured at chest height of the worker. To measure surface temperature (e.g., skin surface) or a remote temperature, a thermoelectric thermometer, thermocouple (Figure 13–2), or thermistor should be used.

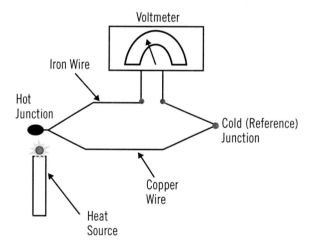

Figure 13–2. A thermocouple.

AIR SPEED/WIND SPEED

Wind (air movement) can be produced by body movement or air movement. Directional air speed is measured by rotating a vane anemometer (Figure 13–3) or a swinging vane velometer (Figure 13–4).

Nondirectional air speed is measured using instruments that depend on the rate of cooling of a heated element. The resulting reading is a factor of the moving air's cooling power. Instruments used to measure nondirectional speed include thermoanemometers (Figure 13–5), anemotherms, and kata thermometers.

Figure 13–3. A rotating vane anemometer.

Figure 13–4. A swinging vane velometer (no individually calibrated fittings required)

Air speed (V) can also be estimated as follows:
- No sensation of air movement
  V < 0.2 m/s
- Sensation of light breeze
  0.2 < V < 1.0 m/s
- Sensation of moderate breeze
  1.0 < V < 1.5 m/s
- Sensation of heavy breeze
  V > 1.5 m/s

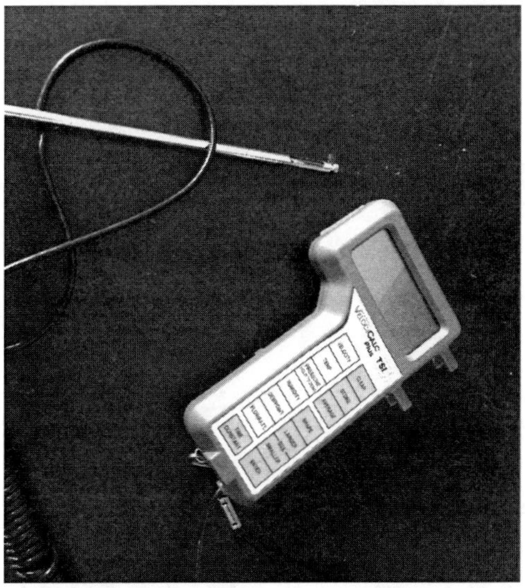

RADIANT HEAT

Radiant heat sources are classified as artificial (e.g., infrared radiation in iron and steel industries, the glass industry, and foundries) or natural (e.g., solar radiation).

Figure 13–5. A thermoanemometer (thermal anemometer).

In general, black globe thermometers or radiometers may be used to measure occupational radiation, and pyrheliometers or pyranometers may be used to measure solar radiation.

A net radiometer consists of a thermopile with the sensitive elements exposed on the two opposite faces of a blackened disc. These radiometers, such as infrared pyrometers, are used to measure surface temperatures ranging from –30°C to 3,000°C.

Direct solar radiation is measured with a pyrheliometer. A pyrheliometer consists of a tube that can be directed at the sun's disc and at a thermal sensor. Generally, a pyrheliometer with a thermopile as the sensor and a view angle of 5.7 degrees is recommended. A pyranometer is used to measure diffuse and total solar radiation or other radiation between 0.35 and 2.5 mm, which includes the UV, visible, and infrared ranges.

However, the black globe thermometer is the most commonly used to measure the thermal load of solar and infrared radiation affecting workers and the tr. This thermometer consists of a 6-in.-diameter thin-capper sphere shell that is painted matte black. A thermometer is inserted through a rubber stopper in a hole drilled at the top of the shell, and the bulb of the thermometer is located in the center of the globe. The black globe absorbs the radiant heat, and the thermometer reaches equilibrium after about 25 min.

Radiant heat is represented by the tr, which is calculated by measuring the globe temperature (tg [in °C]) and other environmental factors.

The tr (in °C) is calculated as follows:

$$tr = tg + (1.8V^{0.5})(tg - ta)$$

The ta (in °C) and V (in m/s) in the formula account for the heat energy exchanged by convection around the globe.

Example

Find the mean radiant temperature (tr) when the globe thermometer reads 34°C, the dry-bulb temperature (ta) is 30.5°C, and air velocity (V) is 1.3 m/s.

Solution:

$tr = tg + (1.8V^{0.5})(tg - ta)$
$tr = 34 + (1.8 \times 1.3^{0.5})(34 - 30.5)$
$tr = 41.2°C$

RELATIVE HUMIDITY

Humidity is the amount of water vapor in the air. Relative humidity, expressed as a percentage, is the amount of moisture in the air compared to the total amount that the air could contain at saturation at the same temperature. Given the relative humidity and the air temperature, water vapor pressure can be determined.

The psychrometric chart (Figure 13–6) is designed to indicate the relationship between ta, wet-bulb temperature (tw), relative humidity, water vapor pressure, and dew point. Although the psychrometric chart is constructed assuming standard barometric pressure, conversions should be performed when nonstandard conditions exist.

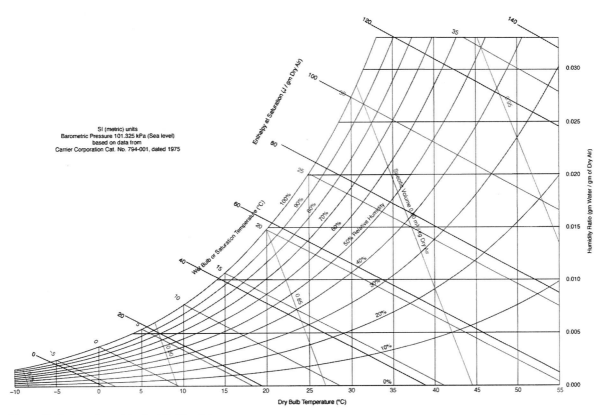

Figure 13-6. A psychrometric chart (in SI units).

A sling psychrometer (Figure 13-7) is the most common instrument used to measure relative humidity.

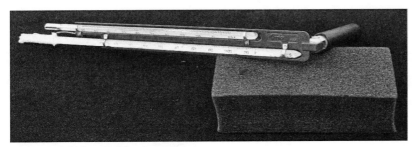

Figure 13-7. A sling psychrometer.

THERMAL STRESS EVALUATION

WET-BULB GLOBE TEMPERATURE INDEX

Wet-bulb globe temperature (WBGT) is a simple and suitable technique to measure the combined effects of environmental factors related to heat stress (Figures 13-8).

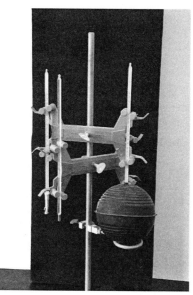

Figure 13–8. A setup to determine WBGT: (from left to right) wet bulb thermometer, dry bulb thermometer, and globe thermometer.

See Table 13–2 for recommendations from the American Conference of Governmental Industrial Hygienists (ACGIH®) for permissible heat exposure Threshold Limit Values (TLV®).

By using the ACGIH recommendations and estimating the worker's metabolic rate, the job's hazard ranking can be determined.

WBGT values (°C or °F) are calculated using the following equations:

(1) Outdoors with solar load
$$WBGT = 0.7tnw + 0.2tg + 0.1ta$$
(2) Indoors or outdoors with no solar load
$$WBGT = 0.7tnw + 0.3tg$$

where

tnw = Natural wet-bulb temperature (°C or °F)

Table 13–2. TLV Chart Recommended by the ACGIH (2009). WBGT is Given in Degrees Celsius.

Work Demands	Acclimatized				Unacclimatized			
	Light	Moderate	Heavy	Very Heavy	Light	Moderate	Heavy	Very Heavy
100% Work	29.5	27.5	26		27.5	25	22.5	
75% Work; 25% Rest	30.5	28.5	27.5		29	26.5	24.5	
50% Work; 50% Rest	31.5	29.5	28.5	27.5	30	28	26.5	25
25% Work; 75% Rest	32.5	31	30	29.5	31	29	28	26.5

Example

In a foundry, the natural wet-bulb temperature is 19°C, the globe temperature is 36°C, and the ambient temperature is 27°C. What is the WBGT index inside the shop?

Solution:

WBGT = 0.7tnw + 0.3tg

WBGT = (0.7 × 19) + (0.3 × 36) = 24.1°C

ADJUSTED AIR TEMPERATURE

In workplaces where each worker wears a vapor- and air-impermeable encapsulating ensemble, the WBGT is not the appropriate measurement of environmental heat stress. Instead, the adjusted air temperature (tadj) must be measured and used.

$$tadj = (tr + ta)/2, \text{ in } °C$$

- When tadj exceeds 20°C, physiological monitoring (of oral temperature and pulse rate) should be performed.
- The suggested frequency of physiological monitoring during moderate work varies from once every 2 hours at a tadj of 24°C to once every 15 minutes at a tadj of 32°C.

Figure 13–9. A natural wet-bulb thermometer.

WINDCHILL INDEX

The windchill index (WI) indicates the cooling effect of any combination of temperature and wind velocity (air movement). The WI does not take into account the body part exposed to the cold, the level of activity, or the amount of clothing worn. (See Table 13–3.)

Table 13–3. Windchill Index Chart

Air Speed (mph)	Temperature (°F)																	Air Speed m/s
Calm	40	35	30	25	20	15	10	5	0	-5	-10	-15	-20	-25	-30	-35	-40	-45	Calm
5	36	31	25	19	13	7	1	-5	-11	-16	-22	-28	-34	-40	-46	-52	-57	-63	2
10	34	27	21	15	9	3	-4	-10	-16	-22	-28	-35	-41	-47	-53	-59	-66	-72	4
15	32	25	19	13	6	0	-7	-13	-19	-26	-32	-39	-45	-51	-58	-64	-71	-77	7
20	30	24	17	11	4	-2	-9	-15	-22	-29	-35	-42	-48	-55	-61	-68	-74	-81	9
25	29	33	16	9	3	-4	-11	-17	-24	-31	-37	-44	-51	-58	-64	-71	-78	-84	11
30	28	22	15	8	1	-5	-12	-19	-26	-33	-39	-46	-53	-60	-67	-73	-80	-87	13
35	28	21	14	7	0	-7	-14	-21	-27	-34	-41	-48	-55	-62	-69	-76	-82	-89	16
40	27	20	13	6	-1	-8	-15	-22	-29	-36	-43	-50	-57	-64	-71	-78	-84	-91	18
45	26	19	12	5	-2	-9	-16	-23	-30	-37	-44	-51	-58	-65	-72	-79	-86	-93	20
50	26	19	12	4	-3	-10	-17	-24	-31	-38	-45	-52	-60	-67	-74	-81	-88	-95	22
55	25	18	11	4	-3	-11	-18	-25	-32	-39	-46	-54	-61	-68	-75	-82	-89	-97	25
60	25	17	10	3	-4	-11	-19	-26	-33	-40	-48	-55	-62	-69	-76	-84	-91	-98	27
Time to Frostbite							30 min		10 min			5 min							Time to Frostbite
Calm	4	2	-1	-4	-7	-9	-12	-15	-18	-21	-23	-26	-29	-32	-34	-37	-40	-43	Calm
	Temperature (°C)																		

Note: Cross-reference to metric units for air speed and temperature

> **Example**
>
> According to Table 13–3, if the actual air temperature is 20°F and the estimated wind speed is 25 mph, the equivalent windchill temperature will be 3°F.

PHYSIOLOGICAL MONITORING

Physiological monitoring includes determining heart rate, body temperature, sweat rate, and metabolic rate.

Heart Rate

To measure heart rate (HR), count the radial pulse during a 30-second period as early as possible in the rest period. The carotid pulse (in the neck) also can be used to indicate the HR. Measuring heart rate may be easier if the employee wears multiple layers of gloves.

HR increase capacity (CHR) is defined by the following equation:

$$CHR = 40\% \times (220 - Age) + 60\% \times Resting\ HR$$

If the HR exceeds 40% of CHR (in beats/min) at the next rest period, shorten the next work cycle by one-third and keep the rest period the same length. The Resting HR should be measured at the beginning of the workshift. If the heart rate still exceeds CHR beats per minute at the next rest period, shorten the following work cycle by one-third again.

Body Temperature

Temperatures vary in each section of the body. Use a clinical thermometer (3 minutes under the tongue) or similar device to measure the oral temperature or tympanic temperature at the end of a work period (before drinking water). An infrared thermometer can also be used to measure tympanic temperature. If the oral temperature exceeds 37.6°C (99.6°F), shorten the next work cycle by one-third without changing the rest period. If the oral temperature still exceeds 37.6°C (99.6°F) at the beginning of the next rest period, shorten the following work cycle by one-third again.

The World Health Organization recommends a limit of 38°C for internal body temperature under conditions of prolonged daily work and heat.

Semipermeable or impermeable clothing should not be worn when the oral temperature exceeds 38.1°C (100.6°F) or the heart rate exceeds the adjusted maximum rate:

$$0.7 \times (220 - age)$$

Sweat Rate

Workers should weigh themselves before and after exposure to check for weight loss that might have occurred from progressive dehydration. To determine sweat rate, various references may be consulted, such as the Predicted 4-hour Sweat Rate (P4SR) developed by the National Hospital for Nervous Diseases in London in 1947.

Subject's Physical Activity

The environmental heat and the heat produced by the body together determine the total heat affecting the individual.

Thus, methods for evaluating heat stress usually require assessing the heat produced by the body, or the metabolic rate. The metabolic rate can be determined either by measuring metabolic rate directly while the worker performs the job or by estimating it. The metabolic rate of a given task may be estimated by adding basal metabolism, posture, degree of body involved in the activity, and vertical motion (see Table 13–1).

Clothing

Some methods of heat stress monitoring assume the workers wear a particular type of clothing, whereas other methods have provisions for acknowledging different types.

For example, the permissible heat exposure TLVs recommended by the ACGIH are valid for the light summer clothing customarily worn by workers in hot environmental conditions.

For each job category where special clothing is required, the permissible heat exposure TLVs should be established by an expert in thermal hazards.

THERMAL ENVIRONMENT CONTROL METHODS

ENGINEERING CONTROLS

Heat stress can be controlled by applying and modifying one or more of the following factors:
- *Metabolic heat production*
  - Provide powered assistance for strenuous tasks.
  - Reduce physical demands of the work.
- *Radiant heat load*
  - Interpose line-of-sight barriers.
  - Provide furnace wall insulation, metallic reflecting screens, and heat-reflective clothing.

- Cover exposed parts of the body.
- *Convective heat load*
  - If the air temperature is above 35°C (skin temperature), workers should do the following:
    - Reduce air temperature.
    - Reduce air speed across the skin.
    - Wear more clothing.
  - If the air temperature is below 35°C, workers should do the following:
    - Increase air speed across the skin.
    - Wear less clothing.
- *Evaporative cooling*
  - Decrease humidity.
  - Increase air speed.

WORK AND HYGIENE PRACTICES AND ADMINISTRATIVE CONTROLS FOR HOT AREAS

Limit exposure times and temperature
- Schedule hot jobs for the cooler part of the day.
- Schedule routine repair and maintenance work for cooler seasons.
- Design work-rest regimens and provide a cool area for rest and recovery.
- Add extra personnel to reduce the exposure time of each member of the field crew.
- Allow freedom to interrupt work when a worker feels severe heat discomfort.
- Enforce water and rest breaks and encourage workers' water intake.
- Segregate (by location and time) hot operations from other operations.
- Allow for acclimatization and physical conditioning.

Reduce metabolic heat load
- Mechanize physical components of the job.
- Reduce exposure time (reduce workday, increase rest time, restrict overtime work).
- Increase work force.
- Consider the effects of drugs, alcohol, and obesity.
- Reduce the metabolic heat load by minimizing employees' unnecessary activities.

ADMINISTRATIVE AND SAFE WORK PRACTICES FOR COLD AREAS
- Substitute, isolate, or redesign the equipment and processes to minimize cold stress.
- Use general or spot heating.

- Minimize air velocity, reduce it to less than 1 m/s (200 ft/min), and shield the work area.
- Cover metal handles of tools and control bars with thermal insulating material.
- Do not use unprotected metal seats.
- Use mechanical lifting aids to reduce manual workload.
- Provide heated warming shelters and encourage workers to use them.
- Provide personal protective equipment and warm clothing.
- Monitor the workplace and design work-rest regimens accordingly.
- Schedule and enforce rest and drink breaks and provide heated rest areas.
- Schedule the work for a warmer time or a warmer work area.
- Assign extra workers to do the job and allow the workers to pace themselves.
- Provide workers with training and orientation on the basic principles of cold stress.
- Allow new workers to become acclimatized to the cold environment.
- Minimize sitting still or standing for long periods of time.
- Maintain safety supervision and a buddy system.
- Establish a medical surveillance program.
- Establish emergency and first-aid procedures.
- Pay special attention to older workers and those with circulatory or other chronic illnesses.
- Encourage workers to stay physically fit, get sufficient sleep, and eat nutritious meals.
- Provide emergency supplies in areas where storms are frequent.

ENHANCE TOLERANCE TO HEAT

Acclimatization

On the first exposure to a hot environment, workers typically experience the following:
- distress and discomfort
- increased core temperatures and heart rates
- headache, giddiness, and nausea
- symptoms of early heat exhaustion

Eventually, working in heat produces a phenomenon called acclimatization. In heat-stressful situations, a person acclimatized to heat will have the following characteristics:
- lower heart rate
- lower body temperature

- higher sweat rate
- more dilute (containing less salt) sweat than a person who is not acclimatized at the start of exposure to excessive heat

Both physical activity and heat stress must take place to initiate the bodily changes that result in acclimatization. Working in heat for about 2 hours per day for one or two weeks will result in essentially complete acclimatization to that work-stress condition. Once attained, acclimatization is lost slowly. A measurable amount, however, can be lost in a few days. Similarly, with sufficient (usually repeated uncomfortable) exposures to the cold, the body may undergo acclimatization that increases comfort and reduces the risk of cold injuries. Workers who are physically unfit, older, obese, or taking medications; using alcohol; or using drugs may not acclimatize easily.

PHYSICAL FITNESS

Obesity

- Obesity predisposes workers to heat disorders.
- Additional weight requires a large expenditure of energy to perform a given job because the body-surface-to-body-weight ratio is less favorable to heat dissipation.
- Poor lower physical fitness and decreased maximum cardiovascular capacity are frequently associated with obesity.
- The fat layer theoretically reduces the direct transfer of heat from the muscles to the skin.

Age

- The aging process leads to a sluggish response of the sweat glands, which in turn leads to less effective regulation of body temperature.
- Aging results in an increased level of skin blood flow associated with exposure to heat.
- Total body water decreases with age, which may be a factor in the higher incidence of heat disorders in older workers.

Skin

- The skin is the body's natural barrier against heat and cold. Thus, maintaining healthy skin and controlling skin temperature are important factors in the body's thermal control.

Dehydration, Salt/Electrolytes
- Working in areas with extreme temperatures causes significant water loss through the skin and lungs as a result of the dry air and sweating.
- Dehydration affects the flow of blood to the extremities and increases the risk of thermal strain.
- Increased fluid intake is essential to prevent dehydration.
- Sweet, nonalcoholic, caffeine-free drinks should be available at the worksite for fluid and calorie replacement.
- Workers should drink water in small quantities of about 150–200 mL (5–7 oz) every 20 minutes to prevent excessive dehydration.
- Because the normal thirst mechanism is not sensitive enough to ensure sufficient water intake, workers should regularly drink water or low-sodium noncarbonated beverages.
- The fluid should be as palatable as possible at temperatures of 10–15°C (50–60°F).
- Two hormones are important in thermoregulation: the antidiuretic hormone, which is released from the pituitary gland, and aldosterone, which is released from the adrenal glands.
  - Changes in plasma volume, changes in the sodium chloride concentration in the plasma, and so forth encourage the release of these hormones.
  - The antidiuretic hormone reduces the amount of water lost from the kidneys but has no effect on the amount of water lost from the sweat glands.
  - Aldosterone reduces the amount of salt lost from both the kidneys and the sweat glands.

Diet
- The importance of replacing lost water and maintaining salt balance cannot be overemphasized because of the importance of water and salt in sustaining the body's necessary physiologic functions.
- People who work in extreme temperatures should have a well-balanced diet.
- A very-high-protein diet might increase urine output to remove nitrogen and thus increase water intake requirements.
- Supplementing the diet with potassium to maintain normal electrolytic levels and functions might be necessary.
- Supplementing the diet with vitamin C might enhance heat acclimatization and thermoregulatory functions.

- Alcohol is a drug that interferes with the functions of the central and peripheral nervous systems and is associated with hypohydration because it suppresses antidiuretic hormone function.
- Therapeutic and social drugs that affect central nervous system activity, cardiovascular reserve, or body hydration could potentially affect heat tolerance.

Screening for Heat Intolerance
- Individuals with low physical work capacities are more likely to develop elevated body temperatures.
- Tolerance of physical work in a hot environment is related to physical work capacity; therefore, heat tolerance might be predicted from physical fitness tests.
- It has been shown that heat-acclimatized workers with a maximum work capacity (Vo2max) of 2.5 L/min oxygen or greater are more tolerant of heat than those with a low work capacity.
- Workers who have experienced a heat illness may be less heat tolerant.

CLOTHING

Hot Environments
- Loose-fitting clothing should be worn in highly humid areas.
- With moderate radiant heat loads, the amount of exposed skin must be minimal. Jobs with high radiant heat loads often require wearing reflective garments.
- Extreme radiant and convective heat exposure may require special insulation or mechanically cooled suits.
- Protective clothing that is adequate for a job varies from simple head cooling to essentially complete isolation of the worker from the environment in an encapsulating suit.
- An encapsulating suit is usually required when a worker must remain in a very hot environment and would experience unacceptably high heat strain without the protection.
- Auxiliary body-cooling systems may include water-cooled garments, air-cooled garments, and (wet or dry) ice-packed vests.

Cold Environments
- In cold environments, the most important parts of the body to protect are the feet, hands, head, and face.
- Workers should wear several layers of clothing instead of single heavy outergarments.

- The outer layer should be windproof and waterproof.
- Recommended clothing includes the following:
  - A cotton T-shirt and shorts or underwear under cotton and wool thermal underwear (two pieces of underwear are preferred)
  - High-wool-content socks (two pairs of socks, with the inside pair cotton)
  - Wool or thermal trousers lapped over boot tops
  - Felt-lined, rubber-bottomed, leather-topped waterproof boots with a removable felt insole
  - A wool shirt or wool sweater over a cotton shirt
  - An anorak, snorkel coat, or arctic parka
  - A hood around the neck that extends past the face
  - A wool knit cap or a hardhat with a liner
  - Wool mittens over gloves
  - A face mask or scarf (a ski mask with eye openings gives better visibility than a snorkel hood)

TRAINING

- Teach workers to recognize the signs/symptoms of various types of heat-induced illnesses.
- Include thermal stress prevention in basic and refresher health and safety training courses.
- Instruct workers in the proper use and care of personal protective devices.
- Instruct workers on the effects of drugs, alcohol, and obesity on the body's tolerance of thermal stress.
- Instruct workers in establishing a buddy system to observe signs/symptoms of thermal stress in each other.

MEDICAL SURVEILLANCE

In order to ascertain a worker's fitness for being placed in and for continuing to work in a particular hot environment, the characteristics of the individual worker, such as age, gender, childbearing potential and pregnancy in women, weight, social habits, chronic or irreversible health status, and acute medical conditions, must be assessed in the context of the extent of heat stress imposed in a given work setting.

The ability of a worker to tolerate elevated heat stress requires the following:

- the cardiac, pulmonary, and renal systems functioning
- the seating mechanism functioning
- the body's fluids and electrolytes being balanced
- the central nervous system's heat regulatory mechanism functioning

HEAT-ALERT PROGRAM TO PREVENT EMERGENCIES

A written heat-alert program should be developed for implementation whenever weather services forecast a heat wave. A heat wave is present when the daily maximum temperature exceeds 35°C or when the daily maximum temperature exceeds 32°C and is 5°C or more above the maximum reached on the preceding days.

SUMMARY

Temperature extremes can occur in a broad range of work settings and can lead to serious health effects and death in a matter of moments. Occupational health practitioners need to know what these health effects are and be able to communicate the hazards to the workers. The abilities to recognize the potential risks of hot and cold environments, accurately evaluate the thermal conditions, and take timely and effective control or response actions are essential skills of industrial hygienists.

REVIEW QUESTIONS

1. A worker's metabolic heat (M) is 61 kcal/h. He gains 135 kcal/h by radiation (R) and 47 kcal/h by convection (C). He loses 229 kcal/h by sweating (E). How much is his body heat storage rate (dH)?
2. In a workplace, ambient temperature (ta) is measured at 39°C and wind speed is measured at 2.3 m/s. Calculate heat exchange by convection. (Assume skin temperature is 35°C.)
3. If mean radiant temperature (tr) in a glass-manufacturing shop is 54°C, what is the radiant heat exchange (R)? (Assume skin temperature is 35°C.)
4. Water vapor of ambient air is 19 mmHg, and wind speed is 1.7 m/s. Calculate the maximum evaporative heat loss (Emax). (Assume skin temperature is 35°C; vapor pressure at 35°C is 42 mmHg.)
5. In a shop, a globe thermometer reads 34.0°C, air velocity is 1.3 m/s, and ambient temperature is 30.5°C. Determine the mean radiant temperature (1°C = 5/9[°F − 32]).
6. Using a sling psychrometer, ta and tw are determined to be 33.9°C (93°F) and 25.6°C (78°F), respectively. Using Figure 13–6, find relative humidity, dew point temperature, and water vapor pressure.

7. Inside a smelting shop, the tg is measured at 33.0°C and the tnw is measured at 28.0°C. Calculate the WBGT in degrees Fahrenheit.
8. The heat stress exposure profile of a glass factory worker, in terms of WBGT, was as follows: 2.7 h at 97°F, 3.2 h at 101°F, and 2.1 h at 76°F. What was this worker's time-weighted average exposure?
9. Consider a tw of 65°F (18.3°C) and a tg of 90°F (32.2°C) in a work environment in which the air is moving at 100 ft/min (0.51 m/s). What is the workers' feeling of warmth in this workplace in terms of *corrected effective temperature* (CET)?

REFERENCES

American Conference of Governmental Industrial Hygienists. *2013 TLVs® and BEIs®: Threshold Limit Values for Chemical Substances and Physical Agents and Biological Exposure Indices*. Cincinnati, OH: ACGIH, 2013.

———. *Industrial Ventilation—A Manual of Recommended Practice*. 21st ed. Cincinnati, OH: ACGIH, 1996.

DiNardi, S. R., ed. *The Occupational Environment—Its Evaluation and Control*. 2nd ed. Fairfax, VA: American Industrial Hygiene Association, 2003.

National Institute for Occupational Safety and Health. *Heat Stress*. Washington DC: NIOSH, 2013. http://www.cdc.gov/niosh/topics/heatstress/.

Occupational Safety and Health Administration. *OSHA Fact Sheet: Protecting Workers from the Effects of Heat*. Washington DC: U.S. Department of Labor, 2011. http://www.osha.gov/OshDoc/data_Hurricane_Facts/heat_stress.pdf.

———. *OSHA Technical Manual: Section III: Chapter 4: Heat Stress; Appendix III: Measurement of Wet Bulb Globe Temperature*. Washington DC: OSHA, 1999. http://www.osha.gov/dts/osta/otm/otm_iii/otm_iii_4.html#iii:4_3.

Parsons, K. C. *Human Thermal Environments: The Effects of Hot, Moderate and Cold Environments on Human Health, Comfort and Performance: The Principles and the Practice*. London: Taylor & Francis, 1993.

Plog, B. A., J. Niland, and P. J. Quinlan, eds. *Fundamentals of Industrial Hygiene.* 4th ed. Itasca, IL: National Safety Council, 1996.

Talty, J. T., ed. *Industrial Hygiene Engineering—Recognition, Measurement, Evaluation and Control.* Park Ridge, NJ: Noyes Data Corporation, 1988.

14
Ergonomics

LEARNING OBJECTIVES

After completing this chapter, readers should be able to do the following:
- Describe the study and goals of occupational ergonomics.
- Identify the signs and symptoms of musculoskeletal injuries.
- Recognize and describe the different types of common musculoskeletal injuries and illnesses.
- Define various aspects of anatomy and anthropometry as they relate to the study of ergonomics.
- Identify categories of contributing risk factors for musculoskeletal hazards.
- Understand various methods for assessing and quantifying musculoskeletal risks and hazards in working environments, activities, and conditions.
- Develop controls that can minimize or reduce workplace musculoskeletal hazards.
- Use the National Institute for Occupational Safety and Health (NIOSH) lifting equation to analyze working conditions and compare them with lifting guidelines.
- Establish an ergonomics workplace safety program and associated training materials.

Photo credit: nndemidchick/iStock

INTRODUCTION

Ergonomics is the study of the relationship among the human body, the human mind, and the physical environment. Ergonomists evaluate a person's size, shape, and neural/cognitive abilities and match them to the materials, equipment, structures, tools, and systems the person interacts with.

Ergonomics builds upon several core sciences, including anatomy, physiology, mathematics, and physics and incorporates anthropometry, psychology, statistics, biomechanics, and engineering.

The primary goals of ergonomics are to help humans perform tasks more efficiently, be more comfortable, and have fewer injuries. By studying how people's bodies interact with the objects and systems around them, ergonomists can design equipment, systems, tools, and machines to work more effectively with those people.

Ergonomics can be used to design simple objects like the handle on a suitcase by identifying the hardness, shape, and size of a handle that would be comfortable to most people who would use that handle. In order to make these identifications, ergonomists would look at the size of most adults' hands to see how wide to make the handle. They would also perform grip tests to see which material is easiest to grip, survey users to see which grips they liked best, and eventually come up with the best design.

Industrial hygienists should be familiar with ergonomics so that they can match the characteristics of workers to the work environments around them and thus minimize or prevent injuries and illnesses. In doing so, they should consider the capabilities of the workers and their interactions with the machines, equipment, tools, and systems. This consideration could be as simple as analyzing whether a worker can lift a box safely or as complex as designing the control panel of a nuclear reactor to be sure the operator will notice important alarms or indicators.

In addition to making a workplace safer, ergonomics typically makes the work environment more efficient. By improving systems for workers, other discrepancies and problems might be corrected or improved, which often leads to fewer mistakes, less rework, and ultimately higher-quality products.

Other efficiency benefits include fewer lost workdays, lower turnover, and reduced costs because there is no need to hire and train temporary or permanent replacement workers. Ergonomics and associated programs also tend to improve worker morale.

In broad terms, ergonomics is divided into three major areas of study. The first area, and the one most commonly practiced by industrial hygienists, is physical ergonomics. This area is associated with the anatomical, physiological, and biomechanical abilities of a person. In industrial hygiene, the workplace layout, equipment, tools, tasks, and general environment are analyzed to determine whether they are consistent with the physical and psychological limitations of the workers.

Cognitive ergonomics is the study of how the mind relates to the work environment in terms of memory, perception, response time, and reasoning. Work load, stress, and hand-machine interaction affect decision-making accuracy and speed, performance of complicated tasks, and reliability. The effects of different types of training on performance are included in cognitive ergonomics.

The study of sociotechnical systems and structures and how they affect the individual is called organizational ergonomics. This is a somewhat specialized field in which industrial hygienists are less likely to be involved. It includes studying organizational behavior and psychology, management systems, communication, and stress management. However, organizational ergonomics is an important area for industrial hygienists to be aware of because of the close relationship between workplace stress and its physical effects on the worker and relationships to safety, accidents, and workplace efficiency.

INJURIES AND ILLNESSES

According to the Occupational Safety and Health Administration (OSHA), 30 percent of all workplace injuries and illnesses in 2011 were musculoskeletal in nature. The industries with the highest musculoskeletal disorder rates that year were health care, transportation and warehousing, retail and wholesale trades, and construction. About 30 percent of all workers' compensation claims are for overexertion and repetitive-motion injuries, and claims for these injuries typically occur more than twice as often as those for other workplace injuries.

SIGNS AND SYMPTOMS

The signs and symptoms associated with musculoskeletal injuries can be difficult to interpret and diagnose. While many symptoms obviously derive from a particular workplace activity, such as a sore finger from pressing a button all day, often it is difficult to pinpoint what in the work environment causes particular injuries or pain. For example, although a worker presses a button continuously with a finger, pain occurs in the elbow. This symptom could be related to multiple ergonomic issues in the work environment.

Likewise, two workers doing the same tasks might experience different musculoskeletal health effects. Both might turn a bolt in place, but due to differences in their anatomies, ages, preexisting conditions, or individual techniques, one worker may develop a sore elbow, whereas the other may develop numb fingers.

Ergonomic musculoskeletal injuries are also difficult to identify because they are often similar to the muscle pain or tenderness that people commonly get from doing other everyday activities. Because the symptoms often aren't evident at work when the injury occurs, they become difficult to associate with the workplace. Workers may not associate pain in the elbow with their jobs, but may instead think they injured themselves playing volleyball, when in actuality, the pain could be a combination of the two, or the pain after volleyball could be due to injuries already incurred at work.

Joint pain can result from numerous workplace conditions and activities, particularly activities with repetitive motions. Swelling can also develop in various parts of the body under stress. Extended standing and walking on hard surfaces have been associated with swelling in the legs and ankles.

Numbness or tingling, particularly in the upper extremities and fingers, is associated with tendinitis and other forms of compression or impacts on nerves and blood vessels. Many workplace activities involving excessive force or pressure over extended periods have been associated with loss of circulation in an anatomical area.

In some particularly difficult cases to diagnose, the injury may not involve pain or swelling but only restricted motion. These cases are difficult to assess because of the normal anatomical differences among individuals and the lack of a baseline to compare the motion loss with. It is also sometimes difficult for people to notice their own loss of motion or other physical capabilities over long periods, and they may incorrectly associate this loss with the normal aging process.

NUMBERS/COSTS

The injuries that genuinely result from poor ergonomic conditions are broadly and commonly known as **musculoskeletal disorders (MSDs)**. Other terms that may be used are **cumulative-trauma disorder**, which indicates that the injury developed over a long period from numerous micro-injuries, and repetitive-strain injury, which describes an injury that develops over time from continuous stressful work activities or conditions.

Medical diagnosis has identified a variety of specific conditions that are musculoskeletal injuries. These medical conditions are not particular to ergonomic issues but can result from other types of injuries and accidents. Medical treatments for these disorders can involve surgeries to repair anatomical damage or pharmaceuticals to relieve swelling or pain. Therapeutic treatments such as physical therapy, massage, chiropractic, and acupuncture are also used to treat some musculoskeletal disorders.

MUSCULOSKELETAL DISORDERS

CARPAL TUNNEL SYNDROME

When the median nerve, which runs through the carpal tunnel from the forearm into the palm of the hand, is pressed or squeezed at the wrist, the nerve can become inflamed within the small passageway of ligaments and bones and lead to tingling, loss of sensation, and pain in the hand and fingers. This is known as carpal tunnel syndrome (Figure 14–1). Sometimes the pain radiates up from the wrist, through the forearm and elbow, and all the way to the shoulder. Carpal tunnel syndrome is one of the most common musculoskeletal workplace injuries.

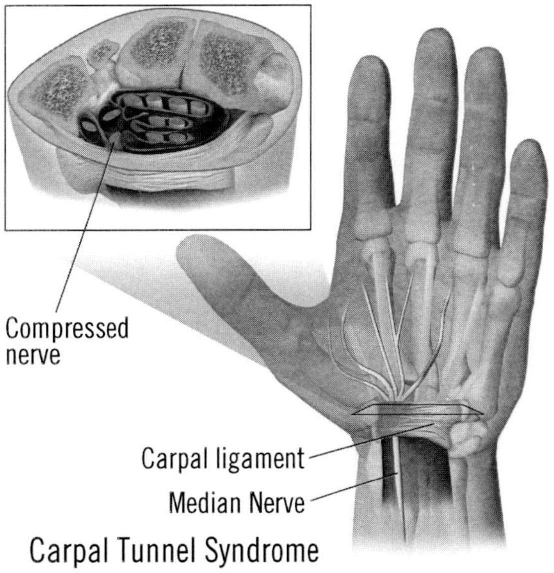

Carpal tunnel is not typically associated with problems involving the median nerve itself but with problems involving the anatomy surrounding the nerve. It is closely associated with individual characteristics and the type of activity being performed. Individuals with small wrists tend to be affected more often. For this reason, women are more likely to suffer from carpal tunnel syndrome than men.

Figure 14–1. Anatomical cross section of a wrist.

(Source: Blausen.com staff. "Blausen Gallery 2014." Wikiversity Journal of Medicine. DOI:10.15347/wjm/2014.010.)

Other contributing factors include trauma or injury to the wrist that causes swelling, such as sprain or fracture; hyperactivity of the pituitary gland; hypothyroidism; rheumatoid arthritis; mechanical problems with the wrist joint; work stress; repeated use of vibrating hand tools; fluid retention during pregnancy or menopause; and the development of a cyst or tumor in the tunnel. In some cases, no cause can be identified.

TENDINITIS

Tendinitis, the swelling of a tendon, is one of the most common injuries associated with overuse. When tendons have a low blood supply, they do not receive enough oxygen and nutrients and are thus easily damaged by overuse and by rest periods that are not sufficient to allow for full healing. Symptoms include swelling, pain during muscle movement, and general tenderness.

THORACIC OUTLET SYNDROME

Thoracic outlet syndrome occurs when the nerves and blood vessels that pass through the space between the upper ribs and the clavicle are compressed (see Figure 14–2). The compression leads to pain, numbness, weakness, and swelling in the arms and hands. This condition is often associated with tasks that require raising the arms or hands over the shoulders for prolonged periods.

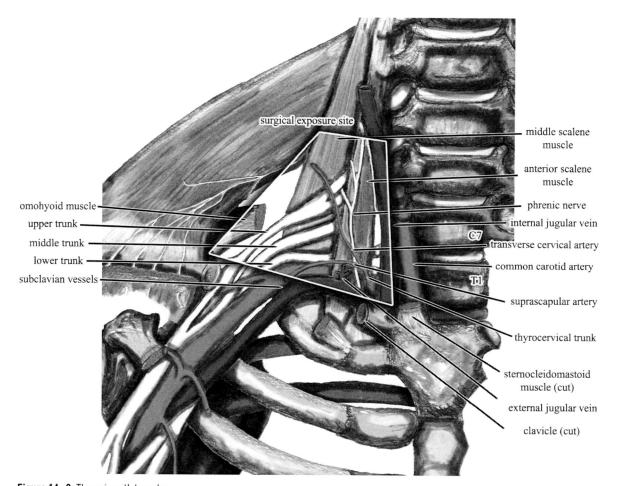

Figure 14–2. Thoracic outlet syndrome.

(Source: Nicholas Zaorsky, M.D.)

LOWER BACK STRAIN

Lower back strain is one of the most common and serious musculoskeletal injuries. When the lower back is stressed, the muscles and ligaments that hold the bones and spinal column in place are stretched too far, causing tiny tears in the tissue. Over time, this repetitive stress can cause the spinal column to go out of line, resulting in lower back pain and pain in other parts of the body.

Back stress can be caused by extreme exertion, bending and crouching repeatedly, and lifting objects that are beyond the weight appropriate for a person. It can also be caused by sitting in the same position for extended periods.

HAND-ARM VIBRATION SYNDROME AND WHITE FINGER SYNDROME

Hand-arm vibration syndrome involves the loss of sensory perception, numbness in fingers, and muscle weakness due to excessive exposure to vibrating tools or equipment such as power drills, chainsaws, and power presses. In advanced cases, the vibration affects the blood vessels in the area and reduces the circulation, causing the fingers to turn white, which is known as Raynaud's syndrome or white finger syndrome.

TRIGGER FINGER

Trigger finger often occurs when the hand or fingers are extended and the flexor tendon becomes irritated and momentarily stuck at the mouth of the tendon sheath tunnel. At some point the tendon is freed, and there is a noticeable pop as the finger is extended. As the tendon becomes increasingly irritated over time, movement within the tunnel becomes increasingly strained and difficult. The tendon and tendon sheath thicken, and the opening of the tunnel becomes smaller, causing even more friction and swelling.

ROTATOR CUFF DAMAGE

The shoulder is a ball-and-socket joint consisting of the humeral head (the upper end of the bone of the upper arm) fitting into the glenoid fossa of the scapula (shoulder blade). This anatomy allows the arm to move in a variety of directions. The humeral head is held in place by bands of cartilage and is connected to the shoulder and scapula by rotator cuff muscles, which both control movement and stabilize the humerus.

The rotator cuff and its associated muscles, tendons, and ligaments can all be damaged by excessive use or acute injury. In the workplace, continuous lifting and shoulder movement, particularly moving objects above the shoulder, can lead to accelerated degeneration of these muscles and tendons. This can cause pain and significantly limit the shoulder's range of motion and use. Acute injury can result from sudden force on the arms or shoulders, as when extending the arms to brace for a collision. Micro-injuries from repeated wear and tear can also cause rotator cuff injuries, including bone spurs and tendinitis in the area.

EPICONDYLITIS

Lateral epicondylitis is sometimes called tennis elbow. It is an inflammation of the tendons that connect to the forearm muscles around the lateral elbow. These tendons are easily damaged by overuse and repetitive motions, and the continuous microscopic tears that develop in the tendon lead to pain and tenderness.

This type of tendinitis also involves the muscle that is responsible for stabilizing the wrist when the elbow is straight. Painters, plumbers, carpenters, auto workers, cooks, and butchers commonly have this injury.

OTHER HEALTH EFFECTS

Regardless of the symptoms, disorders, and medical treatments, it is important to understand that the underlying cause is poor ergonomic workplace conditions. These working conditions need to be improved in order to protect the worker from reinjury and to prevent other workers from experiencing the same injuries.

ANATOMY

In order to be able to apply some of the fundamental concepts of ergonomics, a few basic anatomy terms must be understood. When the human body is flat on its back in a supine position, with the palms facing upward, and divided into quarters, the body can then be segmented into three hypothetical planes. These are called the frontal (coronal), medial (sagittal), and transverse (horizontal) planes (Figure 14–3).

Any body part in front of the frontal plane is considered frontal or anterior, and parts that fall behind this plane are called dorsal or posterior. Also, any parts that are respectively closer to the front of the body than other parts are considered frontal. Similarly, in terms of the transverse plane, the terms superior and inferior describe position, such as the superior vena cava and the inferior vena cava of the circulatory system. The terms upper and lower may also be used in this respect, as in the upper lobe of the lung.

The medial plane divides the body in half vertically and creates a left side and a right side. The part of the

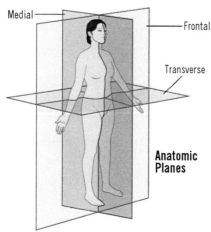

Figure 14–3. Anatomical planes of the body.

body toward the center is called medial; the part on the outside is called lateral.

The different positions or movements of the body are also important to understand when studying ergonomics. When a body part moves away from the medial plane (to the left or right), the movement is called abduction. A motion that moves a body part toward the body is called adduction. So in moving the right leg out to the right side laterally, one is abducting the leg. When one moves the right leg toward the center, one is adducting the leg to the left.

Flexion results when a body part is bent forward or backward perpendicular to the frontal plane. In contrast, stretching a body part toward the frontal plane is called extension. Bending the elbow is a good example of flexion, and moving the elbow back and straight down by one's side is a good example of extension. Other examples are pulling the lower leg and foot back to kick a ball (flexion) and then moving it forward again to kick the ball by extending the leg (extension).

In order to describe twisting, we use the terms pronation and supination. To visualize these movements, think of the frontal plane as dividing the hand and fingers of a person lying on a table in a neutral position (with the palms up). If the hand is turned so that the palm faces backward or down, the hand is moved into pronation. When the hand is turned back to its normal position, with the palm facing forward or up, the hand is in supination; if the hand goes beyond the neutral position in the other direction, the hand becomes pronated (Figure 14–4).

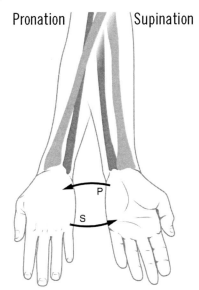

Figure 14–4. Pronation and supination.

(Source: OpenStax College. "Anatomy & Physiology." Connexions Web site. http://cnx.org/content/col11496/1.6/, June 19, 2013.)

These basic anatomical terms can be used to describe a worker's various motions or positions. Being able to accurately and consistently describe and measure these movements is necessary to understand how some adverse health effects occur and to determine which of the positions or behaviors contributed to the injuries and illnesses.

If workers are screwing plastic caps on bottles all day, they may continually abduct and adduct their wrists thousands of times a day. If the workers flex their wrists each time, it can make the effects even worse. Part of the ergonomist's task is to measure the levels of abduction, adduction, and flexion; measure the number of times the activities are performed; and make correlations between these measurements and the injury and illness data collected on the work force.

ANTHROPOMETRY

According to the Centers for Disease Control and Prevention, "Anthropometry is the science that defines physical measures of a person's size, form, and functional capacities." Of the potential ergonomic risk factors, one of the easiest to observe is when the physical dimensions of a work task are designed in such a way that the body dimensions, or anthropometry, of the workers are not accommodated. This mismatch of dimensions results in the workers' being unable to perform the task or needing to assume a stressful or awkward posture while working. These anthropometric measures are closely related to the interactions of workers in their surrounding work spaces and their interactions with tools and equipment.

It is relatively easy to measure the various human dimensions. Anthropometers can measure body dimensions such as arm length or the ranges of reaches of different-sized individuals. Newer technologies such as three-dimensional scanners also aid in gathering anthropometric data. An example of a static anthropometric data set is provided in Table 14–1.

Simple design criteria should be considered to accommodate workers performing their responsibilities. For example, shorter workers may have trouble reaching equipment to perform their tasks properly, and larger/taller employees may have difficulty fitting into a confined space. When cost and function allow, adjustability should be included in workspace design so that workers can modify their setups to accommodate their individual anthropometric measurements and easily and safely perform their tasks. Unfortunately, it is not always cost-effective to design for every single worker who is not part of the user population, which is why data such as the dimensions in Table 14–1 are very useful.

DESIGN FOR CLEARANCE

In general, when designing workspace entrances, use the measurements for a 90th-percentile male and the static anthropometric dimension of interest. The assumption is that if the entrance is big enough to accommodate a larger male, it can also accommodate anyone smaller. For example, to design a door height using the average male height in Table 14–1, the following empirical equation would be used:

$$Door\ height = x + s \times z_{90} = 1{,}756 + 67 \times 1.28 = 1{,}842\ mm$$

where x = The average/mean stature
  s = Standard deviation
  z_{90} = Empirically derived safety factor

Table 14-1. Example Anthropometric Data for U.S. Adults (19–60 years old)

Measurement	Men				Women			
	5%	Mean	95%	Standard Deviation	5%	Mean	95%	Standard Deviation
Heights (Standing)								
1. Height	1,632	1,746	1,860	67	1,525	1,628	1,732	64
2. Shoulder height standing	1,342	1,443	1,546	62	1241	1,334	1,432	58
3. Waist height	853	928	1,009	48	789	862	938	45
4. Fingertip height, standing	591	653	716	40	551	610	670	36
Heights (Sitting)								
5. Sitting height	855	914	972	36	795	852	910	35
6. Sitting eye height	735	792	848	34	685	739	794	33
Lengths (Reach)								
7. Shoulder-elbow length	340	369	399	18	308	336	365	17
9. Overhead grip reach, sitting	1,221	1,310	1,401	55	1,122	1,208	1,295	51
10. Overhead grip reach, standing	1,958	2,107	2,260	92	1,808	1,947	2,094	87
11. Forward grip reach	693	751	813	37	632	686	744	34
12. Downward grip reach	612	666	722	33	557	700	664	33
Hands								
13. Hand length	179	194	211	10	165	181	197	10
14. Hand breadth	84	90	98	4	73	79	86	4
15. Weight (kg)	58	78	99	13	39	62	85	14

Empirical equations are based on measurements, observations, and experience rather than on mathematical theories. As a result, these equations often include parameters with hidden units and factors, which means that sometimes the units on the right side of the equation do not match the units on the left side. In some cases, as in the equation above, factors are added to provide an additional safety factor to broaden the acceptable range.

If we assume that all female workers in the population are shorter than a 90th-percentile male, then a height of 1,842 mm would accommodate 95% of the total population.

DESIGN FOR REACH
Designing for reaches can be a bit more complicated depending on the static dimension of interest in the design, but the general design objective in this case is to accommodate the 10th-percentile female worker. The assumption is that

if shorter workers can reach the equipment, taller workers can as well. For example, to design a switch so that no workers have to reach above shoulder height to operate it, apply the appropriate values from Table 14–1 to the same empirical equation as above:

$$\text{Switch height} = x - s \times z_{10} = 1{,}334 - 58 - 1.28 = 1{,}259 \text{ mm}$$

If we assume that all the male workers in the population are taller than a 10th-percentile female, then a height of 1,259 mm would accommodate 95% of the total worker population.

Using static anthropometry to design dynamic work tasks is merely a first step. User trials and other testing should be performed to assess the adequacy of any task design before putting it into practice in the workplace. The comfort of the workers and the usability of the design are important factors in keeping workers healthy and happy on the job.

CAPACITY FOR WORK

Using muscles to perform work tasks takes energy. The heart and lungs meet these energy demands physiologically. When the demands are high, as they are in strenuous work carried out over a long period of time, a worker may experience general body fatigue. Table 14–2 provides examples of metabolic demands for certain activities, expressed as an amount of energy consumed by a person per hour. NIOSH recommends that the energy demands of a task not exceed one-third of the short-term maximum capacity of a worker over the course of an eight-hour workday. That limits the energy demands to 300 kcal/h and 210 kcal/h for men and women, respectively.

Table 14–2. Example Metabolic Energy Costs

Task	kcal/h
Resting, prone	80–90
Resting, seated	95–100
Standing, at ease	100–110
Driving automobile	170
Walking, casual	175–225
Pushing wheelbarrow	300–400
Shoveling	235–525
Climbing stairs	450–775

There are two ways to avoid exceeding long-term work capacity. The first is to lower the demands of the task, although that is not always practical or possible. The second approach allows workers to rest throughout the day so that the cumulative demand is effectively lowered. For example, consider the wheelbarrow task in Table 14–2, and assume the energy demands of the task are at a high level (400 kcal/h) and male workers are performing the activity. If rest and hydration are provided to the workers, the following equation may be used to determine the amount of rest needed. In this case, we will assume the rest occurs while seated.

$$\text{Rest} = (\text{Capacity} - E_{Job}) / (E_{Rest} - E_{Job}) = (300 - 400) / (100 - 400) = 33\%$$

In this scenario, 33% of the workday, or 2 hours 40 min, should be provided to the workers as rest time. It is important to note that rest time is more effective if it is spread out over the day in short periods rather than provided as one long rest period.

ASSESSING AND CONTROLLING ERGONOMIC WORKPLACE HAZARDS

ANTICIPATION: HOW, WHY, AND WHERE INJURIES HAPPEN

OSHA and the Bureau of Labor and Statistics provide a considerable amount of data identifying risky jobs and tasks, in terms of musculoskeletal injuries and illnesses. Most injuries are associated with highly repetitive tasks, awkward postures, or the use of excessive force to perform a task. These parameters and others related to ergonomic injuries are described in the following sections.

Repetitive Hazards

In general, musculoskeletal disorders associated with poor ergonomic workplace design can be related to two distinct causes. First, worker tasks are excessively repetitive, causing continued stress and strain on body parts. In addition, the duration of stresses and strains can be excessive, even if the pressure or force is not high.

An example of a repetitive hazard is placing a product into a specific position numerous times a day, which may entail repeating some motions hundreds of times per day or even per hour. The constant stress on individual muscle and tendon cells can lead to microtears and other damage, and tissue, skin, and bones can become inflamed and swollen.

Body Angles and Awkward Postures

Another significant risk factor for musculoskeletal injuries is awkward worker postures in relation to the work space. Awkward postures include flexion and extension of wrists or arms and, basically, any body angles outside of the normal ranges of motion.

The best worker positions are those where body angles are neutral or not in flexion or extension. The extremities should also not stay in abduction or adduction.

Work tasks that require excessive reaching have been associated with musculoskeletal injuries. A good rule of thumb for a suitable workplace setup is that a worker's reach should extend to 95 percent of the maximum reach only occasionally during the workshift. Most of the workday, the reach should be within a range of about 65 cm for the average woman and 80 cm for the average man. Reaching beyond these safe distances more than occasionally can put workers at risk for injury.

Twisting

Twisting of the body and body parts while performing work tasks is clearly associated with musculoskeletal injuries and illnesses. Extreme or continuous pronation and supination—and the movement back and forth between the two—have been shown to be hazardous and are associated with tendinitis and swelling.

Bending

Bending the body excessively often leads to injuries. The farther and longer a body part is bent, the greater the likelihood of injury. Sitting on the floor with crossed legs for four hours is a greater strain on the knees than sitting in a chair for the same amount of time.

Static Postures

Sometimes keeping the body in a static position for extended periods can be just as bad as engaging in repetitive movements. Holding a heavy tool over the head can put excessive pressure and force on muscles and tendons, which will soon become tired and then damaged. Blood flow may be restricted and nerves compressed during these excessive durations, leading to musculoskeletal injury.

Force

It is obvious that acute excessive force can cause injury, such as a heavy tool falling and breaking a bone in the foot. But even moderate force, if spread out over time, can result in musculoskeletal damage. Most people have used one of their hands in place of a hammer at some point, usually without any noticeable damage. However, if that same action and force were performed 100 times daily, the cumulative effects could eventually cause a musculoskeletal injury.

Force is also involved in the weight of a part or equipment that needs to be lifted—that is, a worker must exert force to pick up a part or piece of equipment. Typically, the heavier the object, the more force a worker will have to exert and the more hazardous the lift is.

Vibration

Although not a common ergonomic concern, excessive vibration can also have significant negative effects. A work environment with vibration should be evaluated to ensure worker safety.

Temperature Extremes

Because research has shown that cold and wet environments can exacerbate ergonomic hazards, special analyses should be conducted on those who work in these conditions. In cold and wet work environments, the body parts are stressed, and circulation often decreases. Tense muscles are more likely to be strained under these conditions.

Cold temperatures force the body to work harder and expend more energy and have been associated with an increase in musculoskeletal injuries. Because the cold reduces circulation to the extremities, the muscles have a harder time continuing to function and repair micro-injuries.

RECOGNITION: JOB SITE EVALUATIONS AND WALKTHROUGHS

Perhaps more than for any other area of industrial hygiene, a large amount of information about the ergonomic conditions and risks at particular worksites can be determined by reviewing injury and illness reports. Analyzing the locations, jobs, and workers associated with different types of injuries can provide useful information about where initial, detailed evaluations should be made.

Another useful source of information is employee surveys seeking feedback about perceived injuries or illnesses. Employees may be asked to indicate and rate on a pictorial diagram whether and where they have pain or discomfort. This information can then be analyzed to see whether any particular jobs or locations of the facility affect workers' health and require evaluation. Employees may also be asked to provide detailed information about the jobs they perform and any extreme ergonomic activities they engage in throughout the day.

As industrial hygienists tour a facility, they should observe the flow of materials and become familiar with processes and associated jobs and tasks while looking for ergonomic risk factors. They should make notes about workers' repetition of tasks, forces workers use to perform activities, workers' postures, and work task duration. Much of the information, such as the weight of boxes being lifted, needs to come from a knowledgeable manager of the work area, such as the supervisor.

The use of tools, particularly those rigged up by the workers, should be noted during the recognition phase. These tools often represent ergonomic stressors that the workers are aware of and are trying to alleviate.

The industrial hygiene recognition phase should also consider other environmental conditions in the work space such as extreme temperatures, noise levels, lighting, vibration, and chemical or biological hazards. The psychological or organizational stressors should also be documented in the recognition phase if possible.

In addition to having access to the supervisor of a given area, it can sometimes be helpful to have access to the workers performing the jobs and tasks. Answers to the industrial hygienist's short questions can go a long way in orienting a study and future evaluations of the workplace.

EVALUATION

Evaluating ergonomic hazards can be done in a variety of ways. Measurements can be made of physical conditions such as force, distance, posture, pressure, and acceleration. Data regarding factors such as duration, frequency, and repetition can also be measured and recorded.

Repetition

Job observations should be made and documented on data sheets. For example, the number of times that a worker lifts a box, the distance the box is moved, and the box's weight should all be noted. These data can then be compared with injury and illness reports to see whether there is a relationship between lifting a box and injuries. Carrying out repetitive activities can be compared with recommended levels to determine whether a workplace hazard exists.

Video recordings of workplace activities can also be used to make detailed records of worker activities.

Duration

In addition to repetition, the duration that a posture is held can be a risk factor. Holding a heavy box for five minutes can significantly contribute to the wear and tear on muscles, ligaments, nerves, and blood vessels, especially if workers must hold this position several times a day. Holding a heavy tool, especially when it is not being used, can be exhausting. Pressing a button down for a few seconds might not be strenuous, but holding it down for five minutes at a time could become painful. It is important to identify activities for which the worker is in a static posture and must remain in that position several minutes. Although these conditions can be difficult to identify, it is important to do so because these job tasks can be just as hazardous as seemingly more strenuous job tasks.

Force

Dynamometers

A dynamometer measures the force or power in the muscular effort needed to perform an activity. This device usually contains a spring to be compressed or a weight to be sustained by applying force, combined with an index, or automatic recorder, to show the force necessary to perform the work (as shown in Figure 14–5). In ergonomics, a dynamometer is used to measure forces involved in tasks such as pushing or pulling.

Dynamometers can measure small forces (1–3 lb) like those used in pinching motions, while others can measure more than 100 lb. Units are converted from force units. In addition to documenting a force, certain musculoskeletal disorders are associated with a reduction in grip strength, which can be measured with this device.

Dynamometers can have a meter-type display or be digital. Digital models often provide such data management features as input memory/recall and the ability to download data to a PC. Other features of dynamometers are a peak-hold needle on a dial display, adjustable grip handle, and dual-scale readout. Most devices come with a variety of attachments (as shown in Figure 14–6) to allow the forces needed for pushing, pulling, lifting, lowering, holding stationary, and other actions to be measured.

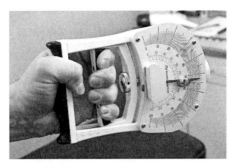

Figure 14–5. A dynamometer to measure hand grip strength.

(Source: BanksPhotos/iStock)

Vibration

Accelerometers

Workplace vibration levels can be measured using accelerometers. Results can be compared with recommended vibration levels for a variety of tasks and jobs. Levels are typically given in units of millimeters per second squared because it is the acceleration or quick and continuous change in direction that is believed to be most associated with the biological hazards of vibration.

Accelerometers can be designed to fit on specific parts of the body to measure local vibration levels. A common type of accelerometer to measure wrist vibration is shown in Figure 14–7. Accelerometers can be placed on equipment, such as a chair or a vehicle, to evaluate the overall vibration levels that the worker's entire body is exposed to.

Figure 14–6. A dynamometer with a variety of attachments.

(Source: Used with permission from 3B Scientific.)

Posture

Goniometers

Goniometers measure joint angles and range of motion. In ergonomics, goniometers are used to evaluate the postures and functional reaches of workers in workplace analysis and design. The various postures needed to complete tasks can be quantified and analyzed scientifically. Knowing the posture and joint angles used in tasks can assist the ergonomist in making workplace improvements to reduce risks of injury or illness.

Figure 14–7. A portable accelerometer for measuring wrist vibration.

(Source: Agricultural Technology Research Program at Georgia Tech)

Figure 14–8. A typical handheld goniometer for measuring body joint angles

(Source: EmmaLDavis/Wikimedia Commons)

A traditional goniometer looks like a protractor with arms, as shown in Figure 14–8. To use a goniometer, (1) align the fulcrum of the goniometer with the fulcrum of the joint to be measured, (2) align the stationary arm with the limb being measured, and (3) hold the arms of the goniometer in place as the limb is moved through its range of motion to provide a reading in degrees.

Many jobs do not lend themselves to quantitative analysis; luckily, posture may be evaluated using less quantitative methods. For example, someone who works in a confined space may not have the ideal ergonomic working conditions. Because it can be challenging to evaluate these spaces for hazardous conditions and activities, it is best to build upon the available information on any one of the

> **Case: Accommodating All Workers**
>
> Bill was the industrial hygienist at a local packaging center, and he was also responsible for workplace safety, including ergonomics. He noticed that the injury and illness logs showed unusually high absences from the small-parcel department, where 100 employees packaged small products into boxes and then shipped them to off-site locations via tractor-trailer trucks loaded at the center. The absence rate due to injury in that department was 10.2 days per 100 days, as compared to a rate of 4.2 days per 100 days for the rest of the company. When Bill looked further into the available information, he saw that many of the injuries causing the absences were sore shoulders and thoracic outlet syndrome, and one rotator cuff replacement.
>
> Bill went to the department to look more closely at some of the workers' activities to try to better understand what might be associated with or causing the injuries. As Bill walked through the department, he observed the activities, processes, repetitive actions, and body postures of the workers. He noticed that in one area, the workers had to lift up the filled packages to place them on the assembly line so that they could roll on bearings to the next workstation, where they would be taped shut and a label would be applied.
>
> Bill used a goniometer to measure the angles of the shoulders of some of the workers and noticed that the taller workers (over 5 ft. 10 in.) had to lift the box only to chest height to be able to place it on the conveyor. But anyone under that height, some of whom were much shorter, had to reach up significantly higher, placing additional stress on the shoulder joint. When he compared the recorded injuries to the workers' heights, he noticed that all the injuries had occurred in the shorter workers.
>
> It would have been too expensive to move the conveyor line already built into the production process, but it was possible for Bill to have a small platform built to raise the shorter workers up to the level where lifting the parcels did not put their shoulders into the hazardous range of motion. Injury rates in this department quickly fell to the company's average, and absenteeism in the department decreased dramatically.

potential hazards to identify possible improvements or controls. Beyond that, ergonomists need to be creative and build upon their knowledge of ergonomic principles to minimize awkward postures, repetition, long durations, and other work stressors.

CONTROL OF MUSCULOSKELETAL HAZARDS

Musculoskeletal hazard prevention can be approached in much the same way that most other hazard prevention methods are approached: anticipating hazardous conditions in the industries and jobs that have the highest musculoskeletal risks; recognizing during site observations and walkthroughs potentially hazardous areas and activities; evaluating the hazards using standardized and scientific methods; and controlling the hazards through elimination/substitution, engineering, administrative, and personal protective equipment approaches.

WORKPLACE DESIGN

Fit the Job to the Worker
The key feature of ergonomics is designing the workplace to fit the worker, rather than the other way around. The initial design phase is the best time to fit the workplace to the worker. The design should consider process flow and movements of workers throughout the production process.

If the workplace already exists, the first control, or method, to be used to improve the ergonomic environment, is to redesign and change the work space to fit the workers. Physically changing the process flow, machine-worker interfaces, and physical structures of the work process can make the space and the required activities less hazardous.

Adjust the Work Space
Whether in the initial design phase or after the workplace has already been built, the work space can be adjusted to optimally fit the worker. On the basis of anthropometry and anatomy, some key aspects can be focused on to improve workplace design.

Same Level
Whenever possible, it is usually advantageous to place the materials being handled or the work being done at the level of the workers' hands or forearms. Needing to lift or lower materials from one level to another not only requires additional work, but also increases the risk factors of musculoskeletal injuries.

Boxes that travel from a conveyor belt to a workstation ideally arrive at the workstation at the same level as the worker, requiring a worker to merely slide them off the belt rather than lift them up or down.

Forward Facing

When work is to be done on a product, it is best if the product is placed directly in front of the worker, which minimizes turning or twisting to perform work activities. In addition, the product should be within easy reach of the worker's arms and hands and not require leaning over or extending the arms for long periods or repetitively.

Reduce the Force

The forces used to perform work should be controlled so that they fall within the safe levels of the workers expected to perform the tasks. These forces include the weight of the product or boxes, the distance or height the product has to be moved up or down, and the grip needed to grasp the material.

Pushing or pulling a heavy cart can require excessive force from a worker. Possible ways to reduce the force needed include reducing the weight of the cart, improving the traction of the wheels, and mechanically powering the cart.

Machines should be designed so that they do not require workers to use excessive force to operate them by turning valves, working levers, steering, moving machine parts, and pushing buttons.

Improve Grip Strength

Materials or containers with handles are generally easier to lift and carry than materials and containers without handles. Even when they have the same amount of weight, handled containers tend to require less grip strength to lift. Boxes with handles or openings on the sides are easier to lift and carry than boxes with smooth sides, which require force to lift or hold, tend to be difficult to handle, and require the worker to expend a lot of energy. If a box without handles must be lifted, grabbing the opposite corners improves the grip.

The sizes and shapes of handles can be designed to provide maximum control with minimal force or energy. Handle material can also be designed and selected for maximum comfort and minimal damage to tissues.

Use Tools to Support the Task

Ergonomic tools and aids that are already in place should be evaluated for their appropriateness and effectiveness. It is not uncommon for workers, and even management, to address musculoskeletal issues by purchasing or creating tools or equipment. It is easy to look at an online catalog and purchase "ergonomically designed" office furniture; unfortunately, sometimes the furniture is ergonomic in name only.

Tools and aids already present at the workplace may actually provide relief from musculoskeletal stressors, however, and should not be discarded thoughtlessly. Cushioned mats in a hospital dietary department could be a great idea, but they must be of the appropriate design and material so that they do not absorb moisture, can be cleaned properly, and allow water to flow freely away from the mat to nearby drains.

Tools can often be used to improve the ergonomic conditions of a workstation or job. Large tools can provide leverage to move parts; they can also reduce how far workers need to reach. In addition, even small tools can improve leverage, reduce forces and pressure on body parts and eliminate the awkward body angles or postures needed to perform a task.

Electric or motorized power tools can greatly increase the amount of physical work that can be done. It is difficult to remember a time when it was necessary to perform all drilling tasks by hand. The same is true for saws, staplers, electric knives, and a host of other power tools.

Despite the benefits of power tools, they may pose their own musculoskeletal risks. The vibration and weight of a power hand drill can pose musculoskeletal stress and strain when a worker drills for extended periods. The worker's awkward posture when holding the drill may lead to a musculoskeletal injury. The grip strength required to hold a power tool for long durations may also be hazardous. It is thus important to analyze the work conditions of power tool use and to implement controls such as designing the tool to minimize musculoskeletal hazards, reducing the tool's weight, improving the grip, reducing vibration forces, and reducing the noise and heat generated. Heavy hand tools can be hung from the ceiling with flexible cords to reduce the length of time workers need to hold the tools and reduce the force necessary to position the tools. Other features that help improve hand tool use include

- curved handles
- extensions
- product supports
- step stools
- pliable buttons

Use Special Equipment

When handling large products or machines, using motorized and powered equipment can reduce musculoskeletal loads (see Figure 14–9). Motorized lift pallets, for example, can be

Figure 14–9. A motorized material lift device.

(Source: Electro Kinetic Technologies)

used to lift heavy products up to the work zone, eliminating the worker having to lift the material.

Antifatigue mats for workers to stand on can reduce stress on the lower extremities, back, and feet. They can also improve safety in the workplace and make workers more productive.

Antivibration mounts can be added to heavy equipment to reduce vibration and noise. They can also be added to seats and vehicles to reduce whole-body vibration. Hand tools, especially powered ones, can be equipped with cushioned grips and support devices to reduce the vibration the worker experiences.

Ergonomically Design Facilities

Workplace structures can be designed, built, and renovated to reduce ergonomic risks in a variety of ways. By maximizing product and process flow in a facility, it is often possible to reduce the musculoskeletal stressors on workers that are associated with walking distances and changes in elevation and to control other environmental conditions such as lighting, noise, temperature, and exposure to hazardous chemicals or biological agents.

Processes can be designed to maximize workflow and reduce the need for workers to stand. The heights of work processes can be designed and constructed for maximal worker efficiency and comfort. Ideally, facilities can be designed and built with movable features to allow for product and process design changes and different sizes of workers.

Individual work spaces can be designed with adjustable work surface heights, seating, and lighting; appropriate footrests and armrests; and antifatigue mats on the floor.

Lighting can play a role in the ability to perform many tasks, particularly those that require skilled or detailed work. Lighting not only directly affects safety in a workplace but, if not corresponding to the tasks at hand, can also add stresses to muscles and eyestrain.

ADMINISTRATIVE ERGONOMIC CONTROLS

An effective and comprehensive ergonomics program can substantially reduce the number and severity of workplace musculoskeletal injuries and illnesses. Applying ergonomic evaluations and controls consistently can result in significant reductions in lost workdays, absenteeism, and turnover. Ergonomics programs can make the operation of any organization more efficient.

As with any other safety or industrial hygiene program, in order to be fully effective, an ergonomics program should be written down, communicated effectively and thoroughly to workers, and reviewed and revised frequently to ensure its continuous improvement. Much like a lockout/tagout program, in which specific directions and written procedures are provided for each individual machine, specific ergonomics procedures should be prepared for each individual workstation (or, at a minimum, each job or task) that include the conditions, hazards, possible symptoms of overexposure, and controls that should be in place to protect workers.

An effective ergonomics program is a continuous process of design, evaluation, and control of the workplace to minimize hazardous worker exposures. According to OSHA, an effective ergonomics program (1) includes strong management support, (2) involves workers in worksite assessments and solutions, (3) provides appropriate training, (4) identifies hazards before they cause harm, and (5) encourages workers to report hazards and symptoms promptly.

A variety of support information and tools that can provide guidance and help develop an effective ergonomics program is available. Existing OSHA guidelines for ergonomic assessment and control include the following, which are available on the OSHA website:

- Ergonomics Program Management Guidelines for Meatpacking Plants
- Guidelines for Foundries: Solutions for the Prevention of Musculoskeletal Injuries in Foundries
- Guidelines for Nursing Homes: Ergonomics for the Prevention of Musculoskeletal Disorders
- Guidelines for Shipyards: Ergonomics for the Prevention of Musculoskeletal Disorders
- Guidelines for Retail Grocery Stores: Ergonomics for the Prevention of Musculoskeletal Disorders. Updated Guidelines: Prevention of Musculoskeletal Injuries in Poultry Processing

Eliminate or Reduce Stressful Noise

Hazardous noise levels over an 8-hour day are about 85 dBA. This is about the noise level of a standard lawnmower. In expectedly noisy jobs, like landscaping or manufacturing, a hearing conservation program allows for occupational noise measurement and the implementation of audiograms, hearing protection, and training.

But depending on the work location, even noise levels below 85 dBA can contribute to poor ergonomic conditions. Hospital laboratories now use many large automatic machines to do much of the analytical work, but one result is that noise levels in these labs have increased dramatically. With the noise contributions of these machines, of telephones, and of normal conversation in the area, the noise levels can approach 80 dBA. Although still below hazardous levels, this background noise adds to worker stress in an already stressful job. Noise control methods should be considered to reduce the noise levels in these work locations.

Identify and Reduce Organizational Stressors

To use the hospital laboratory as an example again, such areas require a fast pace, high levels of concentration, long work hours, and shiftwork (nights and evenings); and the workers often have to deal with staffing shortages, poor morale, and lack of empowerment to change their working conditions. These factors can contribute to poor ergonomic working conditions and increases in musculoskeletal injuries. Again, these factors are difficult for an industrial hygienist to quantify and change. However, it is important to understand that these factors play a role in the ergonomics of the area. Policies and procedures can be reviewed to ensure that the introduction of stress into the work process is minimized and that work is being administered fairly.

Minimize and Control Chemical, Biological, and Radiological Hazards

The addition of other stressors, such as toxic chemicals, radioactive materials, and biological agents, can contribute to musculoskeletal injuries. The laboratory workers mentioned previously have to protect themselves from bloodborne pathogens and toxic chemicals in addition to all their other stressors. They may need to don special personal protective equipment to perform their tasks and utilize other special equipment.

Develop and Implement Ergonomics Training Programs

Training programs should include the signs and symptoms of musculoskeletal health effects. Workers should be encouraged to report these signs and symptoms promptly so that hazardous workplace conditions can be identified and further damage to their health and the health of others can be prevented.

Specific administrative controls that are typically used to provide effective ergonomic controls include the following activities:

- Reduce shift length or curtail overtime.
- Rotate workers through several jobs with different physical demands to reduce the stress to limbs and body regions.
- Schedule more breaks to allow for rest and recovery.
- Broaden or vary job content to offset certain risk factors (e.g., repetitive motions or static and awkward postures).
- Adjust the work pace to relieve repetitive-motion risks and give workers more control over the work process.
- Train workers to recognize risk factors for musculoskeletal hazards and work practices that can reduce the risks.
- Teach workers about the benefits and methods of physical conditioning, exercising, and stretching to reduce musculoskeletal injuries.

RECOMMENDED LIMITS FOR LIFTING

The NIOSH Work Practices Guide for Manual Lifting provides detailed recommendations for manual lifting activities. The Guide's basic advice is to avoid doing any lifting activities that involve a high degree of repetition, twisting, or reaching; involve high forces or heavy loads; or require impulsive (quick) movements during the lift. In 1991, NIOSH revised the 1981 lifting equation so that it now calculates a recommended weight limit (RWL) for an individual task. The RWL is divided into the actual task load to calculate a lifting index. On the basis of the biomechanical, epidemiological, physiological, and psychophysical data used to develop the lifting equation, lifting index values greater than 1.0 present some degree of increased risk to workers.

The RWL is calculated by multiplying a load constant by a set of multipliers that are based on the physical dimensions of the work task. Worker size is not considered in the set of multipliers. The formula for calculating the recommended weight limit is as follows (see Table 14–3 for definitions of the elements of the equation):

$$RWL = LC \times HM \times VM \times DM \times AM \times FM \times CM$$

The horizontal multiplier is calculated using the distance from the center of the ankles to the location of the hands during the lift, and the vertical multiplier is calculated using the distance from the bottom of the feet to the location of the hands during the lift. The distance multiplier is calculated using the vertical distance the load travels during the lift. The asymmetry multiplier is based on the amount of twisting at the torso during the lift. Table 14–3 provides the equations for calculating these multipliers. The coupling and frequency multipliers are provided in Tables 14–4 and 14–5, respectively.

Although the RWL and load index can be used to make decisions about lifting tasks, they are based on several assumptions that may not apply to all lifting tasks. These assumptions are as follows:
- The equation and index assume a balanced, two-handed lift in which the worker is not slipping or falling.
- The equation and index do not apply to one-handed tasks performed while sitting or kneeling and do not apply to lifts made in excessively constrained or awkward postures.
- The equation and index assume that other manual handling activities, such as pushing, pulling, carrying, walking, or climbing, or static efforts such as holding represent no more than 20% of the total work activity for the workshift. These activities are not considered in the calculation of the RWL.

Table 14–3. Example Equations for Calculating Multipliers

	Metric	U.S. Standard
LC = Load constant =	23 kg	51 lb
HM = Horizontal multiplier =	25/H	10/H
VM = Vertical multiplier =	$1 - (0.003 \times (V - 75))$	$1 - (0.0075 \times (V - 30))$
DM = Distance multiplier =	$0.82 + (4.5/D)$	$0.82 + (1.8/D)$
AM = Asymmetry multiplier =	$1 - (0.0032A)$	$1 - (0.0032A)$
CM = Coupling multiplier (see Table 14–4)		
FM = Frequency multiplier (see Table 14–5)		

Table 14–4. Examples of Coupling Multipliers

	Metric	U.S. Customary
Couplings	V < 75 cm (30 in.)	V = 75 cm (30 in.)
Good	1.00	1.00
Poor	0.90	0.90

Table 14–5. Examples of Frequency Multipliers

	Work Duration (Continuous)			
	≤ 8 h		≤ 2 h	
Frequency (Lifts/min)	V < 75	V ≥ 75	V < 75	V ≥ 75
0.2	0.85	0.85	0.95	0.95
0.5	0.81	0.81	0.92	0.92
1	0.75	0.75	0.88	0.88
2	0.65	0.65	0.84	0.84
3	0.55	0.55	0.79	0.79
4	0.45	0.45	0.72	0.72
5	0.35	0.35	0.60	0.60
6	0.27	0.27	0.50	0.50
7	0.22	0.22	0.42	0.42
8	0.18	0.18	0.35	0.35
9	0	0.15	0.30	0.30
10	0	0.13	0.26	0.26
11	0	0	0	0.23
12	0	0	0	0.21
13	0	0	0	0
14	0	0	0	0
15	0	0	0	0
> 15	0	0	0	0

PERSONAL PROTECTIVE EQUIPMENT

Generally speaking, personal protective equipment used for ergonomic purposes is designed to support muscle function and movement or to reduce fatigue. Common examples of this type of PPE include wrist braces, back belts, and antifatigue working surfaces. Unlike a respirator or face shield, which attempts to provide a protective barrier between the worker and the hazard, ergonomic equipment supports only the activity of the worker. Consequently, ergonomic personal protective equipment is vulnerable to behavioral misuse, even when the worker does not intend to misuse the PPE. Misconceptions about the level of protection that a back belt provides, for example, have led some workers to increase the amount of lifting they perform, putting themselves at additional risk. There are also examples of ergonomic hazards that arise as a result of using other forms of personal protective equipment as protection from nonergonomic hazards. Workers who grip a tool with extra force because they are wearing gloves is an example. Another is an awkward head or neck posture that results from trying to accommodate eyewear or head protection. Ergonomic risk is mitigated most effectively through the redesign of jobs and equipment.

SUMMARY

Musculoskeletal injuries comprise a significant number of workplace injuries and occur in nearly every industry and work setting. Industrial hygienists use ergonomics to evaluate workplace musculoskeletal risks and to design and implement controls that minimize the risks. The key is to fit the job and workplace to the size and abilities of the worker, rather than force the worker to do a task that is too difficult and hazardous and that will result in injuries.

REVIEW QUESTIONS

1. What is ergonomics?
2. Why is ergonomics important to the successful operations of, say, an airline company?
3. If an ergonomist walked into a factory, what worker activities would he or she scrutinize to identify musculoskeletal hazards?
4. What is carpal tunnel syndrome, and why are women more likely to experience it?
5. Which plane divides the body into front and back halves?

6. If workers turn a dial clockwise with their right hand, what type of motion are they performing? If they perform the same activity with their left hand, what type of action are they performing?
7. What is anthropometry, and why is it important in ergonomics?
8. How many kilocalories per hour does a person's body use while driving an automobile?
9. What are three main working conditions or activities associated with musculoskeletal injuries?
10. How can the force needed to perform a task be measured?
11. What does an accelerometer measure?
12. How does a goniometer work, and what does it measure?
13. What is a key aspect of workplace design in terms of ergonomic performance?
14. What are some ways that facility design can reduce musculoskeletal hazards?
15. Why should workers be involved in developing ergonomic programs or improvements?
16. What are some organizational stressors that can increase the risk of musculoskeletal injuries?

REFERENCES

Bureau of Labor Statistics. *Report on the Days of Job Transfer or Restriction. Pilot Study—New Data on Case Circumstances and Worker Characteristics*, 2011. Washington DC: BLS, 2013. stats.bls.gov/iif/oshwc/osh/case/djtr2011.pdf.

Centers for Disease Control and Prevention. "Anthropometry." Accessed April 27, 2015. www.cdc.gov/niosh/topics/anthropometry/.

Cohen, A. L., C. C. Gjessing, L. J. Fine, et al. *Elements of Ergonomics Programs: A Primer Based on Workplace Evaluations of Musculoskeletal Disorders*. DHHS (NIOSH) Publication No. 97-117. Washington DC: Department of Health and Human Services, National Institute for Occupational Safety and Health, 1997.

eMedicineHealth.com. *Rotator Cuff Injury*. San Clemente, CA: eMedicineHealth.com, 2014. www.emedicinehealth.com/rotator_cuff_injury/article_em.htm.

Gradjean, E. *Fitting the Task to the Man*. P. 49. London: Taylor Francis, 1993.

National Institute for Occupational Safety and Health. "Anthropometry." Washington DC: NIOSH, 2013. www.cdc.gov/niosh/topics/anthropometry/.

National Institutes of Health. "Carpal Tunnel Syndrome Fact Sheet." Bethesda, MD: NIH, 2014. www.ninds.nih.gov/disorders/carpal_tunnel/detail_carpal_tunnel.htm.

Occupational Safety and Health Administration. *Prevention of Musculoskeletal Disorders in the Workplace*. Washington DC: OSHA, 2014. www.osha.gov/SLTC/ergonomics/.

Utterback, D. F., and T. M. Schnorr, eds. *Use of Workers' Compensation Data for Occupational Injury & Illness Prevention*. DHHS (NIOSH) Publication No. 2010-152. Washington DC: Department of Health and Human Services, National Institute for Occupational Safety and Health, 2010.

Waters, T. R., V. Putz-Anderson, and A. Garg. *Applications Manual for the Revised NIOSH Lifting Equation*. DHHS (NIOSH) Publication No. 94-110. Washington DC: National Institute for Occupational Safety and Health, 1994.

WorkSafeBC. *MSI Prevention Guidance Sheet: Tools for Ergonomic Assessments*. Vancouver, BC, Canada: WorkSafeBC, 2014. www2.worksafebc.com/PDFs/ergonomics/ErgoToolsGuidanceSheet.pdf.

15
Biological Hazards

LEARNING OBJECTIVES

After completing this chapter, readers should be able to do the following:

- Identify the industries where workers are at risk of exposure to hazardous biological agents and infectious diseases.
- Describe the common pathways and routes of exposure of hazardous biological agents.
- Identify different characteristics of diseases that affect their abilities to harm workers.
- Understand the human defense systems against infectious diseases.
- Describe the workplace controls that can minimize and eliminate the spread of infectious agents.
- Identify types and levels of cleaning, disinfection, and sterilization.
- Describe the different levels of biological laboratory safety and equipment and handling practices.

Photo credit: dra_schwartz/iStock

INTRODUCTION

Although not previously thought of as relevant to industrial hygiene, biological safety and control of hazardous infectious materials have grown in importance and awareness, and therefore the role of the industrial hygienist has grown as well. The increased awareness of biological hazards stemmed from the Occupational Safety and Health Administration's (OSHA's) Bloodborne Pathogens Standard (29 CFR 1910.1030). **Bloodborne pathogens** are infectious microorganisms in the blood that can cause diseases in humans. Workers exposed to bloodborne pathogens are at risk for serious or life-threatening illnesses, such as hepatitis B virus (HBV), hepatitis C virus (HCV), and human immunodeficiency virus (HIV). The standard's requirements indicate what employers must do to protect workers who can reasonably be expected to come into contact with blood or other potentially infectious materials (OPIM) while performing their job duties.

The education and tools that an industrial hygienist brings to the area of infection control and biological safety are not usually included in the training that many of the professionals in medical or biological research receive. Ventilation, air monitoring, exposure and risk assessment, and contamination control are all part of routine industrial hygiene training and procedures. Industrial hygiene is making significant contributions to the understanding and control of hazardous biological agents in a variety of industries.

BIOLOGICAL HAZARDS

Microorganisms comprise a broad range of biological species including bacteria, protozoa, fungi, algae, viruses, and prions. In addition to causing diseases, many biological materials, such as certain species of mold and fungi, animal dander, and agricultural dusts, can cause allergic or inflammatory reactions in exposed individuals. Some biological agents release endotoxins through their natural processes. Fungi often emit mycotoxins, which have been shown to cause allergic reactions in humans. Organisms that cause such diseases in humans, animals, or the environment are of primary interest to industrial hygienists.

Biological hazards associated with agriculture and animal husbandry, particularly research and veterinary care, require the expertise of an industrial hygienist to evaluate and control exposures within facilities and to ensure hazardous agents are not released into the surrounding community. Industrial hygiene methods of control in the research laboratory are particularly important in agricultural and other environmental research facilities.

INFECTIOUS DISEASES

Infectious diseases result when microorganisms enter and multiply in the body of a host and cause a pathogenic effect. These agents may directly harm tissues, or they may cause the body to react to their presence or to the toxins they generate. Some of these diseases can spread, directly or indirectly, from one person to another. Zoonotic diseases are infectious diseases in animals that can cause illness when transmitted to humans.

PATHWAYS OF EXPOSURE

Pathways of exposure, the means by which hazardous infectious agents move through the environment to reach the target, include air, water, people, surfaces, and inanimate objects like dust. While air is considered one of the most common environmental pathways, it often does not provide a viable environment for many living biological materials. If the temperature, humidity, or even UV light levels are not within a specific range, many infectious agents cannot live for more than a few minutes. Tuberculosis is an exception because the disease can live several minutes in the air at room temperature and in a broad range of humidity levels. As a result, tuberculosis is fairly transmissible via an airborne pathway.

Pathogenic agents also can be transmitted in mucous droplets in the air that are emitted when an infected person sneezes or coughs. Even normal breathing can release a small amount of these droplets into the air.

Some diseases are transmitted by direct contact when a person touches a surface that contains the infectious agent. This surface could be another person, an animal, a contaminated object, or even him- or herself. For example, when people do not wash their hands after using the restroom, they risk accidently ingesting an agent (e.g., *Escherichia coli*) from their own feces, resulting in gastroenteritis. Similar to a disease's ability to survive in the air, survival on surfaces depends on temperature, availability of moisture, and the presence of UV radiation. (See Chapter 1, Introduction to Industrial Hygiene, for more information on pathways of exposure.)

ROUTES OF EXPOSURE

Routes of exposure, the ways infectious organisms enter the body, include inhalation into the respiratory tract, dermal absorption, or injection through the skin. Many infectious agents can be absorbed into the body through ingestion and by contact with mucous membranes, such as the eyes. Infectious pathogens on surfaces or food can be unknowingly ingested and absorbed directly into the gastrointestinal tract and cause disease.

As mentioned earlier, inhalation is a route of exposure to organisms traveling in the air or those that are contained in droplets coughed or sneezed out by a nearby infected person. These particles are breathed in through the nose and pharynx and pass through the trachea and bronchioles to the alveoli deep in the lungs.

Injection into tissue or direct contact with open tissue through cuts or open wounds is another way that infectious agents can enter the body. HIV and AIDS are well-known infectious diseases that can be contracted if blood from an infected person enters the bloodstream of another person. For example, if a person's skin is punctured by a needle already used by an infected person, the first person is exposed to the disease and at risk for infection.

In humans, the inborn immune system is the first line of defense against pathogenic agents. The body's immediate responses typically include immune cells moving to the site of infection and white blood cells removing dead cells.

Anatomical barriers to pathogens include the skin, gastrointestinal tract, respiratory tract, and eyes. The skin provides a physical barrier for sensitive tissues and can remove pathogens from its surface through sweating, which washes contaminants away. The skin can also slough off epithelial cells in a process called desquamation, which releases any pathogens that may have been absorbed into the skin surface.

The gastrointestinal barrier has a variety of defense mechanisms that are active against infectious agents. Gastric acids, digestive enzymes, bile, and the normal gut flora all act as deterrents because pathogens cannot survive in that environment.

One feature of the respiratory system that can fight pathogens is the mucociliary escalator, which is where invasive particles are caught in mucous and then pushed by pulsating cilia toward the throat to be swallowed. A variety of enzymes and surfactants in the lung reduces the survival of infectious pathogens. The nasopharynx secretes lysozymes, which kill microorganisms, and produces mucous and saliva to capture and move the microorganisms out of the body.

Finally, the eyes actively work to eliminate pathogens by tearing.

The human body also has several defense mechanisms that fight pathogens. Phagocytosis is a process in which cells of the immune system kill and remove invading microorganisms. Macrophages move outside the capillary system to invade and destroy infectious agents. Neutrophils contain toxic substances that kill or inhibit the growth of bacteria or fungi.

LIKELIHOOD OF INFECTION

Infection begins when an agent colonizes in the body by growing and multiplying successfully. Individuals who are weak, malnourished, or already sick or who have depressed immune systems are especially susceptible to infection. Once inside the body, pathogenic agents can travel to and accumulate in specific organs or can cause systemic (throughout the body) infections. The likelihood and severity of infection vary among exposed individuals and based on the type of pathogenic agent.

The viability of an agent is its ability to survive in the source, in the environment, or while in its vector of transmission. If an agent lives only a few seconds in UV light, it may not be sufficiently viable to move farther into the environment, which reduces its overall risk.

Infectiveness is a measure of an agent's ability to colonize and cause an infection. This concept also relates to the agent quantity necessary to inoculate the host—the infectious dose. A pathogen that needs only a few bacteria or cells is more infectious than one that needs several hundred or thousand cells to cause infection in a healthy person. For many pathogenic agents, the infectious dose is not known or well documented. Most procedures for infection control are thus not based on solid evidence about what works for different levels or concentrations of exposure but rather on anecdotal evidence concerning which procedures are effective and which are not. Infectiveness is an important factor in how many people are likely to get a given disease and, therefore, how likely the disease is to spread.

Virulence is a measure of how aggressive the infectious agent is and indicates the speed with which victims become ill and how rapidly they succumb to the disease. Ironically, some of the more virulent diseases have limited viability because the victims become ill so quickly and severely. These outbreaks are more likely to be contained geographically because the victims quickly become ill and die. Historically, the Marburg virus and Ebola fever fell in this category of diseases, and outbreaks were usually contained to a given village. With the advent of air travel, however, the possibility of these types of agents reaching populated areas poses a serious threat to public health. The Ebola outbreak in 2014, which primarily affected countries in West Africa, illustrates the potential of the diseases to spread by air travel, as the only person to die of the disease in the United States contracted it before leaving Liberia. Luckily, he did not infect anyone else during his flight to the United States, but two health care workers who treated him at a Dallas hospital became ill. Both health care workers recovered (CDC 2014).

EPIDEMIOLOGY OF BIOLOGICAL AGENTS

Epidemiology is broadly defined as the study of the transmission and spread of disease. A variety of statistical and analytical techniques are used to study a given population. Descriptive statistics and studies can be used to determine how disease is distributed throughout the group, and analytical statistics can then be used to observe relationships and causes of diseases.

An **epidemic** is the widespread occurrence of a disease in a particular geographical area that affects an unusually large number of population members and is beyond what is normally expected or encountered (as in endemic diseases). However, epidemics are not only related to infectious or contagious diseases. Rather, the obesity epidemic in the United States and the global tobacco epidemic are examples of noncontagious health issues that can be measured and analyzed using epidemiology.

Pandemics are epidemics that spread to geographical areas where the disease is not typically expected to occur. Pandemics usually affect much larger numbers of people than epidemics do.

Epidemiological statistics and principles are commonly used to evaluate health outcomes in workplace populations. By surveying workplace injuries and illnesses, it is possible to gather information about the number and severity of events. Data can be gathered to indicate where accidents and illnesses occur and what types of workers are affected.

Slightly more advanced analytical epidemiological statistics can be used to identify correlations between worker health outcomes and numerous variables in working conditions and other factors. For example, aggregated exposure data from air monitoring of a number of chemicals may be compared with data on which workers are getting a certain type of lung disease; this information can then be used to help determine which chemical might be causing the disease.

In response to the hepatitis B pandemic in the United States, OSHA issued its Bloodborne Pathogens standard (discussed in more detail later in this chapter), which includes a requirement that employers offer free vaccinations and post-exposure evaluations to workers who might be occupationally exposed to the disease. The standard also requires employers to provide personal protective equipment (e.g., gloves, gowns, masks) at no cost to employees (OSHA 2012).

WORK EXPOSURES AND CONTROLS

HEALTH CARE

According to the Centers for Disease Control and Prevention (CDC), health care is the fastest-growing sector of the U.S. economy, employing over 18 million workers (CDC 2014). Many of these workers are routinely exposed to infectious agents. Typical infectious pathogens in U.S. hospitals include tuberculosis; influenza; *Staphylococcus aureas* (staph); methicillin-resistant *Staphylococcus aureas* (MRSA); HIV; and hepatitis A, B, and C.

Despite the potential of occupational exposure to many of these agents, there is often little emphasis placed on protecting workers. It is widely thought that health care worker exposures and subsequent negative health outcomes are underreported and not clearly understood. During the 2003 Severe Acute Respiratory Syndrome (SARS) epidemic in Toronto, for example, inadequate precautions were taken, and 40% of all cases of the disease were the health care workers providing patient care (Liu et al.).

The primary activities for infection control in hospitals revolve around patient care. However, statistics show that approximately 13% of all patients entering a hospital for treatment will develop an infection while in the hospital (WHO 2006). These are known as **hospital-acquired infections (HAIs)**. In a Canadian study, deaths caused by Clostridium difficile alone accounted for 5.7% of HAI deaths. In the province of Quebec, this number was 14.9% (Gravel, Miller, and Simor 2009).

Primary responsibilities for infection control in a hospital rest with the hospital's Infection Control Department or Infection Control Committee. This committee develops and implements a hospital infection control program and maintains records to track the various infection rates in the hospital. Members of this committee include representatives from other hospital departments, such as surgery, nursing, management, quality, occupational health, and environmental health and safety. It is recommended that the environmental health and safety representative be skilled in industrial hygiene concepts and methodologies.

BLOODBORNE PATHOGEN STANDARD

OSHA created the Bloodborne Pathogen standard (29 CFR 1910.1030) as a way to provide added protections to workers who may be exposed to human blood or **other potentially infected materials (OPIM)**. The standard requires employers to identify workers potentially exposed to blood and OPIM, identify and implement engineering and administrative controls, provide training and required personal protective equipment, initiate medical surveillance, and ensure the availability of certain vaccines. The standard also includes specific reporting requirements for exposures and sticks from sharps/needles. Prophylactic treatment must also be available for workers who have been exposed to blood or OPIM. This regulation covers not only health care workers but also potentially exposed individuals such as law enforcement officers, prison workers, school teachers, body modification professionals, and athletic trainers.

In all cases, **universal precautions** must be followed. Universal precautions are preemptive in nature and stipulate that, in cases where uncertainty exists as to whether or not materials are contaminated with potentially infectious bodily fluids, the assumption should be that the materials are contaminated, and workers should take appropriate precautions and handle the materials as though they are contaminated.

Exposure Control Plan

As a way to ensure that all of OSHA's required activities are in place in a given company, the standard requires workplaces where exposure is reasonably anticipated to develop an **exposure control plan** (ECP). A model ECP appears in the example below.

Example

Model Exposure Control Plan

The _(Company Name)_ is committed to providing a safe and healthful work environment for our entire staff. In pursuit of this goal, the following exposure control plan (ECP) is provided to eliminate or minimize occupational exposure to bloodborne pathogens in accordance with OSHA standard 29 *CFR* 1910.1030, "Occupational Exposure to Bloodborne Pathogens."

The ECP is a key document to assist our organization in implementing and ensuring compliance with the standard, thereby protecting our employees. This ECP includes:

- Determination of employee exposure
- Implementation of various methods of exposure control, including:
  - Universal precautions
  - Engineering and work practice controls
  - Personal protective equipment
  - Housekeeping
- Hepatitis B vaccination
- Post-exposure evaluation and follow-up
- Communication of hazards to employees and training
- Recordkeeping
- Procedures for evaluating circumstances surrounding exposure incidents

(Source: OSHA 2003)

Ventilation Systems

Ventilation systems are a particularly important aspect of infection control in health care, and the industrial hygienist plays an important role in ensuring their proper operation. For example, troubleshooting ventilation deficiencies is a responsibility of the industrial hygienist.

Many hospital areas, such as surgery, special procedures, endoscopy, oncology, and sterile instrument preparation areas, require outdoor air to be filtered. The filtration systems should be evaluated at installation and regularly thereafter to ensure they are performing correctly.

In many areas of a hospital, a specific balance is needed among the different ventilation functions. Most patient rooms, for example, are designed so that the air supply pressure is positive compared to that in adjacent areas, with the goal of preventing potentially contaminated air in hallways and washrooms from being pulled into the patients' rooms. But in some rooms, such as one in which the patient has an infectious disease like tuberculosis, the air in the room must be at negative pressure compared to that in adjacent areas. This ensures that the air in the tuberculosis patient's room doesn't move into the public spaces or other patients' rooms.

Exposure risk assessments can often identify where airborne infectious agents might be present and allow hospital employers to determine the appropriate protection for their employees. For example, N95 respirators have been shown to be effective in preventing transmission of tuberculosis to workers who enter infected patients' rooms. To ensure that workers are protected against exposure, hospitals should have written respiratory protection programs and associated annual respiratory fit-testing and training for all affected staff. See Chapter 8, Ventilation, and Chapter 9, Respiratory Protection, for more information on these topics.

Medical Waste

The majority of waste from a hospital goes into the same waste-handling stream as that of the households in a city, county, or other area and ultimately ends up in a municipal landfill or incinerator. However, certain types of medical infectious waste are required to be handled and disposed of differently.

OSHA identifies several requirements for the handling, storage, and disposal of regulated medical wastes. In general, all contaminated materials or potentially contaminated materials must be placed in containers designed to be sealable and leakproof. The containers must be constructed of materials capable of securely confining the items placed inside (e.g., sharps containers) and must remain closed except when materials are being placed inside (Figure 15-1). Additionally, all containers used for storing blood or other potentially infectious materials must be color-coded and exhibit the bio-hazard warning label. This labeling requirement also extends to all containers used to store or transport blood or potentially infectious materials, including refrigerators,

Figure 15–1. Medical waste container.

freezers, and portable containers. Because requirements for the disposal of regulated medical waste vary according to geographic location, it is important to be aware of all federal, state, and local requirements prior to disposing of these wastes (29 CFR 1030(d)(4)(iii)).

Medical Surveillance

Medical surveillance is a preventive practice used to detect and eliminate exposures to hazardous agents. Medical screening is a subset of medical surveillance and involves early diagnosis and treatment of individuals with symptoms or illnesses associated with hazardous agent exposures. Most health care workers fall under varying levels of medical surveillance, depending on their jobs. For instance, workers receive hepatitis evaluations and are required to be offered vaccines to prevent diseases that could result after exposure to them. Annual required health care worker tuberculosis evaluations are an example of medical screening designed to identify workers who have become infected.

The OSHA medical surveillance requirements set forth in the Bloodborne Pathogen standard are generally clinically focused (screening), with the information on medical and work histories, physical fitness, and biological testing obtained during the monitoring and analysis elements of medical surveillance. The initial screenings are used to establish a baseline of the employee's health, which is then used to monitor the employee's future health, with the goal of identifying (and subsequently controlling) health issues that arise because of the employee's exposures to hazardous agents in the workplace (OSHA n.d.; Wesdock and Sokas 2000).

Cleaning, Disinfection, Sterilization

In health care, industrial hygienists are often involved in selecting the products to clean, disinfect, and sterilize tools, equipment, and rooms. (These activities are covered in more detail in the Review of Terms section later in this chapter.) Because these products are typically dermal or respiratory irritants and sensitizers, every opportunity should be taken to eliminate or substitute each product with a less hazardous version whenever possible. Often, an infection control representative's interest in using the most effective product must be weighed against the potential negative health effects on the workers who use the product and on the patients in the rooms where the products are used.

Case: Health Care and Infection Control

A large hospital's endoscopy department had two automatic machines for cleaning and high-level disinfection with glutaraldehyde of the endoscopes between uses. The machines were side by side on a benchtop in a small room that had a dedicated exhaust system. In addition, the disinfection machines were completely enclosed and included an exhaust system that drew vapors from each device directly into the room's exhaust ducts, protecting workers from exposure.

One day the hospital's industrial hygiene manager received a call from the occupational medicine physician reporting that an endoscopy technician had been seen for respiratory irritation and dizziness. The technician believed that the ventilation in the disinfection room was not working and that he had breathed in hazardous vapors.

A subsequent interview with the worker revealed that the two disinfection units had broken and had been taken out of service earlier in the week. Until repairs could be made, the technician was cleaning the endoscopes by hand in the sinks of a room immediately adjacent to the disinfection room. The technician was working seven hours each day washing and rinsing the endoscopes by hand using the glutaraldehyde.

However, air samples indicated that the contaminant's concentration levels were well below the NIOSH recommended exposure limit, and measures of the room's ventilation system indicated it was operating to standards.

Further interviews with the worker brought to light that he had been wearing the same gloves several days in a row after the automatic machines went down. In addition, on the day he noticed symptoms, the worker had accidentally splashed some of the glutaraldehyde on his forearm, and it trickled down his arm into his glove. Rather than remove the glove and wash his arm and hands, he had continued to work another two hours until his lunch break. At that point, he noticed he was having difficulty breathing and felt dizzy, so he reported to the Occupational Medicine Department.

This case highlights several key points that pertain to health care and infection control. First, the chemicals used to kill infectious agents can be extremely hazardous to humans, and the industrial hygienist has an important role in protecting workers from those types of chemicals.

Second, workers need to understand the pathways, routes, and symptoms of exposure to the chemicals they work with. In this case, everyone initially assumed the symptoms developed from the airborne route because they were respiratory. The concept that respiratory symptoms can arise from a dermal exposure is not intuitive; however, that was clearly what occurred in this case.

Third, proper and timely maintenance of systems and equipment is closely linked to good worker health and safety. As seen in this situation, broken equipment and delaying repairs can expose workers to illnesses.

Germicidal UV Radiation

UVC radiation (230–290 nm) has been shown to be an excellent germicide against a wide variety of infectious agents. Unfortunately, it is also a probable human carcinogen, as indicated by the International Agency of Research on Cancer (IARC).

In recent years, using germicidal UV radiation in health care has been promoted as another tool against infectious agents. UV radiation has been used inside ventilation systems, in water filtration devices, and even to disinfect surfaces.

The effectiveness of UV radiation in large-space disinfection is still being debated. Nevertheless, the prime concern of the industrial hygienist is the safety of workers and members of the public who may be exposed to this radiation. Recommended allowable levels are provided in the American Conference of Governmental Industrial Hygienists' (ACGIH®) Threshold Limit Values (TLVs®) and are the industry standard for allowable worker exposures. At levels that are germicidal, the ACGIH TLVs® are exceeded in only a few seconds. Therefore, workers should not be exposed to UV radiation under any circumstances. Anyone operating or maintaining systems that use UV radiation should receive training on the hazardous effects of exposure and on how to properly control his or her own exposure.

> **Case: UV Radiation and Surgery**
>
> A large urban hospital decided to use germicidal lights in its orthopedic operating rooms to "kill the germs" during surgical procedures. After several years and numerous complaints from staff that the UV radiation was giving them sunburns, the nurses' union negotiated for workplace monitoring. When measurements were subsequently taken, the germicidal UV levels were found to be 150 times the ACGIH recommended levels.
>
> After additional lengthy negotiations among the union, hospital management, and surgeons, the hospital finally decided to discontinue using UV radiation in the surgical suites. Oddly enough, during all the years the UV had been used, no data were ever provided that the use of the radiation reduced surgical infection rates. Workers had thus been exposed for decades, with no demonstration that there was any benefit.

BIOLOGICAL LABORATORIES AND INFECTIOUS DISEASE RESEARCH

Occupational health and safety regulations and guidelines for laboratories are developed by a number of agencies in the United States. Although OSHA has several requirements that apply specifically to laboratories, the requirements are not comprehensive. However, OSHA has allied with the American Biological Safety Association to expand its focus on worker safety for workers in biological laboratories and formalize an outreach program.

The Centers for Disease Control and Prevention (CDC) is another federal agency that has staff and resources available to provide guidance and support to protect workers who handle hazardous biological and pathogenic materials. Although the CDC is not a regulatory or enforcement body, it creates numerous standards for and guidelines of practice in biological and medical laboratories that handle infectious agents. The CDC also provides a registration system for institutions that handle and store biological materials and maintains a national database of these institutions.

According to the Bureau of Labor Statistics, approximately 325,800 people work in medical and research laboratories and facilities that deal with hazardous biological and infectious agents (BLS 2014). To ensure these workers' safety, several practices have been put in place.

Wherever workers use hazardous biological agents, a biological safety officer should be on the premises. This individual has the training and experience to ensure that facility activities are conducted with controls that minimize worker exposure to hazardous agents and prevent the agents' release into the work environment or surrounding community.

In addition to the biological safety officer, each organization is required to create an institutional biosafety committee comprised of representatives of different departments of the organization. This committee should meet periodically to review and decide on various project and process proposals on how workers should handle hazardous biological or pathogenic materials.

RISK ASSESSMENT

So that employees can work safely in laboratories, the CDC recommends that companies perform a risk assessment made up of five steps. In the first step, the hazardous agents to be used are identified and ranked in terms of their hazard levels. Guidance on appropriate hazard level ranking is provided by the CDC and other reputable health sources. When a hazard rating is not available for an infectious agent, a conservative approach to working with the agent is recommended. Federal and other support organizations can be consulted for detailed advice and information when necessary.

The second step in risk assessment is to clarify the safeguards available to protect workers and ensure that the hazardous agent can be handled safely. The agent's suspension medium, physical form, concentration, and volume should be considered. If possible, procedures that may generate small-particle aerosols should not be used. The amounts that need to be handled and how they will be used are also reviewed.

Third, a biosafety level is determined; and additional precautions to be used with the agent, such as safety equipment or personal protective equipment, are reviewed. Specific characteristics of the agent should be considered when deciding on the best means of control in the workplace.

In the fourth step of risk assessment, the levels of training and experience required to safely handle the agent are identified. Also included in this step is an evaluation of the competencies of the staff and the capabilities of the equipment to safely handle the agent.

The last step is to go over the assessment results and present them to the institutional biosafety committee for its review and approval. Depending on the work being performed and the jurisdiction, it may also be necessary to have external funding agencies and state and local government authorities review the assessment.

In general, the risks of infectious agents used in laboratories fall in the basic categorizations determined by expert groups and standards bodies. Table 15–1 presents some of the basic risk group classifications (and descriptions of the applicable agents) used by the U.S. National Institutes of Health and the World Health Organization.

Table 15–1. Classification of Infectious Microorganisms by Risk Group

Risk Group Classification	NIH Guidelines for Research Involving Recombinant DNA Molecules, 2002	World Health Organization, Laboratory Biosafety Manual, 3rd Edition, 2004
Risk Group 1	Agents not associated with disease in healthy adult humans.	(No or low individual and community risk) A microorganism unlikely to cause human or animal disease.
Risk Group 2	Agents associated with human disease that is rarely serious and for which preventive or therapeutic interventions are often available.	(Moderate individual risk; low community risk) A pathogen that can cause human or animal disease but is unlikely to be a serious hazard to laboratory workers, the community, livestock or the environment. Laboratory exposures may cause serious infection, but effective treatment and preventive measures are available and the risk of spread of infection is limited.
Risk Group 3	Agents associated with serious or lethal human disease for which preventive or therapeutic interventions may be available (high individual risk but low community risk).	(High individual risk; low community risk) A pathogen that usually causes serious human or animal disease but does not ordinarily spread from one infected individual to another. Effective treatment and preventive measures are available.
Risk Group 4	Agents likely to cause serious or lethal human disease for which preventive or therapeutic interventions are not usually available (high individual risk and high community risk).	(High individual and community risk) A pathogen that usually causes serious human or animal disease and can be readily transmitted from one individual to another, directly or indirectly. Effective treatment and preventive measures are not usually available.

Biosafety Levels

Closely related to the designated risk levels, the CDC has identified biosafety levels (BSLs) that describe the safety controls for and the techniques, facilities, and equipment to be used when handling the various types of hazardous infectious agents (Table 15–2). Biosafety levels are based on the hazardous characteristics of an organism, such as its virulence, environmental viability, infectivity, and pathogenicity, and on the typical routes of exposure to the organism. The designations are also based on the availability of vaccines and the effectiveness of treatments. An agent that is commonly lethal, with no protective vaccine or treatment, has a higher biosafety level.

Facility Design

In the field of biosafety, protective controls are separated into primary barriers and secondary barriers. Primary barriers are the first control or contact with the infectious agent and are designed to protect the workers from exposure. Primary safety equipment includes enclosed containers, biological safety cabinets, and safety centrifuge cups. Personal protective equipment is also considered a primary barrier and includes gloves, lab coats, shoe covers, respirators, safety glasses or goggles, and face shields.

Secondary barriers are designed and constructed to provide additional protections to workers in the laboratory and also to provide protection to people outside the laboratory and to the local environment and community around the facility. Appropriate secondary barriers are determined by the level of hazard of the agents being used in the laboratory and typically become more stringent as the biosafety level designation rises. Secondary barriers commonly include access control, self-closing doors, decontamination devices (autoclaves), and designated hand-washing facilities. As hazard levels increase, secondary barriers may be supplemented with specially designed and filtered ventilation systems, airlocks at laboratory entrances, and various levels of laboratory isolation.

Biological Safety Cabinets

A variety of biological safety cabinets (BSCs) has been designed to protect workers, the environment, and the products that are being manipulated and used in a hood. The various hood designs provide varying levels of protection, so a thorough understanding of each design is necessary to ensure workers are protected from whatever organism they are dealing with in the hood.

Class 1 BSCs

Class 1 BSCs provide protection to the workers and the environment but no protection to the product. Similar to chemical fume hoods, these cabinets pull

Table 15–2. Recommended Biosafety Controls

Biosafety Level	Agent Descriptions	Recommended Controls
BSL-1	Practices and procedures are suitable for work involving agents of no known or of minimal potential hazard to laboratory personnel and the environment. Work is performed with defined and characterized strains of viable microorganisms not known to cause disease in healthy individuals, although there are some agents in this category, which are termed opportunistic and may cause disease in compromised individuals (e.g., in immunosuppressed individuals, in the aged, or in infants).	(1) The laboratory is not separated from the general traffic patterns of others, (2) work is generally performed on open bench tops, (3) special containment equipment and devices are not generally needed, and (4) laboratory personnel have specific training in the procedures conducted in the laboratory and are supervised by personnel with general training in microbiology or related field.
BSL-2	Practices and procedures are suitable for work involving agents of moderate potential risk to personnel and the environment. These agents can cause disease in healthy individuals and pose a moderate risk to the environment.	(1) Access to the laboratory is limited when work with these organisms is being performed; (2) the use of biological safety cabinets or protective equipment is recommended when performing work that can cause the potential for generation of aerosols (pipeting, centrifugation procedures, vortexing, etc.); and (3) laboratory personnel have specific training in handling pathogenic materials, are familiar with the hazards associated with the specific agents they are using, and are directed by scientists who are competent and familiar with good microbiological practices.
BSL-3	Practices and procedures are suitable for work involving indigenous or exotic agents where the potential for infection is real and the disease may have serious or lethal consequences. Work with these agents is performed in special containment facilities.	Precautions for use of these agents require BSL-1 and BSL-2 practices plus the following. (1) Access to the laboratory is limited to those individuals performing the work. (2) All work with these agents is performed in biological safety cabinets with special laboratory practices and procedures. The laboratory has special engineering and design features, such as an airlock entrance zone, sealed floor and wall penetrations, and directional airflow (negative pressure to the surrounding areas). (3) Laboratory personnel must have specialized training to handle pathogenic and potentially lethal agents and are supervised by a competent scientist. They must adhere strictly to special practices and procedures.
BSL-4	Practices and procedures are required for work with dangerous and exotic agents that pose a high individual risk of life-threatening disease. This is the highest level of containment (maximum containment) and requires a containment facility that is generally a separate building or completely isolated zone with complex, specialized ventilation requirements and waste management systems to prevent release of viable agents to the environment.	Specialized training of all laboratory workers is required, and strict adherence to appropriate specialized practices and procedures is mandated. Precautions for use of these agents include BSL-1, BSL-2, and BSL-3 practices and procedures plus specialized BSL-4 procedures listed in the CDC Guidelines (includes entrance only through a clothing change room, removal of street clothes, and donning of complete laboratory clothing along with showering upon leaving the containment area, complete isolation from agents used, and other specialized practices). Personnel must be specially trained and shown to be proficient in the use of the agents at this containment level.

(Source: ABSA/OSHA 2009)

in air through the sash opening in the front, eject it out the back of the hood, and directly exhaust it to the building exhaust system, which has the necessary negative pressure to pull air through this system. The exhaust system in this type of hood often includes a high-efficiency particulate air (HEPA) filter to clean the air before exhausting to the environment.

Class 2 BSCs
Class 2 BSCs have a special laminar airflow design that reduces the turbulence of the air moving across the surface of the hood. The input air is also filtered through a HEPA filter before entry. Both of these features not only protect the worker but also protect the product from contamination. There are four different types of Class 2 BSC hoods, each with particular design features and particular purposes.

Class 2 Type A1 and A2 hoods are designed so that the air supply comes from the lab environment itself. Air is drawn in through a HEPA filter and passed over the working surface, away from the breathing zone of the worker. As the air is drawn out of the hood, 30 percent is passed through a HEPA filter and released back into the laboratory; the other 70 percent is passed through a HEPA filter and recirculated back through the hood. Because these hoods do not usually exhaust to the outdoors but vent back into the laboratory, they should not be used when experiments involve volatile toxic chemicals, including most pharmaceuticals used in clinical labs.

Class 2 Type B1 hoods are hard-ducted and partially exhaust to the building's exhaust system. However, in this type of hood, 30 percent of the air is recirculated through a HEPA filter and reintroduced into the hood, which means that if this type of hood is used for anything except minute quantities of toxic volatile materials, a worker could inhale the chemical in the 30 percent reintroduced air.

The air in Class 2 Type B2 hoods is 100% exhausted through a HEPA filter and through the building's exhaust system to the outdoors. Air taken into the hood is also passed through HEPA filters. These hoods offer good protection for the products being worked on, offer the best protection for workers using volatile hazardous materials, and also ensure the safety of other occupants of the laboratory.

Class 3 BSCs
Class 3 cabinets provide maximum protection for workers and the environment. These cabinets are typically used for working with highly infectious microbiological agents and involve an enclosed system under negative pressure. Workers insert their hands into long, heavy-duty gloves attached to the device, and they look through the viewing window to manipulate the experiments. Materials are brought into and out of the cabinet through sealed antechambers and dunk

tanks. Exhausts from these hoods pass through either two HEPA filters or one HEPA filter and an incinerator before they are released into the environment.

Vertical Flow Clean Benches

Vertical flow clean benches, often found in hospital pharmacies, provide a clean area for preparing intravenous solutions. Because these are not biological safety cabinets, they meet only the minimum standards for product protection and offer no worker protection from infectious or toxic agents. Therefore, these benches should never be used to manipulate potentially infectious or toxic materials or to prepare antineoplastic agents.

CLEANING AND DECONTAMINATION

Developing methods to control the spread of hazardous microbiological agents in any facility can be a daunting proposition for an industrial hygienist. Workers and the environment need to be protected from inadvertent exposure to pathogenic agents. In addition, the methods and materials commonly used to clean and decontaminate surfaces, work spaces, and wastes often involve extremely dangerous activities and toxic chemicals or physical agents. The industrial hygienist thus needs to be aware of all the associated hazards and the best means to eliminate or reduce them from the decontamination process.

REVIEW OF TERMS

Bloodborne pathogens include infectious microorganisms present in the blood that can cause disease in humans. They include, but are not limited to, hepatitis B virus (HBV), hepatitis C virus (HCV), and human immunodeficiency virus (HIV), the virus that causes AIDS.

Cleaning is the removal of superficial substances from the outer surfaces of equipment and materials. To attain effective disinfection and sterilization, surfaces must first be cleaned effectively.

Decontamination is a general term that refers to procedures from the simple washing of surfaces up to the thorough disinfection and sterilization of surfaces, tools, and equipment. It typically involves removing biological agents to reduce the spread and growth of microbes.

Disinfection involves reducing the amount of pathogenic microorganisms on surfaces; it does not eliminate all microbiological agents, such as bacterial spores. Disinfection is typically designed to reduce the pathogenic agents of interest in controlled situations. The three levels of disinfection are based on clinical or laboratory effectiveness needs:

- *Low-Level Disinfection* reduces or eliminates most vegetative bacteria and fungi. It can also inactivate some viruses.
- *Intermediate-Level Disinfection* kills vegetative microorganisms and fungi and can inactivate most viruses. This level of disinfection is typically used in laboratories and some health care areas.
- *High-Level Disinfection* eliminates vegetative microorganisms and viruses. It can reduce, but not entirely eliminate, bacterial spores. Because high-level disinfection can be effective within 10 to 30 minutes, it is especially suited for use on medical devices and instruments.
- *Sterilization* completely eliminates all viable microorganisms, bacterial spores, and viruses. Dry heat, ethylene oxide, ultrasonic and sonic vibration, hydrogen peroxide, vaporized peracetic acid, gaseous chlorine dioxide, formaldehyde steam, ozone, filtration, ethylene oxide, and ionizing or nonionizing radiation are typically used in sterilization processes (CDC 2008).

The descending order of resistance to germicidal chemicals is presented in Table 15–3.

Table 15–3. Descending Order of Resistance to Germicidal Chemicals

Bacterial Spores
Bacillus subtilis, Clostridium sporogenes

Mycobacteria
Mycobacterium tuberculosis var. *bovis*,
Nontuberculous mycobacteria

Nonlipid or Small Viruses
Poliovirus, Coxsackie virus, Rhinovirus

Fungi
Trichophyton spp., *Cryptococcus* spp., *Candida* spp.

Vegetative Bacteria
Pseudomonas aeruginosa, Staphylococcus aureus, Salmonella choleraesuis,
Enterococci

Lipid or Medium-Sized Viruses
Herpes simplex virus, Cytomegalovirus, Respiratory syncytial virus, Hepatitis B virus, Hepatitis C virus, HIV, Hantavirus, Ebola virus

HAZARDS AND CONTROLS FOR THE USE OF DECONTAMINATION METHODS AND MATERIALS

It should not be surprising that products that are effective at killing microorganisms can also be hazardous to humans. Even most cleaning products can cause dermatitis, sensitization, and irritation to the skin, eyes, and respiratory tract. Workers who use these products often show signs and symptoms of exposure and are sometimes harmed.

Disinfectants and sterilants can be even more harmful to workers when used incorrectly. Pathways of exposure include evaporation of liquids and vapors and dermal and respiratory means involving splashes, personal protective equipment failure, and inhalation of vapors or gases. Some toxic agents, such as ethylene oxide and formaldehyde, may be lethal if certain concentrations are inhaled or carcinogenic.

Controls need to be developed for each type of cleaner, disinfectant, and sterilant and take into account the specific location and use. Exposure assessment and environmental monitoring may need to be conducted. When available, the results should be compared with occupational exposure limits.

When health effects information is unavailable, the industrial hygienist needs to research potential health effects from measured chemical levels and determine the necessary controls. Workers handling hazardous and toxic chemicals need specific training to ensure safe use of these agents, understanding of controls, awareness of symptoms of exposure, appropriate emergency response, and pursuit of medical treatment if exposed.

DECONTAMINATION OF LARGE SPACES

The decontamination of large spaces is an extremely hazardous procedure because it usually involves high levels of toxic chemicals or UV germicidal radiation. BSL-3 and BSL-4 laboratory areas should have walls, floors, and ceilings that are sealed, water resistant, and impermeable to prevent microbial growth and simplify decontamination.

Fumigation is often used to ensure that all microbes in the cracks and crevices of a facility have been eradicated. For example, fumigation is used in BSL-3 and BSL-4 laboratories, animal care facilities, and health care settings. Chemicals commonly used in fumigation include formaldehyde, paraformaldehyde, hydrogen peroxide vapor, and chlorine dioxide gas.

Facilities where fumigation may be used should be constructed with the openings in walls, floors, and ceilings for electrical, plumbing, heating, and cooling sealed tightly to prevent the escape of the fumigant vapors during and after fumigation procedures.

The industrial hygienist plays an integral role in fumigation operations by providing support in developing the fumigation procedures, employee training, area monitoring during the event, emergency response, and record keeping. Because concentrations of chemicals used in these operations typically exceed occupational exposure limits and sometimes exceed levels that are immediately dangerous to life and health, it is imperative that areas adjacent to, and well beyond, the fumigation area are monitored for chemicals that have leaked from the fumigation area. Even small amounts of these potent gases or vapors can be harmful if inhaled, especially for people who already have respiratory problems.

WASTE HANDLING

A method for decontaminating all biological wastes (e.g., autoclave, chemical disinfection, radiation, incineration, or other validated decontamination method) should be available in the facility, preferably within the laboratory. When biological wastes cannot be decontaminated in the laboratory, strict labeling, handling, and transport protocols must be followed by specially trained personnel.

MEDICAL SURVEILLANCE

All lab workers and animal handlers should be evaluated by a licensed health care practitioner as part of a medical surveillance program. Workers should be trained on the signs, symptoms, and incubation periods of exposures to the agents with which they are working. Working with certain agents requires blood samples to be taken from workers and stored for future reference.

SUMMARY

Biological hazards in the occupational setting have become more important in the past couple decades, and the role of the industrial hygienist in providing methods to control the exposures has also expanded. OSHA originally created the Bloodborne Pathogen standard, with an emphasis on PPE and administrative controls, to protect workers from AIDS and HIV. However, more recently the occupational exposures of workers to other diseases such as tuberculosis, SARS, and Ebola have demonstrated that it is important for industrial hygienists to understand the pathways and routes of exposure to continually improve PPE designs, ventilation controls, and disinfection/decontamination techniques.

REVIEW QUESTIONS

1. What are some factors that affect the viability of infectious agents in the environment?
2. Describe one of the body's natural defenses that keeps infectious agents out or eliminates them once they get into the body.
3. What are some of the factors that affect the likelihood of becoming infected?
4. What is the difference between infectiveness and virulence?
5. How is a pandemic different from an epidemic?

6. UV germicidal radiation can be a useful tool to control infectious agents in health care settings. What challenges does the use of UV radiation present to the industrial hygienist?
7. What does a biological safety officer do?
8. According to the World Health Organization, what types of agents fall into Risk Group 3?
9. What types of agents can be handled in a BSL-2 laboratory?
10. What are four examples of secondary barriers in biological safety facilities?
11. What type of Class 2 hood offers the best protection to both workers and the product?
12. What is the main difference between disinfection and sterilization?

REFERENCES

ABSA/OSHA [American Biological Safety Association/Occupational Safety and Health Administration]. ABSA/OSHA *Alliance Fact Sheet, Biosafety Levels (US)*. 2009. Accessed June 16, 2014. www.absa.org/pdf/OSHABSLFactSheet.pdf.

Bureau of Labor Statistics. *Occupational Outlook Handbook: Medical and Clinical Laboratory Technologists and Technicians*. Washington DC: BLS, 2014. www.bls.gov/ooh/healthcare/medical-and-clinical-laboratory-technologists-and-technicians.htm.

Centers for Disease Control and Prevention. *Biosafety in Microbiological and Biomedical Laboratories*. HHS Publication No. (CDC) 21-1112. 5th ed. Atlanta: CDC, 2009.

———. "Cases of Ebola Diagnosed in the United States." www.cdc.gov/vhf/ebola/outbreaks/2014-west-africa/united-states-imported-case.html, 2014.

———. "Guidelines for Disinfection and Sterilization in Healthcare Facilities." 2008. Accessed July 11, 2015. www.cdc.gov/hicpac/Disinfection_Sterilization/13_10otherSterilizationMethods.html#a1.

———. *Guidelines for Environmental Infection Control in Health-Care Facilities: Recommendations of CDC and the Healthcare Infection Control Practices Advisory Committee (HICPAC)*. Atlanta: CDC, 2003. www.cdc.gov/hicpac/pubs.html.

———. *Healthcare Workers*. Atlanta: CDC, 2014. www.cdc.gov/niosh/topics/healthcare/.

Gravel, D., M. Miller, and A. Simor. "Health Care-Associated Clostridium Difficile Infection in Adults Admitted to Acute Care Hospitals in Canada: A Canadian Nosocomial Infection Surveillance Program Study." Clinical Infectious Diseases 48 (2009): 568–76

Liu, W., F. Tang, L. Fang, S. de Vlas, H. Ma, J. Zhou, C. Richardus, and W. Cao. "Risk Factors for SARS Infection among Hospital Workers in Beijing: A Case Control Study." Tropical Medicine and International Health 14, Supplement 1 (November 2009): 52–59.

National Institutes of Health Office of Biotechnology Activities. *NIH Guidelines for Research Involving Recombinant DNA Molecules*. Bethesda, MD: NIH, 2002.

Occupational Safety and Health Administration. Bloodborne Pathogens, 29 CFR 1910.1030 (2012). www.osha.gov/pls/oshaweb/owadisp.show_document?p_table=STANDARDS&p_id=10051.

———. "Medical Screening and Surveillance." www.osha.gov/SLTC/medicalsurveillance/, n.d.

———. "Model Plans and Programs for the OSHA Bloodborne Pathogens and Hazard Communications Standards." www.osha.gov/Publications/osha3186.pdf, 2003.

———. "OSHA Fact Sheet. Hepatitis B Vaccination Protection." www.osha.gov/OshDoc/data_BloodborneFacts/bbfact05.html, n.d.

Wesdock, James C., and Rosemary K. Sokas. "Medical Surveillance in Work-Site Safety and Health Programs." *American Family Physician* 61, no. 9 (May 1, 2000): 2785–90.

World Health Organization. *Laboratory Biosafety Manual*. 3rd ed. Geneva: WHO, 2004.

———. World Alliance for Patient Safety. *Manual for Observers*. Geneva, Switzerland: WHO, 2006.

Index

A

Abduction, 327
ABET (Accreditation Board for Engineering and Technology), 4
Absolute respirator filters, 198
Absorbed dose, 268–269
Absorption
    in noise reduction, 254
    skin and, 216–220, 227
    of toxic agents/chemicals, 72, 75
    vapor- and gas-removing respirators and, 200
Accelerometers, 335
Acceptable levels of risk, 9
Acclimatization, 311–313, 315
Accreditation Board for Engineering and Technology (ABET), 4
Acetone, calculating concentrations of, 115, 116, 119–120
ACGIH. *See* American Conference of Governmental Industrial Hygienists
Acne, 222–223
Acoustical absorption, 254
Acrocynosis, 301
Acrylates, 224
Actionable goals, 54
Action level, 107–108
Action Level (trade publication), 19
Action limit, 108
Activated charcoal, 125, 126, 200
Active transport, 75
Activity (radioactive decay rate), 265–266
Acute health effects, 8
Acute radiation syndrome, 271
Adduction, 327
Adenocarcinoma, 85
Adhesives and skin, 221
Adjustable work surfaces, 340
Adjusted air temperature, 308
Adjusting work space, 337–338
ADME (Absorption, Distribution, Metabolism and Elimination), 72. *See also* Toxicology
Administrative controls, 15–16, 35. *See also* Engineering controls; Management systems
    for bloodborne pathogens, 355–356
    for dermal exposure, 230
    ergonomics and, 340–342
    for noise reduction, 254–255, 341–342
    for radiation, 283, 286
    for thermal stress, 311–312
Adsorption, 125

vapor- and gas-removing respirators and, 200
Adverse reactions, 66, 68
Advisory Committees (OSH Act, Section 7), 32
Aerodynamic diameter (AD), 141–142
Aerodynamics and noise reduction, 251
Aerosols, 137–158. *See also* Respiratory protection
    aerosol deposition, 141–144
    air-sampling pump calibration, 145–150
    capture velocity and, 171
    control of, 156–157
    definition, 138, 195
    health effects from inhalation, 139–140
    monitoring, 144–150
    respirable particulate concentration, assessing, 152–156
    respirators for, 201–202
    in respiratory tract, 73, 82
    sizes of various types, 143
    total particulate and mist concentration, assessing, 150–152
    types of, 138–139
Age of workers, 294, 313
Agreement States, 28, 276
Agriculture industry, 17
    biological hazards in, 350
    dermal hazards in, 224
    noise in, 238
    thermal stress in, 294
Agrochemicals and skin, 221
AIHA. *See* American Industrial Hygiene Association
AIHce (American Industrial Hygiene Conference and Exposition), 19
Air. *See also* Ventilation
    ambient air and ventilation, 161
    as exposure pathway, 4
    ventilation systems, 15, 35
Airborne concentrations, calculations for, 114–119
Airborne contaminants
    infectious diseases and, 351, 357
    inhalation of, 195
    respiratory protection and, 192, 195
Airborne Occupational Exposure Limits guidelines, 97–98. *See also* Occupational exposure limits
Air changes per hour (ACH), 181
Air cleaning devices, 178–180
Air flow and noise reduction, 252
Air-line respirators, 203–205
Air monitoring and aerosols, 144–150
Air-purifying respirators, 201–202, 204, 205
Air sampling, 12, 116, 119–120, 122–132

Air-sampling pumps, 125, 126, 145–150
Air speed/wind speed, measuring, 296, 302–303
Air supply and exhaust, measuring, 183–184
Air-supply system, 161, 162
Air travel and disease spread, 353
Air velocity instruments, 183
Airways. *See* Respiratory system
Alanine aminotransferase (ALT), 81
ALARA (as low as reasonably achievable), 101, 273–276
Alcohol and drugs, 311, 313, 314
Aldosterone, 314
Allergens and skin, 221, 222
Allergic contact dermatitis, 87–88, 221
Allergies, 7, 140
Alpha-emitters, 263
Alpha-particles, 263, 268–270, 275, 276
Alternative solutions, 61–62
Alveoli, 139, 193, 194, 352
Alveoli and alveolus gas exchange, 82, 83
Ambient aerosol, 209
Ambient air temperature, 219, 294, 296, 302
Ambient noise, 254–255
Amendments (U.S. Constitution), 25, 28
American Biological Safety Association, 360–361
American Conference of Governmental Industrial Hygienists (ACGIH), 19
    Biological Exposure Indices (ACGIH BEIs), 102
    Threshold Limit Values (ACGIH TLV), 98, 99–100, 102, 227, 246–247, 307–308, 360
American Industrial Hygiene Association (AIHA), 18–19, 54, 58, 154
    on Workplace Environmental Exposure Levels (WEELs), 98
American Industrial Hygiene Conference and Exposition (AIHce), 19
American National Standard for Criteria for Safety Symbols (ANSI Z535.3), 54
American National Standard for Environmental and Facility Safety Signs (ANSI Z535.2), 53
American National Standard for Product Safety Information in Product Manuals, Instructions, and Other Collateral Materials (ANSI Z535.6), 54
American National Standard for Product Safety Signs and Labels (ANSI Z535.4), 54
American National Standard for Safety Colors (ANSI Z535.1), 53
American National Standard for Safety Tags and Barricade Tapes (for Temporary Hazards) (ANSI Z535.5), 54
American National Standards Institute (ANSI), 20–21, 53–54, 203, 283
American Society for Quality Control, 55
American Society of Safety Engineers (ASSE), 19
Americium-241, 263, 264
Ames assays, 90
Amplitude, 239

Analytical Method 0500 for Particulates Not Otherwise Regulated, Total (NIOSH manual), 155
Anaphylaxis, 69, 71
Anatomical appendages of skin, 216, 217
Anatomy
    anatomical planes, 326–327
    body positions and movements, 327
    of ear, 242–243
    hand and wrist, 323
    of nerve cell, 84
    of respiratory system, 193–194
    of skin, 214–215
Anemotherms, 303
Animals and lab studies, 69–70, 74, 78, 85, 86, 88, 90, 103
Animal-to-human uncertainty factor, 102–103, 105, 106
Annual occupational dose limits, 272–273
Anode, 267
ANSI. *See* American National Standards Institute
Antennas and RF radiation, 284, 286
Anterior (anatomy), 326
Anthropometry, 328–330
Antibiotics, as sensitizer, 221
Anticipation of hazards, 9–10, 332–334
Antidiuretic hormone, 314
Antifatigue mats, 340
Antihistamines, 218
Antilabor laws, 29
Anvil bone (incus), 242
APF (assigned protection factor), 205
Apoptosis, 88
Appeals Courts, U.S., 27
Aprons, 230
A-range sound frequencies (dBA), 244–247
Area under the dose *versus* time curve (AUC), 78
Arsenic, 83
Artificial radiant heat, 303
Asbestiform silicate minerals, 139
Asbestos
    asbestosis, 6, 8, 140
    dermal hazards and, 224
    fibrosis and, 140
    as pulmonary carcinogen, 83
    safety program types, 49
A-scale devices, 248
As low as reasonably achievable (ALARA), 101, 273–275
Aspartate aminotransferase (AST), 81
Asphyxiant chemicals, 7
Asphyxiation, 6, 116
Aspirin, 78
Assays, 87, 90
ASSE (American Society of Safety Engineers), 19
Assigned protection factor (APF), 205
Assumption of Risk, 29
AST (aspartate aminotransferase), 81

Asthma, 140
Atmosphere-supplying respirators, 201, 203–204
Atomic mass, 262
Atomic number (Z), 262
Atomic structure, 262–263
AUC (Area under the dose *versus* time curve), 78
Audiograms, 341
Audiometric testing, 37
Audits, 52
Autoclaves, 363
Auto-ignition temperature, 116
Autoimmunity, 87
Autonomous body functions, 84
Axial fans, 175, 176, 252
Axons, 84

B

Bacampicillin, 224
Back stress and strain, 325
Backup latches, 229
Backward-curved fan, 175–176
Bacterial disease, spread of, 140
Bacterial spores, disinfection of, 367
Baghouse dust collection, 179
Balance, in ventilation system, 160
Balometer, 183
Banana oil (isoamyl acetate), 208
Barometric pressure, 305
Barriers
    biosafety and, 363–366
    in dermal exposure control, 229
    in heat exposure, 310
    sound barriers, 15, 253–254
Basal cell carcinoma, 279
Basal heat, 296
Basal metabolism, 296, 310
B cells, 86
Becquerel (Bq), 265
Beer's Law, 130
Behavioral changes, 69, 70–71, 299–300
Behaviors of gases, 117–119
BEIs (Biological Exposure Indices), 19
BEIs (Biological Exposure Indices of ACGIH), 102, 227
Benchmark dose (BMD), 102
Bending body, as hazard, 333
Benign cancer, 223
Benign neoplasms, 89
Benign pneumoconiosis, 140
Benzocaine, 224
Beryllium, 140
Beryllium-9, 264
Beta-decay, 264
Beta-emitters, 263
Beta-particles, 263, 268–270, 275
Biliary excretion, 77
Bill of Rights (1789), 25

Bills, in legislative process, 26
Biological Exposure Indices (ACGIH BEIs), 102, 227
Biological Exposure Indices (BEI; of OSHA), 228
Biological Exposure Indices (BEIs), 19
Biological exposure limits, 102
Biological factors in skin absorption, 217
Biological half-life (T½bio), 77–78, 266–267
Biological hazards, 7, 66, 349–371. *See also*
    Respiratory protection; Ventilation
    biological laboratories and infectious disease safety, 360–366
    bloodborne pathogen standard (OSHA), exposure and controls for, 354, 355–356
    cleaning, disinfection, sterilization products, 358
    cleaning and decontamination, 366–367
    decontamination methods and materials, hazards and controls for, 367–369
    epidemiology of biological agents, 354
    exposure control plans (ECP), 356
    germicidal UV radiation, 359–360, 368
    health care industry, exposure and controls in, 354–360
    infection likelihood, 353
    infectious diseases, pathways and routes of exposure, 351–352
    medical and biological waste, 357–358, 369
    medical surveillance, 358, 369
    overview, 350
    risk assessment and lab work, 361–366
    ventilation systems, 356–357
Biological safety cabinets (BSC) and classes of, 363–366
Biological safety officer, 361
Biomarkers of effects, 69–71
Biosafety, 360–366
Biosafety levels (BSLs) and controls, 362, 363–364
Biotransformation, 76
Bitrex, 208
Black globe thermometers, 304
Black light (UV-A), 277, 278
Bleaches and skin, 224
Bloodborne pathogens
    explanation of, 350, 366
    OSHA standard and exposure and control for, 354, 355–356
    (OSHA standards, Subpart Z 1910.1030), 36, 37, 350, 354
    safety program for, 49
Blood cell analysis, 69
Blood pressure, cold disorders and, 300
Blood serum biochemical markers, 69
Blood-to-gas partition coefficient, 73, 77
Blood vessel abnormalities, 300–301
Blue-light hazard, 279–281
BMD (benchmark dose), 102
Body angles and awkward postures, 332–333, 340
Body dimension measurements, 328, 329

Body heat storage load, 295–296
Body temperature and thermal stress, 295, 300, 309, 312
Body weight, 69, 70
Boiling point, 116
Bone marrow, 86
Bone spurs, 326
Booths, as protection, 15
Boots, 315–316
Bowman's capsule (kidney), 81
Boyle's Law, 119
Brain cancer, 285, 286
Breakthrough, 200
Bremsstrahlung x-rays, 267, 275
British Standards Institute, 60
British Thermal Units (BTU), 295
Bronchi and bronchioles, 193, 194
Bronchitis, 140
Brownian diffusion/motion, 117, 141, 142, 144, 197, 198
BTU (British Thermal Units), 295
Bubble tubes, 146–150, 152
Building egress laws, 29
Burns. *See also* Injuries and illnesses
    chemical burns, 87, 88
    radiation and, 271, 280
    UV radiation and, 277
    workplace risks and, 7, 223, 294
Business. *See also* Management systems
    efficiency and, 47, 52, 53, 61
    profitability, 29, 30, 47, 55, 61
    Six Sigma in, 57–58
    total quality management, 55–57
Byproduct materials, 11
Byssinosis, 140

C

Cadmium, 140
Cadmium-109, 264
Caffeine, 78
Calculations and equations
    activity and radioactive decay, 265–267
    airborne concentrations, gas and vapors, 114–119
    air flow capture, 171–172
    capture velocity, 170–171
    daily noise dose (DND), 245–247
    dose equivalent, 270
    fit factor, 209
    flow rates, 145–147
    heart rate increase capacity (CHR), 309
    heat exchange (worker and environment), 295–299
    inverse-square law, 261, 274, 281
    for maximum use concentration, 206
    noise frequency and fan blades, 251–252
    noise reduction and barriers, 253
    radiant heat temperature, 304
    recommended weight limit (RWL) for lifting, 343–344
    rest needs of workers, 331
    RF radiation, 284–285
    sound pressure level (SPL), 240–242, 250, 252–253
    UV dose calculation, 278
    velocity pressure, 167–169
    for volumetric flow rate, 163–165
    wavelength and frequency, 240, 260
    wet-bulb globe temperature, 306–307
    workspace design, 328–330
Calibration, 128, 145–150
Cancer, 6–8
    brain cancer, 285, 286
    cell and tumor types, 89–90
    lung cancer and aerosols, 140–141
    OELs and cancer potential, 101
    radiation and, 272, 273, 277, 286
    skin cancer, 223–224, 279
Cancer notation, 101
Canopy hoods, 174
Capacity for work, 330–332
Capture hoods, 156–157, 174
Capture velocity, 170, 171
Carbon-14, 264
Carbon arc welding, 282
Carbon disulfide, 125
Carcinogenesis and genotoxicity, 88–90, 93
Carcinogens
    chemicals as, 7, 83, 93
    germicidal UV radiation, 360
    low-frequency magnetic fields, 286
    OELs and, 101–102
Carcinoma, 89
Cardiac pacemakers, 286
Cardio-circulatory damage, 7
Cardiovascular system, 313, 314, 316
Careers and jobs, 17–18
Carotid pulse, 309
Cartridges and canisters, respiratory protection and, 195, 199, 200–201, 204
Casino hotels, 38
Catalyst, vapor- and gas-removing respirators and, 200–201
Cataracts, 7, 279, 280, 285
Cathode-ray tubes (CRTs), 267
Cattle ranching and thermal stress, 294
Cause and effect, 56
    toxic agents and, 71–72
CDC. *See* Centers for Disease Control and Prevention
Ceiling limit, 99, 100
Cell-based immunity, 86, 87
Cell damage, 66
Cell sensitivity and radiation, 271
Cellular duplication, 70

Cement and cement industry, 221, 224
Centers for Disease Control and Prevention (CDC), 4, 17, 27, 34, 328, 361, 363–364
Central nervous system, 83–84, 271, 314, 316
Centrifugal fans, 175–176
Cerebral dysfunction/confusion, 300
Cesium-137, 265, 266–267
CFR. *See Code of Federal Regulations*
Chalicosis, 140
Change schedules, 200–201
Charcoal cartridges/filters, 180
Charles' Law, 119
Checklists, 52
Chemical hygiene program, 49
Chemical Safety Board, 17
Chemicals and chemical hazards, 6, 7. *See also*
   Dermal hazards; Gases and vapors; Occupational exposure limits; Toxicology
      absorption, distribution and elimination of in body, 72, 75–78
      airborne OELs, of various organizations, 98
      cancer and, 223
      decontamination and cleaning, 367–369
      elimination of related hazards, 13–14
      EPA and, 26
      factories and, 10
      hazard communication programs on, 67
      industry standard guidance on, 227–228
      mechanisms of action and, 68–71
      personal protective equipment and, 16
      properties of chemicals, 114–117
      recognizing hazards, 11
      Safety Data Sheets (SDSs), 11, 67
      skin exposure and, 214, 216–220, 224, 225–227
      Threshold Limit Values, 19, 98
      toxic agent examples, organ system and effects on, 92–93
      toxicokinetics and, 72
Chemical-specific adjustment factors, 103
Chemisorption, 200
Chilblain, 301
Child labor, 29
Chloracne, 223
Chlorine dioxide, as disinfectant, 367
Chloroform, 73
Chromates and skin, 221
Chronic exposures, 79
Chronic health effects, 8
Chronic obstructive pulmonary diseases, 140
Chronic obstructive pulmonary diseases (COPD), 140
Cilia, 194, 352
Circulatory rates and skin absorption, 218
Cirrhosis, 80
Citations (OSH Act, Section 9), 32, 39
Civil penalties (OSHA Act, Section 17), 33
Classification of hazards, 67
Cleaning, 229, 232, 358, 366–369

Cleaning industry, dermal hazards and, 224, 225
Cleanliness requirements and ventilation, 161
Clearance, design for, 328–330
Clients, responsibility to, 16
Climate requirements and ventilation, 161
Clinical measures, 69
Closed reactors, 229
Closing conference (OSHA), 39
Clostridium difficile, 355
Clothing, 296–298, 315–316. *See also* Personal protective equipment (PPE)
Coal tar, 221, 224
Coal worker's pneumoconiosis, 140
Cochlea, 242, 243
Code of Federal Regulations (CFR), 27, 36, 43
    construction industry standard, 192–193
      respiratory protection standard, 192–193, 204–206
Cognitive ergonomics, 321
Coherence and plausibility (Hill's criteria), 72
Cold disorders, 300–301
Cold environment clothing, 315–316. *See also* Thermal stressors
Collection
    of gases and vapors in air, 119–120
    theories, methods and media, 122–126
Combination respirators, 204
Combustible chemicals, 7
Combustible-gas/multiple-gas monitors, 128
Combustion, heat of, 295
Comet assay, 90
Committees, health and safety, 16
Communications, 16, 56, 67
Communication systems, 28
Compensation, 29
    labor unions and, 29
Compliance. *See also* Regulations; Standards
    dermal hazards, 227–228
    noise levels and, 249
    radiation and, 276
    safety equipment and, 17
    surveillance and, 52
Compliance officers (OSHA), 37–39
Comprehensive inspections (OSHA), 38
Compressed Gas Association standards (ANSI), 203
Concentration
    aerosols and, 139
    chemicals and rate of absorption, 218
    gas or vapor concentration in air, 114–115
    OELs and concentration units, 97–99, 103–104
    particulates and mists, 150–152
    respirable particulates, 152–156
Conductimetric DRI, 129
Conferences, 19
Confidentiality, 16
Confined spaces, identification and control of (OSHA standards, Subpart J), 36, 37

Congress
    Constitutional amendments and, 25
    legislative process and, 26
    OSH Act and, 30–31
    OSHA rulemaking and, 35
Congress & Expo (National Safety Council), 20
Consensus-building, 53
Consistency (Hill's criteria), 71
Constant-flow air-line respirator, 203–205
Constitution, U.S., 24–28
Construction industry, 17
    dermal hazards in, 224
    musculoskeletal injuries in, 321
    noise in, 238
    OSHA inspections and, 38
    respirator standards in, 192
    thermal stress in, 294
Construction site hazards, 10
Consulting jobs, 18
Consumer Product Safety Commission, 17
Contact dermatitis, 87–88, 217, 220–222, 225, 367
Contact with objects and equipment, fatalities by, 40
Contaminants and ventilation, 160–162, 169–175
Contaminated surfaces, 218, 222, 229
Contamination control and biosafety, 360–366
Contamination control and radiation, 276
Continuity equation, 164–165
Continuous improvement (TQM), 55, 56, 60, 61
Continuous/integrated monitoring, 144–145
Continuous mechanical handling equipment for loose bulk materials-Couplings and hose components used in pneumatic handling-Safety code (ISO 5031:1977), 59–60
Continuous mechanical handling equipment-Safety code-General rules (ISO 1819:1977), 59
Continuous sampling, 124–125
Contributory Negligence, 29
Control of hazards, 13–17. *See also* Administrative controls; Assessing and controlling hazards; Elimination and substitution of hazards; Engineering controls; *specific hazards*
Convection heat exchange, 298
Convective heat load, 310–311
Convulsions, 84
Cooling units, 161
Cornea and radiation, 278, 279
Corrosive chemicals, 7
    burns and, 87, 88
Corti, organ of, 242
Cosmetology, dermal hazards and, 224
Costs
    hearing loss and, 238
    of occupational skin diseases, 214, 220
    respirators and cartridges/canisters, 200, 201
    of skin cancer, 223
Coulomb per kilogram (C/kg)., 268
Coulometric detectors, 129

Counts per minute (cpm), 268
Court of Appeals, U.S., 32
Coverage (OSH Act, Section 4), 31
Cramps and heat disorder, 300
Creatinine, 82
Creosote, 223
Criminal penalties (OSHA Act, Section 17), 33
Cross-contamination, 219
Crystalline silica, 139, 156
C-scale meters, 248
Cumulative effect of noise, 238, 243–244
Cumulative-trauma disorder, 322
Curie (Ci), 265
Current Intelligence Bulletin (CIB) 61: A Strategy for Assigning New NIOSH Skin Notations, 227
Customer-orientation (TQM), 55
Cuts and punctures, 12
Cutting fluids and skin, 221–222, 224
Cyclone filtration system, 179–180
Cyclones and cyclone calibration, 152–154
Cysts, 223
Cytotoxic T lymphocyte (CTL) assay, 87

D

Daily noise dose (DND), 245–247, 249
Dairy farm operations, 38
Damping and noise reduction, 251, 252, 254
Data, as assessment tool, 11, 354
Database insufficiency, 103, 105
Data differences and OELs, 108
dBA (A-range sound frequencies), 244–247
Death of cells/tissues. *See* Necrosis
Deaths. *See* Fatalities
Decibels (dB), 240–241
Decontamination, 366–369
Decontamination devices (autoclaves), 363
Default value, 105
Dehydration, 310, 313–314
Delayed hypersensitivity response (DHR) assay, 87
Demand mode and respirators, 205
Deming wheel, 57
Demonstration-level worksites, 60–61
Dendrites, 84
Density, 116, 117, 275
Department of Energy, U.S., 17, 26
Department of Health and Human Services, U.S., 26, 27
Department of Labor, U.S. (DOL), 25–26, 28, 29, 33, 332
Derivation and occupational exposure limits, 102–106
Dermal contact and exposure, 5, 73–74, 214. *See also* Skin
Dermal exposure and toxicity, 225, 226
Dermal hazards, 213–235
    acne, 222–223
    administrative and work practice controls, 230

compliance and industry standard guidance, 227–228
contact dermatitis, 217, 220–222
controlling skin exposure and proper PPE use, 228–232
elimination and substitution, 228
engineering controls and, 229
exposure assessment, 225–227
factors affecting absorption, 217–220
overview, 214
physical skin damage, 223
PPE, 219, 222
products for skin care, 232
skin absorption process, 216–220
skin anatomy and functions, 214–216
skin cancer, 223–224
skin injury and illness, 220–224
systemic diseases and, 223
urticaria, 222
workplace exposures, 224–225
Dermis (underlying skin layer), 214–215
DES (diethylstilbestrol), 85
Desquamation, 271, 352
Detector tubes, colormetric indicator tubes and pumps, 124
Developmental and reproductive effects/toxicity, 69, 70, 85–86, 94
Diagnoses, difficulties with, 322
Diet and nutrition, 314
Diethyl ether, 77
Diethylstilbestrol (DES), 85
Diffusion, 117. *See also* Brownian diffusion/motion
skin absorption and, 216–217
Dilution, 161
Dioctyl phthalate aerosol, 199
Diplomacy, need for, 17
Direct assessment methods, 227
Direct contact and infection spread, 351
Direct-reading detection devices, 126–132
Disease *vs.* injury, 8
Disinfectants and detergents, 221, 222, 224, 225
Disinfection in health care industry, 358, 360, 366–367
Disintegration (radioactive decay), 263–267
Disposable air-purifying respirators, 202
Disposal regulations, 357–358
Distal tubule (kidney), 81
Distance
in exposure reduction, 229, 252
in radiation protection, 274, 280–281
Distribution of toxic agents/chemicals, 72, 75–76
District Courts, U.S., 33
DNA and carcinogenesis, 89, 90
DNA and radiation, 269, 278
Documentation, 11, 67, 335, 356
DOL. *See* Department of Labor, U.S.
Dorr-Oliver cyclone, 153–154

Dorsal (anatomy), 326
Dose, 103
Dose equivalent, 270
Dose limits (rem), 272–273
Dose rate, 274, 278
Dose-response
OELs and assessment of, 96, 97, 102–104
toxicology and, 78–80
Dose-response relationship (Hill's criteria), 72
Dosimetry, 126–127, 249, 269
Draeger Safety Inc., 124
Dräger X-am 5000 (personal gas detector), 128
Drugs and alcohol, 311, 313, 314
Dry-bulb temperature, 302
Dry-bulb temperature/thermometer, 294, 306
DSEN (expected dermal sensitizers), 101
Ducted return system, 161
Ducts, 161, 165–175
Duct velocities, recommended, 175
Duct velocity and flow, measuring, 184–185
Duration
aerosols and, 139–140
chemical exposure and, 80, 102, 218, 226
glove use and, 230–231
as musculoskeletal hazard, 6–7, 332, 335
noise exposure and, 243, 245–247
Dusts
explanation of, 138
as exposure pathway, 4, 351
in respiratory tract, 82
Duties (OSH Act, Section 5), 31–32
Dyes and skin, 221, 224
Dynamometers, 335–336

E

Ear anatomy, 242–243
Ear drum (tympanic membrane), 242, 243
Earmuffs, 254–255
Ebola, 353
Economic analyses, 35
EC SCOEL (European Commission Scientific Committee on Occupational Exposure Limit Values), 98
Eczema. *See* Contact dermatitis
Education and training
administrative controls and training, 15
careers and jobs, 17–18
college-level programs, 3–4
dermal exposure training and workers, 230, 231
health and safety representatives training, 16
national programs, 4
OSH Act and, 34
OSHA inspections and, 39
on personal protective equipment, 16–17, 231
professional organizations, 19–21
for radiation safety officers, 276
resources for (OSHA), 42–43

safety programs and procedures, 50
types of, 3–4
Effective dose, 76, 79
Effective dose equivalent, 272–273
Effective half-life (T½eff), 266–267
Efficiency in workplace, 320
Efficiency (organizational), 47, 52, 53, 61
Effluent cleaning systems, 179–180
Elastomeric facepieces, 207
Electrical
    general requirements (OSHA standard, 1910.303), 41
    wiring methods (OSHA standard, 1910.305), 41
Electrical attraction, 262
Electrical signals and sound, 242
Electric field (E), 260, 261, 277, 284–287
Electric field meter, 287
Electrochemical detectors, 128–129
Electrolytes, 69, 81, 313–314, 316
    in electrochemical detectors, 128–129
Electromagnetic fields, 284
Electromagnetic radiation, basics, 260–262
Electromagnetic radiation bands, nonionizing, 277
Electromagnetic spectrum, 131, 262
Electromagnetic wave (figure), 261
Electron capture, 264
Electron capture detectors, 130
Electron microscopes, 267
Electrons, 262–263
Electrostatic attractions, 218
Electrostatic precipitation filtration, 197, 198
Electrostatic precipitator dust-collection, 179
Embalmers, dermal hazards and, 224
Embryo and fetal development/toxicity, 85–86
Emergency action plans (OSHA standards, Subpart E), 37
Emergency exit routes (OSHA standards, Subpart E), 37
Emergency preparedness, 28, 49
Emergency response, 230, 368
Emergency response jobs, 294
Emphysema, 83, 140
Empirical equations, 328–330
Employee involvement (TQM), 55
Employees
    communicating with, 67
    hearing protection, 244–247, 254–255
    inspections, right to request, 32
    protections in health care environment, 355–358
    respirators and, 206–210
    responsibilities of, 50
    safety training for, 230, 231, 342, 357, 360, 368
    thermal stressors training, 294–295, 312, 316
    total quality management and, 55
    as whistleblowers, 32
    workplace and work process redesign, 337–340
Employee surveys, 334

Employers. *See also* Management systems
    bloodborne pathogens standard (OSHA) and, 354
    citations and violations, 39
    OSH Act on, 31–32
    respirators and, 204–210
    responsibilities to employees, 48
Encapsulating suits, 315
Enclosed containers, 363
Enclosures and noise reduction, 253, 254
Endocrine hormone systems, 85
Endothelial cells, 84
Endothermic process, 295
Endotoxins, 350
Energy industry, 18
Energy loss in ducts, 168
Energy transfer, 261, 284, 295
Enforcement activities (OSHA), 37–39
Enforcement (OSH Act, Section 10), 32
Engineering controls
    for dermal hazards, 229
    for ergonomics and musculoskeletal hazards, 337–340
    of hazards, 14–15, 35
    inhalation hazards and, 192
    for noise, 238, 250–255
    for radiation, 275–276, 280–281
    for thermal stressors, 310–311
Entertainment industry, noise in, 238
Environmental damage, 47
Environmental factors in workplace, 7
Environmental Management (ISO 14001), 59
Environmental monitoring, 368
Environmental pressure variations, 6
Environmental protection, ANSI and, 53
Environmental Protection Agency (EPA), 17, 26, 28
Epicondylitis (tennis elbow), 326
Epidemics, 354
Epidemiological studies, 69, 225
Epidemiology of biological agents, 354
Epidermis (outer skin layer), 214–215
Epiglottis, 194
Equilibrium, 73, 75–77
Equipment. *See* Personal protective equipment (PPE)
Equipment failures, 6
Ergonomics, 7, 12, 319–348
    administrative ergonomic controls, 340–343
    anatomy, 326–327
    anthropometry, 328–330
    anticipation: how, why and where injuries happen, 332–334
    capacity for work, 330–332
    carpal tunnel syndrome, 323
    controlling hazards, 337–343
    epicondylitis (tennis elbow), 326
    ergonomic hazards, assessing and controlling, 331–338

evaluation of ergonomic hazards, 335–338
hand-arm vibration syndrome, 325
industrial hygiene jobs and, 18
injuries and illnesses, 321–323
lifting, recommended limits for, 343–345
lower back strain, 325
metabolic energy costs of typical activities, 331
musculoskeletal disorders, 323–326
overview, 320–321
PPE for, 345
professional organizations on, 20
recognition: job site evaluations and walkthroughs, 33–334
rotator cuff damage, 325–326
tendinitis, 322, 324
thoracic outlet syndrome, 324
trigger finger, 325
workplace design, 337–340
Ergonomics program design, 341
Ergonomics Program Management Guidelines for Meatpacking Plants (OSHA), 341
Ergonomics training programs, 342–343
Errors in sampling and analysis, 120–121
Erythemal radiation (UV-B light), 277, 278
Erythema (sunburn), 271, 279
Ethanol, 77, 78
Ethics, 47
Ethylene, 73
Ethylene dioxide, as disinfectant, 367
Ethylene oxide, 368
European Commission Scientific Committee on Occupational Exposure Limit Values (EC SCOEL), 98
Evaluation of hazards, 12–13, 306–310, 334–337
Evaporation, 227, 296, 298
Evaporation rate, 115
Evaporative cooling, 311
Evaporative heat loss (E), 298–299
Excretion, 77. *See also* Kidneys
Excursion limit approach, 100
Executive Branch, U.S., 25–26
Exhalation, 77
Exhaust air, 160
Exhaust ventilation system, 161–162
Exothermic process, 295
Expected dermal sensitizers (DSEN), 101
Experimental testing and change schedules, 200
Explosion potential, 138
Explosions, 7
  oxygen displacement and, 6
Exposure. *See also* Pathways of exposure; Routes of exposure; Toxicology
  determining risk, 9
  evaluating, 12–13
  in health care environment, 354–360
  OELs and assessment of, 96
  radiation, detecting and measuring, 267–269
  reducing exposure to aerosols, 156–157
  reducing thermal stress, 311–312
Exposure assessment
  dermal hazards and, 225–227
  radiation and, 267–269, 280–282, 286–287
Exposure assessment and control jobs, 18
Exposure control plans (ECP), 356
Exposure rate (rate of ionization), 268, 274
Extension (body), 327, 332
External respiration, 193, 194
External stimuli and skin, 216
Extremities (body), 300–301, 332
Eyes
  aerosols and, 142
  nonionizing radiation and, 277–279
  pathogens and, 352

F

FAA (Federal Aviation Administration), 31
Fabric filters, 178–179
Facepiece respirators, 202
Facepiece-to-face seal, respirators and, 208
Face shields, 219, 226, 230
Facility design and biosafety, 363
Factories, hazards at, 10
Factory Investigating Commission, 29
Fainting, heat disorders and, 300
Fair Labor Act (1938), 29
Fall protection (OSHA standard, 1926.501), 41
Falls, 10, 40
Family members as employees, 31
Fan blades and noise reduction, 251–252
Fan laws, 175–178
Fans and fan systems, 161, 165–169, 173, 178
Fan total pressure, 176
Far field, 251, 284–285, 287
Farms, 31
Fast-response sound meters, 249
Fatalities
  falls and, 10, 40
  hazard evaluation and, 12
  occupations with high rates, 41
  radiation and, 271–273
  statistics on (OSHA), 40–41, 43
Fat and heat transfer, 313
FCC. *See* Federal Communications Commission
FDA (Food and Drug Administration), 282–283
Feces, as elimination route, 77
Federal agencies and commissions, 27–28
  OSH Act and, 31
Federal Agency Safety Programs and Responsibilities (OSH Act, Section 19), 33
Federal Aviation Administration (FAA), 31
Federal Coal Mine Safety and Health Act (1969), 3
Federal Communications Commission (FCC), 17, 28, 31, 286
Federal Compensation Act (1918), 29

Federal government, 24–27
Federal Railroad Administration (FRA), 31
Federal Register, 26–27, 35
Fellow Servant Rule, 29
Fertility, radiation effect on, 272, 285
Fetal alcohol syndrome, 85
Fetal toxicity, 69, 86
Fiberglass, 224
Fibers, 139
Fibroma, 89
Fibrosis, 80, 83, 140
Field Operations Manual (OSHA), 38
Fifth Amendment (U.S. Constitution), 25
Filter fiber size, 198
Filtering facepiece respirators, 201
Filters, 125, 161, 198–199
    for total particulate sampling, 151
Filtration mechanisms and respirators, 195–198
Fines and OSH Act, 33
Fire, oxygen displacement and, 6
Fire point, 116
Fire prevention plans (OSHA standards, Subpart E), 37
Fire protection and associated systems (OSHA standards, Subpart L), 37
Fires, 7, 28–29
Fires and explosions, fatalities by, 40
First aid and medical services (OSHA standards, Subpart K), 37
First-pass metabolism, 80
First-priority inspections (OSHA), 38
Fishing industry, hazards in, 222, 223
Fish processing, 38
Fit factor, 209
Fit-testing procedures
    (OSHA standards, Subpart I), 37
    respiratory protection and, 193, 207–210
Fitting job to worker, 337
Flame ionization detectors, 129–130
Flammable chemicals, 7
Flanges and air ducts, 170–171
Flash point, 116
Flat meter, 248
Flexion (body), 327, 332
Flexor tendon, 325
Floors (OSHA standards, Subpart D), 36–37
Florists, dermal hazards and, 224
Fluid intake, 314
Fluorescent tracers, 227
Fogs, 138–139
Food and Drug Administration (FDA), 282–283
Food processing hazards, 7
Food service, 224
Footwear, 230
Force, as hazard, 6–7, 322, 325–326, 333, 335, 338
Forestry industry, hazards in, 222
Formaldehyde, 14, 224, 368

Formaldehyde steam, as disinfectant, 367
Formulas and equations
    air changes per hour, 181–182
    Flow rate and fan speed relationships, 176–177
Forward-curved fan, 175, 176
Forward-facing positioning, 338
Foundries, Guidelines for: Solutions for the Prevention of Musculoskeletal Injuries in Foundries (OSHA), 341
Fourth Amendment (U.S. Constitution), 25
FRA (Federal Railroad Administration), 31
Fragrances and skin, 221, 224
Free field, 251
Freezer plants and thermal stress, 294
Frequency (Hertz, Hz)
    hearing loss and, 243–244
    high-frequency *vs.* low-frequency sounds, 250–251
    radiation and, 260–262, 277
    sound and, 239–240, 254
Frictionless bubble tubes, 146
Frontal (coronal) plane, 326
Frostbite and frostnip, 7, 301
Full-shift TWA (time-weighted average), 99–100
Fume hoods, 15, 174
Fumes, 139
Fumigation, 368
Funeral homes, dermal hazards and, 224
Fungi, 350, 367
Furnaces, as hazard source, 280
Fusion, heat of, 295

G

Gamma radiation/rays, 261–264. *See also* Ionizing radiation; Radiation
    health effects of, 271–272
    limiting exposure, 275
    measuring exposure, 268–270
Gases and vapors, 113–135. *See also* Ventilation
    air-sampling instrumentation, 125–126
    behaviors of gases, 117–119
    calculating gas or vapor concentration in air, 119
    chemicals, properties of, 114–117
    collecting gases and vapors in air, 119–120
    collection theories, methods and media, 122–126
    combustible-gas/multiple-gas monitors, 128
    concentration units for, 99
    continuous sampling, 124–125
    coulometric detectors, 129
    detector tubes, colormetric indicator tubes and pumps, 124
    electrochemical detectors, 128–129
    errors in sampling and analysis, 120–121
    as exposure pathway, 4
    in fumigation, 368
    gas laws, 117–118

grab sampling, 122–123
inhalation and, 73
ionization detectors, 129–130
limits of detection (LOD), 120, 122
oxygen displacement and, 6
personal sampling, 126
portable gas chromatography, 132
real-time detection instruments, 126–132
respirators and, 199–201
respiratory system and, 82–83
sampling duration, 122
spectrochemical direct-reading instruments, 130–132
thermochemical detectors, 128
Gastroenteritis, 351
Gastrointestinal tract (GI), 7, 74–75, 352
GDH (glutamate dehydrogenase), 81
Gear for workers. *See* Personal protective equipment
Geiger-Mueller counters, 268
Gene mutations, 88
General Duty Clause (OSH Act, Section 5(a)(1)), 31–32
General environmental controls (OSHA standards, Subpart J), 37
General Industry (OSHA regulations), 36
Generalized gas law, 118–119
General notices, 26
General ventilation systems, 161–162, 180–182
Generated aerosol, 209
Genetic mutations, 7, 269, 272
Genotoxicity, 90
German Research Foundation (DFG), 98, 102
Germ cells, 85
Germicidal chemicals, 367
Germicidal lamps, 278, 279, 282
Germicidal ultraviolet radiation (UV-C light), 156, 224, 277, 359–360, 368
Gestation and radiation exposure, 272
Glomerulus (kidney), 81
Glove boxes, 229
Gloves, 219, 226, 230–231
Glutamate dehydrogenase (GDH), 81
Goal-setting, 15, 48, 53, 54
Goniometers, 335–336
Government, 24–29
Government Accountability Office, 35
Grab sampling, 122–123
Grade D air, 203
Grain-handling facilities, 35
Graphite, 139
Gravimetric analysis, 145
Gray (Gy), 268
Grip strength improvements, 338
Grip tests and measurements, 320, 336
Gross pathology, 69, 70
Guarding Floor and Wall Openings and Holes (OSHA standard, 1910.23), 36
Guide 51-Safety Risk Assessment (ISO/IEC), 59
Guide to Industrial Respiratory Protection (NIOSH), 195

H
Haber, L. T., 106
Hair, as elimination route, 77
Hair dyes and skin, 224
Hair follicles and skin, 214–215, 217
Hair loss, 271
Half-life, 265–267
Half Value Layer (HVL), 275
Hamilton, Alice, 29
Hammer bone (malleus), 242, 243
Hand-arm vibration syndrome, 325
Handheld detection devices, 128–130
Hand tools (ergonomics), 339, 340
Hand-washing facilities, 363
Harmful substances/environments, fatalities by, 40
Hazard assessment and control. *See also* Administrative controls; Engineering controls
broadband optical radiation hazards, 280–282
cleaners, disinfectants and sterilants, 368
ergonomic workplace hazards, 332–338
RF and subradiofrequency fields, 286–287
Hazard Banding Strategies, 110
Hazard communication (OSHA standards, Subpart Z 1910.1200), 36, 37, 41
Hazard communication programs, 67
Hazard elimination and substitution, 13–14
aerosols and, 156
dermal hazards and, 228
ergonomics and musculoskeletal hazards, 340, 341
noise control and, 251
toxic agents/chemicals, 72, 77–78
Hazard notations, 97, 101–102
Hazardous chemical exposure in labs (OSHA standards, Subpart Z 1910.1450), 37
Hazardous energy controls (OSHA standards, Subpart J 1910.147), 37
Hazardous exposure levels, 120
Hazardous spills, 16
Hazardous Waste Operations and Emergency Response (OSHA standards, Subpart H 1910.120), 36–37
Hazard quotient (HQ), 106–109
Hazard ratings (infectious agents), 361–362
Hazard ratio, 205
Hazards. *See also specific hazards*
alternative solutions for, 61–62
anticipation and, 9–10
control of, 13–17
documentation on, 11, 67
evaluation of, 12–13
government resources on, 10
health effects, 8

OELs and hazard characterization, 96
OSH Act on, 31
OSHA inspections, 38–39
OSHA rulemaking for, 34–35
process mapping and, 11–12
recognition and, 10–11
signs and labeling of, 15, 357–358, 369
types, 6–7
HCA (Hearing Conservation Amendment), 245–246
HCP (Hearing Conservation Program), 254
Headaches, heat disorders and, 300, 312
Health and Human Services, U.S. Secretary of, 32
Health and safety, ISO standards, 58–60
Health and safety programs. *See also* Occupational Safety and Health Administration; Toxicology
federal agencies and (OSH Act), 33
SMART Objective approach for, 54
on thermal stress prevention, 316
Value Strategy Manual, 58
Health and safety representatives/teams, 15–16
Health care industry, 17
dermal hazards in, 224, 231
disinfection methods, 358, 360, 366–367
exposure and controls in, 354–360
industrial hygienist jobs in, 17
infectious agents in, 7
musculoskeletal injuries in, 321, 342
noise in, 341–343
Health effects, 6, 8, 67. *See also* Burns; Injuries and illnesses
of aerosol inhalation, 139–140
cleaning and decontamination, 367–368
ionizing radiation and, 271–273
nanoparticles and, 10
nonionizing radiation and, 277, 278–283, 285, 286
of thermal stress, 294, 299–301
Health Physics Journal (trade journal), 20
Health Physics Society (HPS), 20
Hearing. *See also* Noise
conservation programs, 49, 50, 245, 254
protecting, 15, 238, 244–247
Hearing Conservation Amendment (HCA), 245–246
Hearing Conservation Program (HCP), 254, 341
Hearing damage and loss, 6–8, 238, 243–244
Hearing protection devices, 254–255
Heart rate and thermal stress, 309, 312
Heat-alert program, 316
Heat capacity, 295
Heat cramps, 300
Heat (definition), 295
Heat disorders, 299–300
Heat emissions and ventilation, 161
Heat exchange, 295–299
body heat storage load, 295–296
convection heat exchange, 298
evaporative heat loss (E), 298–299
metabolic heat gain, 296–297

radiant heat exchange, 297
terms and definitions, 295
thermal balance, 295
Heat exhaustion, 300
Heating devices and RF radiation, 283, 284
Heating units, 161
Heat lamps, as hazard, 280
Heat of combustion, 128
Heat rash, 300
Heat stress, 10, 294–295, 306–310. *See also* Thermal stressors
Heatstroke, 300
Heat tolerance, 312–315
Hematology, 69
Hematopoietic precursor cells, 86, 87
Hemoglobin, cold disorders and, 301
HEPA (high-efficiency particulate air) filters, 365–366
Hepatitis B pandemic, 354
Hepatitis viruses, 350, 355, 366
Hepatocytes, 80–81
Hertz, Hz. *See* Frequency
Hexavalent chromium, 140
HFES (Human Factors and Ergonomics Society), 20
Hierarchy of controls, 13
Hierarchy of OELs, 110
High-efficiency particulate air (HEPA) filters, 365–366
High-level disinfection, 367
High margin of safety, 108
Highway construction, hazards in, 224
Hill's criteria for causation, 71
Histopathology, 69, 70
History of industrial hygiene, 2–3
HIV and AIDS, 350, 352, 355, 366
Hood entry loss factor, 171
Hoods, types of, 171
Hoods and biosafety cabinets, 363–366
Horizontal standards, 36
Horsepower, fans and, 176, 178
Hospital-acquired infections (HAIs), 355
Hospitality industry, 238
Hospitals
germicidal radiation at, 360
infection control at, 355
lab ergonomics, 341–342
ventilation and air filtration, 357
Hot environment clothing, 315. *See also* Thermal stressors
House of Representatives, U.S., 24, 26
HPS (Health Physics Society), 20
HQ (hazard quotient), 106–109
Human-effects studies, 103
Human Factors and Ergonomics Society (HFES), 20
Human resources professionals, 16
Human variability uncertainty factor, 102–103, 105, 106
Human voice frequencies, 244

Humeral head, 325
Humidity, 200, 294, 305, 311
Humoral immunity, 86, 87
HVL (Half Value Layer), 275
Hydrogen peroxide, as disinfectant, 367
Hydrophilic compounds, 77, 220
Hydrophobic molecules, 75
Hyperplasia, 70
Hypersensitivity, 87
Hypertrophy, 70, 79, 82
Hypodermis, 214–215
Hypothermia, 7, 300

I

IARC (International Agency of Research on Cancer), 360
Ice-packed vests, 315
ICNIRP (International Commission on Non-Ionizing Radiation Protection), 280
Ideal gas law, 117–118
Identification of hazards, 67
IDLH. See Immediately dangerous to life or health
IEC (International Electrotechnical Commission), 282–283
Illnesses. See Injuries and illnesses
Immediately dangerous to life or health (IDLH), respirators and, 201, 203, 204
Imminent danger, 33
Immune suppression, 279
Immune system, 352
Immunological data/studies, 69, 71
Immunotoxicity, 86–87, 93
Impact, in exposure, 218
Impaction, 141–143
Impact response time sound meters, 249
Imprisonment, 33
Inattention and thermal stress, 299
Incentive programs, 60–61
Incident investigation form (ANSI 2005), 54
Incident rates, hazard evaluation and, 12
Inclined manometers, 183, 184
Increased cellular size, 70
Incus (anvil bone), 242
Industrial gauges, 275, 276
Industrial hygiene. See also Management systems; Occupational Safety and Health Administration
    anticipation of hazards, 9–10, 332–334
    approach to, 9–13
    control of hazards, 13–17
    ethics and, 47
    government agencies and, 25–26
    hazard types and health effects, 6–8
    infection control and, 355
    management systems in, 45–63
    occupational exposure, 4–5
    OELs, 95–112 (See also Occupational exposure limits)
    overview and history of, 2–3
    professional organizations, 18–21
    program types, 49–50
    quality assurance audits, 51–52
    recognition of hazards, 10–12
    reducing and controlling thermal stress, 311–312
    risk assessment, 8–9
    sciences and, 3–4
    training in, 4, 350
    U.S. Constitution and, 25
Industrial hygienists. See also Aerosols; Biological hazards; Ergonomics; Gases and vapors; Radiation; Respiratory protection; Risk assessment; Thermal stressors; Toxicology; Ventilation
    aerosol monitoring and analysis how-to, 144–156
    careers for, 17–18
    chemistry understanding, importance of, 114
    cleaning and decontamination, 366–369
    communication skills and, 16
    dermal exposure prevention by, 232
    ergonomics and injury prevention, 320–321
    ergonomics jobs, 18
    infection control, 350, 355, 358
    job-site evaluations and walkthroughs, 334
    in judicial process, 27
    in legislative process, 26–27
    as part of safety team, 16
    radiation safety and, 276
    resources for, 42–43, 91, 109–110, 150, 154, 227–228
    role of in alternative solutions, 61–62
    thermal stress safety and prevention, 294–295
    toxicology as core competency of, 66–68
    ventilation in health care setting, 356–357
Industrial toxicology. See Toxicology
Inertial impaction filtration, 196, 198
Infection and infectious disease. See also Biological hazards
    biological and medical laboratories safety, 360–366
    dermatitis and, 221
    hazardous agents ranking, 361–362
    infection likelihood, 353
    pathways and routes of exposure, 351–352
    toxic agents and, 69, 71
Infectious microorganisms, risk group classifications of, 362
Infectiveness, 353
Inferior (anatomy), 326
Influenza, 355
Infrared (IR) spectrophotometers, 131–132
Infrared pyrometers, 304
Infrared radiation (IR), 15, 260, 262, 277, 280, 282–283, 304
Infrared thermometer, 309
Ingestion, as route of exposure, 5

chemicals and, 74–75, 102
infectious agents, 351–352
Inhalable particulate mass, 142
Inhalation
aerosols and health effects from, 139–140
OELs and, 99
as route of exposure, 5, 70, 73, 102, 352
Initiation (carcinogenesis), 89
Injection, as route of exposure, 5
Injuries and illnesses. *See also* Occupational Safety and Health Act
epidemiology and, 354
hearing loss, 238
injury and illness rates, 30, 38
musculoskeletal disorders, 321–326
OSHA inspections and, 38, 39
SARS, 355
skin and, 220–224
skin care products for, 232
statistics on (OSHA), 39–41, 321
thermal stress and, 294, 299–301
vision damage, 279–280, 282
Injury and illness reports, 334
Injury *vs.* disease, 8
In-line centrifugal fan, 176
Inner ear, 242, 243
Input materials, 11
Inspections
employee rights and, 32
surveillance and, 52
Inspections, Investigations, and Record Keeping (OSHA Act, Section 8), 32
Instantaneous/real-time monitoring, 144, 154
Insurance industry, 17
Integrated systems (TQM), 56
Intercellular lipid pathway, 216
Intercellular respiration, 194
Interception, 141, 143
Interception filtration, 195, 198
Intermediate-level disinfection, 367
Internal dose, 75
Internal respiration, 194
International Agency of Research on Cancer (IARC), 360
International Commission on Non-Ionizing Radiation Protection (ICNIRP), 280
International Electrotechnical Commission (IEC), 282–283
International Labor Organization, 29
International Standards Organization (ISO), on environmental health and safety, 58–60
guides from, 59–60
ISO 14001-Environmental Management, 59
ISO/IEC Guide 51-Safety Risk Assessment, 59
ISO 9001-Quality Management, 59
ISO 31000-Risk Management, 59
International System of units (Système Internationale [SI]), 265

Intrinsic clearance, 76–77
Inverse-square law, 261, 274, 280–281, 287
Ionization detectors, 129–130
Ionization/ion chamber, 267–268
Ionizing radiation, 6–8. *See also* Radiation
activity and radioactive decay calculations, 265–267
atomic structure and isotopes, 262–263
biological effects of, 269–272
detecting and measuring, 267–269
explanation of, 260
limiting dose, 272–276
radioactive decay, modes of, 263–265
regulations and responsibilities for protection, 276
skin cancer and, 224
x-ray-generating machines, 267
Irradiance, 278, 280–281, 284
Irritant contact dermatitis, 221
Irritant smoke, 208
ISO. *See* International Standards Organization
Isoamyl acetate (banana oil), 208
Isotopes, 262–263

J

Janssen, L. L., 200
Jobs in industrial hygiene, 17–18
Job-site evaluations and walkthroughs, 333–334
JOEH *(Journal of Occupational and Environmental Hygiene)*, 19
Joint pain, 322
Joules per square meter (J/m^2), 278
Journal of Occupational and Environmental Hygiene (JOEH; trade journal), 19
Journal of Safety Research (trade journal), 20
Judicial Branch, U.S., 27
Judicial review (OSH Act, Section 11), 32

K

Kata thermometers, 303
Key events, 68
Kidneys
as elimination route, 77
toxicity and, 81–82, 92
Kiloelectron volts (keV), 263
Kinetic energy, 263
Km (enzyme activity), 76–77
Krypton-85, 275–276

L

Labeling hazards, 15, 357–358, 369
Labor, U.S. Secretary of, 32
Laboratories and biological hazards, 360–366
Laboratory analyses, 18, 152, 154–156
Laboratory studies, 69. *See also* Animals and lab studies
Labor organizations, 29–30

Labor Standards, Bureau of, 29
Labor Statistics, Bureau of, 361
Lactate dehydrogenase (LDH), 81
Lactation, as elimination route, 77
Ladders, ANSI and, 20
Ladders (OSHA standards, Subpart D 1926.1053), 36–37, 41
Laminar airflow design, 365
Landscaping, 294, 341
Larynx (voice box), 193, 194
Lasers, 15
    ANSI standard on, 53, 283
    classifications, 283
    hazards and control of, 282–283
    safety program types, 49
Latex, as dermal hazard, 224, 225, 231
Law of conservation of energy, 261
Laws. *See also* Occupational Safety and Health Act
    law-making process, 26–27
    workplace safety, evolution of, 28–29
LDH (lactate dehydrogenase), 81
LD50 (median lethal dose), 78, 79
Lead paint, 13–14
Lead shield, 275
Leaks detection, 126
    respirators and, 209, 210
Legislative Branch, U.S., 26–27
Leukemia, 286
LEV. *See* Local exhaust ventilation
Lifting
    changing work process, 13, 337
    evaluating as hazard, 11, 12–13
    motorized equipment for, 339
    musculoskeletal injury and, 325, 333
    recommended limits, 343–345
Lighting in workplace, 340
Limit of quantitation (LOQ), 120, 122
Limits of detection (LOD), 120, 122
Linear energy transfer (LET), 269–270
Lipid-soluble molecules, 75
Lipophilic chemicals, 220
Liquids, as exposure pathway, 4
Lithium-8, 265
Lithium (Li), 262–263
Liver hypertrophy, 79
Liver toxicity, 71, 75, 80–81, 92
Load index, 343–345
LOAEL. *See* Lowest observed adverse effect level
Local air extraction, 161
Local emphasis programs (OSHA), 38
Local exhaust ventilation (LEV), 161, 162–163, 169, 173
Lockout/tagout (OSHA standards, Subpart J 1910.147), 36, 37, 41
LOD (limits of detection), 120, 122
Logging, 38
Long-term planning, 56

Loop of Henle (kidney), 81
LOQ (limit of quantitation), 120, 122
Lower back strain, 325
Lowest observed adverse effect level (LOAEL), 102, 104, 105
Low-level disinfection, 367
Lumbering and thermal stress, 294
Lung cancer, 140–141
Lungs, 73, 74
    aerosols and, 139–140
    anatomy and function of, 193–194
Lung toxicity, 82–83, 92
Lymphocyte cells, 82, 86
Lysozymes, 352

M

Machine design (ergonomics), 338–339
Machine enclosures, 15
Machine guarding (OSHA standard, 1910.212), 41
Machinery
    dermal hazards and, 224
    as hazard, 10, 11
    noise control in, 251
    OSHA inspections and, 38
Macrophage activation factor assay, 87
Macrophage cells, 82, 86, 87
Macrophages, 352
Macular degeneration, 280
Magnetic field (H), 260, 261, 277, 284, 286–287
Magnetic resonance imagers, 283
Major pneumoconiosis, 140
MAKs (exposure level values of German Research Foundation), 98, 102
Malignant cancer, 223, 224, 279
Malignant neoplasms, 89
Malleus (hammer bone), 242, 243
Management
    industrial hygienists and, 18
    as part of safety team, 16
Management systems, 45–63
    advances in, 53–61
    ANSI, 53–54
    ISO, 58–60
    overview, 46–47
    policies, 48–49
    procedures, 50–51, 230
    programs, 49–50
    quality assurance, 51–52
    regulations, 47
    role of industrial hygienists in alternative solutions, 61–62
    Six Sigma, 57–58
    total quality management, 55–57
    Value Strategy Manual, 58
    Voluntary Protection Programs, 60–61
Manual of Analytical Methods (NIOSH), 150, 154–156

Manufacturer recommendations, 201
Manufacturing industry, 17
　　noise in, 238, 341
　　　Six Sigma and, 58
Manufacturing processes and ventilation systems, 161
Marburg virus, 353
Mass concentration, 150
Mass per unit volume, 99
Material Safety Data Sheet (MSDS), 67. *See also* Safety Data Sheets
Math models, 201
Maximum airborne concentration limits (of German Research Foundation), 98, 102
Maximum sweat rate (Emax), 298–299
Maximum use concentration (MUC), 206
Maximum work capacity, 315, 330–331
Measurable goals, 54
Measurements
　　accuracy and precision in, 121
　　air supply and exhaust, 183–184
　　concentration units and OELs, 99
　　duct velocity and flow, 184–185
　　of electric fields, 287
　　ergonomic hazard evaluation and, 335–338
　　hazard evaluation and, 12–13
　　radiation exposure, 267–269
　　sound level measurement, 248–250
　　thermal measurement, 301–305
　　toxicity and, 67
Meatpacking and thermal stress, 294
Mechanisms of action, 68–72
Medial (sagittal) plane, 326
Median lethal dose (LD50), 78, 79
Median nerve, 323
Medical evaluation, respiratory protection and, 192, 206–207
Medical evaluations (OSHA standards, Subpart I), 37
Medical screenings, for health care workers, 358
Medical surveillance
　　biological hazards, 358, 369
　　　decontamination and cleaning, 369
　　　OELs and, 98, 108
　　　thermal stressors and, 312, 316
Medical waste, 357–358
Megaelectron volts (MeV), 263
Melanoma, 223, 224, 279
Mercury thermometer, 302
Mercury vapor lights, 278
Merit-level worksites, 60–61
Metabolic energy costs of typical activities, 331
Metabolic heat gain, 296–297
Metabolic heat load, reducing, 311
Metabolic heat production, 310
Metabolic products, 69
Metabolic rate, 296, 310
Metabolism. *See also specific organs*
　　liver and, 80
　　skin absorption and, 218–220
　　skin and metabolic functions of, 216, 220
　　toxic agents/chemicals and, 72, 76–77
Metabolites, 76, 220
Metal and Nonmetallic Mines Safety Act (1966), 3
Metal fume fever, 140
Metal fumes, 139
Metals and skin, 221
Metastasis, 88
Meter microphones, 249
Meter weighting, 248
Methicillin-resistant *Staphylococcus aureas* (MRSA), 355
Method differences and OELs, 108
Micronucleus test, 90
Microorganisms, 350
Microphones in sound level meters, 248, 249
Micro-tears and injuries, 322, 326, 332
Microwave ovens, 283
Microwave radiation, 31, 260, 262, 277
Milk (human), 77
Millijoules per square centimeter (mJ/cm^2), 278
Milliwatts per square centimeter (mW/cm^2), 278, 284
Mineral oils, 223
Minerals and skin, 222
Mine Safety and Health Administration (MSHA), 17, 28, 204
Minimum wage, 29
Mining industry, 18, 238
MiniRAE 3000 photo ionization detector, 130
Minor pneumoconiosis, 140
Mists, 138
Mixing factor (air), 182
Mobile phones and RF radiation, 283, 285
Mode of action, 68–72
Modifications, physical, 14–15
Modifying factor, 105
Moisturizers, 232
Molar ratio, 99
Molecular absorption, 131–132
Molecular studies, 69
Molecular weight, 74, 82, 99, 114–115, 117, 219
Monitoring
　　aerosols and, 144–150
　　noise and, 12, 50, 248–250
Monocyte cells, 82, 86
Motion-sensing curtains, 15
Motorized material lift device, 339–340
Mouthpiece respirators, 202
MSHA. *See* Mine Safety and Health Administration
MUC (maximum use concentration), 206
Mucociliary escalator, 352
Mucous droplets and infection spread, 351
Mucous membranes, 194, 352
Mufflers, 253
Multigenerational studies, 70
Muscle function loss, 7

Muscle temperature, 296
Musculoskeletal disorders and injuries, 321–326
    construction site injuries, 10
    development and diagnosis of, 322–323
    ergonomic-related disorders, 323–326
    signs and symptoms of, 321–322
    statistics on, 321
    treatments for, 323
Musculoskeletal hazards, 6–7
    controlling hazards, 337–343
Mutagenicity, 90
Mycobacteria, 367
Mycotoxins, 350
Myeloid precursor cells, 86

N

NaCl aerosol, 198
Nanoparticles, 10
Nasopharyngeal region of respiratory system, 141, 142
National Academy of Sciences (NAS), 96
National Advisory Committee (OSHA), 32
National Commission on State Workmen's Compensation Laws (OSH Act, Section 27), 34
National emphasis programs (OSHA), 38
National Institute for Environmental Health Studies, 17
National Institute for Occupational Safety and Health (NIOSH), 4, 10, 17
    on aerosols and analysis of, 138, 150, 154–156
    on filter classifications, 198–199
    manual lifting guidelines from, 343
    on maximum work capacity, 330–331
    mission of, 27
    as part of OSH Act (Section 22), 34
    recommended exposure levels (NIOSH REL), 98, 101
    on respiratory protection, 195, 204
    skin notation strategy of, 101
    on toxic chemicals and dermal exposure, 227
National Institute of Environmental Health Sciences (NIEHS), 89
National Institutes of Health, 17, 362
National Labor Relations Act (1935), 29
National Safety Council (NSC), 19–20
National Toxicology Program (NTP), 87, 89–90
National Transportation Safety Board, 17
Natural killer (NK) cells, 86, 87
Natural radiant heat, 303
Nausea, heat disorders and, 300, 312
Near field, 251, 284–285, 287
Necrosis, 70, 80, 88, 271
Needles/sharps and infection control, 355
Negative air pressure, 160, 357, 365
Negative-pressure and respirators, 201–202, 207–210
Nelson, T. J., 200
Neoplasia and neoplasms, 88–89
Nerve cell anatomy, 84
Nervous system, 83–84
    skin exposure and, 223
Net radiometer, 304
Neurological toxicity and effects, 69, 70–71, 83–84, 92
Neurotransmitters, 84
Neutron activation, 264
Neutrons, 262–264
Neutrophils, 352
Nickel, 140
    as dermal hazard, 224
Nicotene, 78
NIEHS (National Institute of Environmental Health Sciences), 89
NIOSH. *See* National Institute for Occupational Safety and Health
Nitrogen-14, 264
NK (natural killer) cells, 86, 87
NOAEL (no observed adverse effect level), 102–104
Noise, 7, 8, 237–257. *See also* Hearing
    administrative controls in noise reduction, 254–255, 341–342
    controlling noise levels, 250–255
    ear anatomy, 242–243
    as hazard, 6, 10, 11
    hearing damage and, 243–244
    meter microphones, 249
    meter weighting, 248
    monitoring noise levels, 12, 50
    noise dosimetry, 249
    OSHA standards on (Subpart G), 37
    overview, 238
    personal protective equipment for, 16
    protecting workers' hearing, 244–247
    response time, 248–249
    science of sound creation and propagation, 239–242
    sound barriers and, 15, 253–254
    sound level measurement, 248–250
Noise assessment and control jobs, 18
Noise Reduction Rating (NRR), 254–255
Nonabsolute respirator filters, 198
Nonallergic contact dermatitis, 87
Nonionizing radiation, 7. *See also* Radiation
    broadband optical radiation hazards, assessment and control of, 280–282
    explanation of, 260
    IR radiation sources and hazards, 280
    laser hazards and controls, 282–283
    nonionizing electromagnetic radiation bands, 277
    RF and subradiofrequency fields, assessing and controlling, 286–287
    RF sources and hazards, 283–285
    subradiofrequency fields, 286
    UV radiation sources and hazards, 278–279

visible radiation hazards from nonlaser sources, 279–280
Nonlipid or small viruses, 367
Non-water soluble aerosols, 139
No observed adverse effect level (NOAEL), 102–104
Normal temperature and pressure (NTP), 114
Nose and nasal cavity, 193, 194
Nostrils/nares, 194
Notices of Proposed Rulemaking, 26, 35
NRC. *See* Nuclear Regulatory Commission
NSC (National Safety Council), 19–20
N-series filters, 198
NTP (National Toxicology Program), 87, 89–90
NTP (normal temperature and pressure), 114
Nuclear energy and materials, 28, 31
Nuclear Regulatory Commission (NRC), 17, 28, 31, 270, 272–273, 276
Nucleus, 262–263
Nuclides, 262, 264, 266
Nuisance particles, 82
Numbness, 301, 322
Nurses, 16
Nursing Homes, Guidelines for: Ergonomics for the Prevention of Musculoskeletal Disorders (OSHA), 341

O

Obesity, 311, 313
Occupational Alliance for Risk Science (OARS), Workplace Environmental Exposure Levels, 98
Occupational Ergonomics Technical Group (of Human Factors and Ergonomics Society), 20
Occupational exposure, 4–5
Occupational exposure limits (OELs), 95–112
    airborne OELs, of various organizations, 98
    ceiling limit and, 100
    concentration units and, 99
    decontamination and cleaning methods, 368
    dermal hazards and, 225
    fundamentals of OEL derivation, 102–106
    hazard notations and, 101–102
    industrial hygienists and, 67–68
    key components of, 97–102
    noise and, 244–247
    OEL selection process, 107
    oral-dosing studies and, 74
    radiation and, 273, 286, 287
    resources on, 109–110
    respirator selection and, 205, 206
    short-term exposure limit, 100, 105
    time-weighted average and, 97–100
    use and interpretation, 106–109
    workplace risk assessment and, 96–97
Occupational Health and Safety Management (British OHSAS 18001), 60
Occupational Health and Safety Management Systems (ANSI Z10), 54
Occupational medicine physicians and nurses, 16
Occupational noise exposure (OSHA standards, Subpart G 1910.95), 36–37
Occupational Safety and Health Act (OSH Act; 1970), 3, 24, 30–34
Occupational Safety and Health Administration (OSHA), 30–44
    on aerosols, 150, 151, 154
    Agreement States and, 28
    ANSI and, 20–21, 53–54
    on biological lab safety, 360–361
    on bloodborne pathogens, 36, 37, 350, 354, 355–356
    careers and, 17
    Department of Labor and, 25–26
    on dermal exposure and assessment, 228
    enforcement activities, 37–39
    on ergonomics programs, 341
    Fourth Amendment and, 25
    general industry standards, most frequently accessed, 36
    hazards database, 10
    mission of, 27
    on musculoskeletal injuries/illnesses, 332
    on noise exposure, 244–246
    OSH Act, 30–34
    permissible exposure limits (PELs), 98
    on radiation, 273
    regions of, 38
    resources from, 42–43, 201
    on respiratory protection, 200–201, 204–210
    rule development, 34–35
    rule types, 35–37
    on Safety Data Sheets, 67
    standards, 34–43
        most frequently cited, 41
    statistics on injuries, illnesses, deaths, 39–41, 43, 321
    training by, 4
    Voluntary Protection Programs, 60–61
    website of, 41, 42–43, 341
Occupational Safety and Health Review Commission (OSHRC), 32, 33, 39
Odors, ventilation and, 160
Odor threshold, 120
OELs. *See* Occupational exposure limits
Office of Management and Budget (White House), 35
Office of Workers' Compensation, 29
OHSAS 18001, Occupational Health and Safety Management (British Standards Institute), 60
Oil aerosols, 198–199
Oil/gas well jobs and thermal stress, 294
Oil/grease and skin, 223, 224
Opening conference (OSHA), 39
Optical radiation, 277, 280–282. *See also specific types*
Optical technology, 154

Oral exposure, 74–75, 78
Orbitals, 262
Organ damage, 66, 70
Organic solvents and cleaning, 221–222, 232
Organizational ergonomics, 321
Organizational objectives and goals, 48, 53, 54
Organizational stressors, reducing, 342
Organizational structure, 56
Organophosphate pesticides, 223
OSHA. *See* Occupational Safety and Health Administration
OSHRC (Occupational Safety and Health Review Commission), 32, 33, 39
Ossicles, 242
Other potentially infectious materials (OPIM), 350, 355
Outdoor air intake, 161
Outer ear, 242, 243
Outerwear and thermal stress, 315–316
Oval window (ear), 242
Ovaries, radiation effect on, 272
Overexertion, 321
Overtime pay, 29
Overtime reduction, 311, 342
Ovulation, 85
Oxygen and ventilation, 160
Oxygen displacement, 6
Ozone, 279

P

Painters, dermal hazards and, 224
Pain threshold and hearing, 230
Pandemics, 354
PAPR (powered air-purifying respirator), 202–203, 205
Paracelsus, 66
Partial inspections (OSHA), 38
Particle deposition, 141–144
Particles
 respiratory protection and particle size, 195–198
 respiratory tract and, 73–74, 82–83, 141–144
Partition coefficients, 75–76
Pascal (Pa) and sound, 240–241
Passive diffusion, 74, 75
Pathological measures, 69, 70
Pathways of exposure
 disinfectants and sterilants, 367–368
 infectious diseases and, 351
 overview, 4–5
 skin and, 222, 223, 225–228
PBPK (physiologically based pharmacokinetic) modeling, 102, 103, 106
Peak blood concentration, 75
PELs. *See* Permissible exposure limits
Penalties (OSH Act, Section 17), 33, 39
Percent bioavailability, 75
Percutaneous absorption, 216, 218–219

Performance standards, 35
Peripheral nervous system, 83, 314
Permeation and skin, 217–220
Permissible exposure limits (OSHA standards, Subpart Z 1910.1000), 37
Permissible exposure limits (PELs)
 dermal exposure and, 228
 noise and, 245
 of OSHA, 98, 120
 respirator selection and, 205, 206
Permit-required confined spaces (OSHA standards, 1910.146), 36, 37
Personal dosimetry, 126–127
Personal protective equipment (PPE), 16–17
 biosafety and, 363
 clothing, 315–316
 ergonomics and, 345
 noise control and, 238, 254–255
 OSHA standards and, 35–37
 respirators, 50–51, 192–193, 201–210
 skin and, 219, 222, 225, 230–232
 surrogate skin techniques and, 226
 training on proper use, 231
 for UV radiation, 281–282
Personal sampling, 126
Perspiration. *See* Sweating
Pesticides and skin, 221–224
Petroleum products, 224
Phagocytosis, 352
pH and GI tract, 74
Pharmaceuticals, dermal hazards and, 224
Pharynx (throat), 193, 194
Photochemical eye damage, 279–281
Photo-curing lamps (UV), 278
Photodisruption, 282
Photo ionization detectors, 130
Photokeratitis (corneal inflammation), 279
Photometers, 130
Photons, 261, 282
Photosensitivity (chemical induced), 87, 88
Phototoxic contact dermatitis, 221
Physical abnormalities, neurotoxicity and, 84
Physical activity and thermal stress, 310
Physical barriers, 229
Physical ergonomics, 320–321
Physical fitness of workers, 313–315, 343
Physical hazards, 6, 7, 66
 skin and, 223
Physiologically based pharmacokinetic (PBPK) modeling, 102, 103
Physiological monitoring and thermal stress, 308, 309–310
Pinna, 242, 243
Pitot tubes, 183, 184–185
Plan-do-study-act (PDSA) cycle, 57
Plants and skin, 221, 222
Plasma arc welding, 282

Plasma volume, 314
Plastic manufacturing, dermal hazards in, 224
Plenum return system, 161
Pleura, 194
Pneumoconiosis, 139
Pocket Guide to Chemical Hazards (NIOSH), 227
Point of departure (POD), 102, 104–105
Poison and poison exposure, 8. *See also* Toxicology
Poisonous plants, 5
Policies, management, 48, 51–52
Policies and administrative controls, 15
Policy and regulation development/analysis jobs, 18
Polycyclic aromatic hydrogen compounds, 223
Portable gas chromatography, 132
Portal of entry, 70
Positive pressure and respirators, 202–205, 207, 210
Positron emission tomography (PET) scans, 263
Positrons, 263
Postings in workplace, 32, 39
    OSHA standards (Subpart J 1910.146), 37
Postnatal effects, 69, 70
Posture, as hazard, 6–7, 332–333, 336–338
Posture metabolism, 296, 310
Potassium-40, 265
Potassium cyanide, 78
Power density, 284, 286–287
Powered air-purifying respirator (PAPR), 202–203, 205
Powered industrial trucks (OSHA standards, 1910.178), 36, 41
Power tools (ergonomics), 340
PPE. *See* Personal protective equipment
Precursor cells, 86–88
Predicted 4-hour Sweat Rate (P4SR), 309–310
PremAire Cadet Escape system, 204
Prepackaged substances, 228
Presbycusis (hearing loss), 243
Prescriptive Process-Based OELs, 110
Pressure-demand mode and respirators, 205
Pressure gauges, 184
Pressure-relief valves, 229
Prevention. *See* Administrative controls; Engineering controls
Primary barriers (biosafety), 363
Primary calibration standards, 146
Printing/lithography, dermal exposure in, 224
Prioritizing, evaluating hazards and, 12, 13
Probability and risk, 9
Problem formulation, 102
Procedures for safety programs, 15, 50–52
Proceedings to Counteract Imminent Dangers (OSH Act, Section 13), 33
Process-centered (TQM), 56
Process mapping, 11–12
Productivity, 30
Professional development, 26–27
Professional organizations, 18–21

Professional Safety (trade journal), 19
Profitability, 29, 30, 47, 55, 61
Program administrators, 192
Programmed inspections (OSHA), 38
Program reviews, 51
Programs, in management system, 49–50, 52
Programs and administrative controls, 15
Progression (carcinogenesis), 89
Promotion (carcinogenesis), 89
Pronation, 327, 333
Proposed rulemaking, 26, 35
Protective measures, 14–15
Protons, 262–263
Proximal tubule (kidney), 81
P-series filters, 199
Psychosocial conditions in workplace, 7
Psychrometric chart, 305
Public comment, 35
Public Health Service, 17
Public utilities industry, 17
Pulmonary/alveolar region of respiratory system, 141, 142
Pulmonary function testing, 82, 316
Pump flow rates, 119–122, 145–147, 149
Pumps. *See* air-sampling pumps
Pupil dilation, 84
Purpose (OSH Act, Section 2), 30–31
Pustules, 221, 222
Pyranometer, 304
Pyrheliometer, 304

Q

Qualitative exposure analysis, 226
Qualitative fit-testing, 207–208
Quality assurance, 51–52
Quality assurance jobs, 18
Quality control
    Six Sigma and, 57–58
    total quality management, 55–57
Quality control (TQM), 56
Quality Management (ISO 9001), 59
Quantitative analysis, 337
Quantitative fit-testing, 207, 209
Quantitative health-based OELs, 110
Quantitative sampling and exposure assessment, 226–227
Quiescent cells, 88

R

Rad (absorbed dose measurement), 268, 270
Radar systems and RF radiation, 283
Radial blade fan, 176
Radiant exposure, 278
Radiant heat, measuring, 303–304
Radiant heat exchange, 297
Radiant heat load, 310
Radiant temperature, 296

Radiation, 259–292. *See also* Ionizing radiation; Nonionizing radiation
    ANSI standard on, 54
    atomic structure and isotopes, 262–263
    cancer and, 224, 272, 273, 277, 286
    as disinfectant, 367
    electromagnetic radiation, basics of, 260–262
    fatalities and, 271–273
    federal regulatory agencies and, 28, 31, 270, 272–273, 276
    ionizing radiation, 260, 262–276
    microwave radiation, 31, 260, 262
    nonionizing radiation, 260, 277–287
    overview, 260
    photometers and, 132
    radioactive decay, 263–267
    safety program types, 49
Radiation Safety for Personnel Security Screening Systems Using X-Ray or Gamma Radiation (ANSI N43.17), 54
Radiation safety jobs, 18
Radiation Safety Officer (RSO), 276
Radioactive decay, 263–267
Radiofrequency radiation (RF), 31, 277, 283–287
Radiological/physical half-life (T1/2phys), 265, 266–267
Radiometry/radiometer, 280, 304
Radionuclides, 263–267, 269, 275–276
Radon-222, 264
Radon progency, 264–265
Range of motion, 336
Rashes, 8, 223, 300
Rate of absorption, 218
Rate of work, skin absorption, effect on, 218–219
Rats and chemical exposure studies, 78, 85, 86
Raynaud's phenomenon/syndrome, 301, 325
Reach, design for, 329–330, 339
Reaching, as hazard, 332, 336–337
Reactive chemicals, 7
Realistic goals, 54
Real-time detection instruments, 126–132
Recirculated air, 161
Recognition of hazards, 10–12, 333–334
Recommended exposure levels (RELs), of NIOSH (NIOSH REL), 98, 101, 227
Recommended weight limit (RWL), 343–345
Record keeping
    data sheets, 335
    on fumigation operations, 368
    OSHA inspections and, 39
    on workplace fatalities and catastrophes (OSHA), 43
Recycled air, 162
Regional emphasis programs (OSHA), 38
Registers/grilles, for ventilation, 161
Regulations
    absence of, 9
    ANSI and, 20–21
    on bloodborne pathogens, 355–356
    dermal pathway exposure and, 228
    federal agencies and, 27–28
    hazard evaluation and, 12
    on ionizing radiation, 276
    lasers and, 282–283
    management systems and, 47
    for medical and biological laboratories, 360–361
    on medical waste disposal, 357–358
    new federal regulations, 26–27
    OSHA rule development, 34–35
    program reviews and, 51
    risk and, 9
    ventilation systems and, 161
Regulatory text, 26
Relative humidity, measuring, 305
RELs. *See* Recommended exposure levels
Remote temperature, 302
Removal techniques, 226–227
Renal system functioning, 316
Repetition or rate, as hazard, 6–7, 321, 331, 332, 335, 343. *See also* Musculoskeletal disorders and injuries
Reproductive and developmental effects/toxicity, 69, 70, 85–86, 93
Research
    dermal exposure studies, 226
    on RF radiation, 285
Research and Related Activities (OSH Act, Section 20), 33–34
Research/education industry, 17
Research jobs, 18
Resins and skin, 221
Respirable particulate concentration, 152–156
Respirable particulate mass, 142
Respiration, stages of, 194
Respirator filtration, 195–198
Respirators, 50–51, 192–193, 201–210
    in infection control, 357
Respiratory protection, 191–212
    airborne contaminants, 195
    atmosphere-supplying respirators, 203–204
    combination respirators, 204
    filters, 198–199
    fit-testing procedures, 207–210
    in health care setting, 357
    lungs, anatomy and function, 193–194
    medical evaluation, 206–207
    OSHA standards on, 192–193
    overview, 192
    qualitative fit-testing, 207–208
    quantitative fit-testing, 207, 209
    respiration, 194
    respirator filtration, 195–198
    respirators, types of, 201–204
    respirator selection, OSHA standards on, 204–206

seal check, 210
vapor- and gas-removing respirators and removal mechanisms, 199–201
Respiratory protection (OSHA standard, Subpart I 1910.134), 36, 37, 41, 192–193, 204–206
Respiratory protection programs, 49, 50–51, 192–193
Respiratory system
    aerosols and, 73, 82, 141–142
    anatomy of, 193–194
    hypothermia and, 300
    lung toxicity, 82–83, 92
    particle size and, 73–74, 82–83, 141
    pathogens and, 352
    regions of, 141
    sensitization and, 101
Respiratory tract sensitization (RSEN), 101
Responsibility, determining, 50
Rest for workers, 311, 312, 331–332, 342
Resting heart rate, 309
Restricted access, 230
Retail Grocery Stores, Guidelines for: Ergonomics for the Prevention of Musculoskeletal Disorders (OSHA), 341
Retail Grocery Stores, Guidelines for: Prevention of Musculoskeletal Injuries in Poultry Processing (OSHA), 341
Retail industry, 321
Retina damage, 278–282
Reverberant field, 251
Reviews of Human Factors and Ergonomics (trade publication), 20
Risk
    acceptable levels of, 9, 13
    Six Sigma and, 57
Risk analysis, 58
Risk assessment, 8–9. *See also* Occupational exposure limits
    for biological laboratories, 361–366
    dermal hazards and, 225–227
    in health care setting, 357
    OELs and risk assessment and management, 96–97, 105, 108–109
    physical skin damage and, 223
Risk-based level, 102
Risk characterization, 96, 97, 106–109
Risk group classifications (infectious microorganisms), 362
Risk Management (ISO 31000), 59
Risk-tolerance differences and OELs, 108
Roadway incidents, fatalities by, 40
Roentgen-equivalent man (rem), 270
Roentgen (R) and roentgen per hour (R/h), 268
Rotameter, 148–150
Rotating vane anemometer, 302–303
Rotational speed and noise reduction, 251
Rotator cuff damage, 325–326
Routes of exposure

    controlling skin exposure, 228–232
    infectious agents and, 351–352, 363
    ingestion as, 5, 351–352
    inhalation and, 5, 70, 73, 102, 352
    Safety Data Sheets, 67
    skin as, 5, 214, 223
    toxicology and, 73–75
RSEN (respiratory tract sensitization), 101
R-series filters, 199
RSO (Radiation Safety Officer), 276
Rubbers and skin, 221, 222, 225
Rule development and types (OSHA), 34–37. *See also* Regulations
RWL (recommended weight limit), 343–344

S

Saccharin, 208
SAE (sampling and analytical error), 120–121
Safe concentration, 109
Safe dose determination, 103–105
Safety. *See* Workplace safety
Safety centrifuge cups, 363
Safety culture, 46–47
Safety Data Sheets (SDSs), 11, 67, 335
Safety+Health (trade publication), 20
Safety policy sample, 49
Safety professionals, 16
Safety representatives/teams, 15–16
Safety Technical Group (of Human Factors and Ergonomics Society), 20
Safe Use of Lasers in Manufacturing Environments (ANSI Z136.1-2014), 53
Salt/electrolytes, 313–314, 316
Salt (sodium chloride), 78
Same-level positioning, 339
Sampling, 226. *See also* Aerosols
Sampling and analytical error (SAE), 120–121
Sampling duration, 122
Sampling train, 145, 152
Sanitation worker hazards, 7
Sarcoma, 89
SAR (supplied-air respirator), 205
Saturation, skin and protective clothing, 219
Scaffolding (OSHA standards, Subpart D), 36–37, 41
Scapula, 325
SCBA (self-contained breathing apparatus), 203–205
Schedules, change schedules and respirators, 200–201
Sciences, 3
Scintillation detectors, 268
SDH (sorbitol dehydrogenase), 81
SDSs (Safety Data Sheets), 11, 67, 335
Seal check, 208, 209–210
Sealed sources (radioactive materials), 275
Secondary barriers (biosafety), 363
Secondary calibration standards, 146
Second-priority inspections (OSHA), 38
Sedimentation, 141, 143, 218

Sedimentation filtration, 196, 198
Self-closing doors, 363
Self-contained breathing apparatus (SCBA), 203–205
Semiquantitative methods of exposure assessment, 226
Senate, U.S., 24, 26
Sensation, in air measurement, 303
Sensitization, 69, 71, 83, 101, 221, 367
Sensitizing chemicals, 7
Sensors, electrochemical, 128–129
Service life, 200–201
Severe Acute Respiratory Syndrome (SARS), 355
Severity and risk, 9
Sexual reproduction, 85
SG (specific gravity), 116
Shewhart cycle, 57
Shielding, in radiation protection, 274, 275, 281–282
Shipbreaking, 38
Shipyards, Guidelines for: Ergonomics for the Prevention of Musculoskeletal Disorders (OSHA), 341
Shivering, cold disorders and, 300
Short-term exposure limit (STEL), 99, 105
Siderosis, 140
Sievert (Sv), 270
Signage, hazards and, 15
Silica, 38, 140
Silica gel beads, 125, 126
Silicosis, 8, 140
Sinuses, 193
Six Sigma, 57–58
Sixth Amendment (U.S. Constitution), 25
Skin. *See also* Dermal hazards
    absorption process, 216–220
    aerosols and, 142
    anatomy and functions of, 214–216
    chemical exposure and, 73–74, 101, 214
    controlling skin exposure, 228–232
    damage to, 223
    as elimination route, 77
    as exposure route, 5, 214, 223
    ionizing radiation and, 271
    metabolic functions of, 216
    nonionizing radiation and, 278, 279
    occupational injuries and illnesses, 214, 220–224
    pathogens and, 352
    pathway exposure assessment, 225–228
    rashes, 8, 223
    sample skin surface areas for adult males, 214
    skin care products, 232
    thermal stress and, 301, 313
Skin cancer, 223–224, 277, 279
Skin notations, 101, 227
Skin toxicity, 87–88, 93, 101
Sling psychrometer, 305–306
Slips and trips, 12

Slow-response sound meters, 248
SMART Objective approach, 54
Smokes, 139
Smoke tubes, 183, 184
Soap-film flowmeter, 146
Sodium chloride (table salt), 78
Solar radiation, measuring, 304
Sole proprietors, 31
Solubility, 138, 220
Solvents, as dermal hazard, 221–222, 224
Sorbents, 125–126, 199
Sorbitol dehydrogenase (SDH), 81
Sorption, 199
Sound barriers, 15, 253–254
Sound creation and propagation, 239–242
Sound intensity comparisons, 241
Sound level measurement and meters, 248–250
Sound pressure level (SPL), 240–242, 244–245, 250–253
Sound surveys, 249
Sound waves, properties of, 239
Special emphasis programs (OSHA), 38
Special enforcement areas (OSHA), 41
Special equipment (ergonomics), 339–340
Special fans, 176
Specific absorption rate (SAR), 285, 286
Specification standards, 35
Specific goals, 54
Specific gravity (SG), 116
Specific heat, 295
Specificity (Hill's criteria), 71
Spectral weighting, 280
Spectrochemical direct-reading instruments, 130–132
Speed of light, 260
Sperm, 85, 272
Spider veins and radiation, 271
Spills and cleanup of, 16, 229, 230
Spinal damage, 7, 325. *See also* Neurological toxicity and effects
Spirometry, 82
Splash guards, 229
Spray-painting isolation booth, 229
Squamous cell carcinoma, 279
Stable isotopes, 263
Stairs (OSHA standards, Subpart D), 36–37
Standards
    ANSI, 20–21, 53–54
    on bloodborne pathogens (OSHA), 36, 37, 354, 355–356
    on dermal exposure, 227–228
    ISO, 58–60
    on noise exposure, 244–247
    OSHA, most frequently cited, 41
    on radiation, 273
Stannosis, 140
Stapes (stirrup bone), 242
Staphylococcus aureas (staph), 355

Star-level worksites, 60–61
State government and occupational safety, 28–29
State Plans (OSH Act, Section 18), 33
Static anthropometric data set on U.S. adults, 328, 329
Static pressure (SP), 165–167, 169
Stationary air compressors-Safety rules and code of practice (ISO 5388:1981), 59
Statistical analyses, 57–58
Steatosis, 80
Steiger, William A., 30
STEL (short-term exposure limit), 99, 105
Sterility and radiation, 272, 285
Sterilization (disinfection), 358, 367
Stirrup bone (stapes), 242
Storage equipment for loose bulk materials-Safety code (ISO 8456:1985), 59
Strategic planning, 56
Stratum corneum, 74, 216
    thickness by anatomical site, 218
Strength of association (Hill's criteria), 71
Stress management, 321
Stroke, 7
Strontium-90, 264
Studies, types of, 69
Subcuteneous tissue, 214, 215
Subradiofrequency fields, 277, 286–287
Suction techniques, 227
Suffocation, 7
Summa canisters, 123
sunburn (erythema), 271, 279
Sunlight, 260, 278, 280
Superior (anatomy), 326
Supervisors
    responsibilities of, 50
    total quality management and, 55
Supination, 327, 333
Supplied-air respirator (SAR), 205
Supply air, 160–161
    supply system components, 161–164
Supreme Court, U.S., 27, 32
Surface area and skin, 214, 217, 218
Surface contact and infection spread, 351
Surface contamination, 218, 222, 229
Surface temperature, 302, 304
Surfactants, 228, 232
Surrogate skin techniques, 226
Surveillance, 34, 52
Sweat glands, 214, 215
Sweating
    chemical absorption and, 219
    as elimination route, 77
    heat exchange and, 296, 298
    pathogenic agents in sweat, 352
Sweat rate, 298, 309–310, 312
Swelling (edema), 301, 322, 333
Swinging vane velometer, 302–303
Syncope, 7
The Synergist (trade publication), 18–19
Systematic approaches (TQM), 56
Système Internationale [SI] (International System of units), 265
Systemic diseases and skin exposure, 223
Systemic infections, 353

T

Tanning lamps, 278
Tar and skin, 223
Task load and lifting, 343–344
T cell proliferative response to antigens assay, 87
Technecium-99m, 264
Technical Manual (OSHA), 38, 150, 151, 154
    on skin and contaminated surface monitoring, 228
Technology, nanoparticles as hazard, 10
Tedlar bags, 123
Temperature, 295. *See also* Thermal stressors
    as skin hazard, 223
    ventilation and, 160
Temperature extremes, 6, 333. *See also* Thermal stressors
Tempered air, 160, 162
Temporality of events (Hill's criteria), 72
Tendinitis, 322, 324, 333
Tennis elbow (epicondylitis), 326
Tenth Amendment, 28
Teratogenicity, 69, 70, 86
Terrain, effect on ventilation, 161
Thalidomide, 85
Thermal anemometer, 185
Thermal balance, 295
Thermal conditions, 7
Thermal eye damage, 279–281
Thermal measurement, 301–305
Thermal stressors, 293–318
    adjusted air temperature, 308
    administrative and work practice controls for, 311–312
    air speed/wind speed measurements, 302–303
    ambient air temperature, 302
    body heat storage load, 295–296
    clothing, 315–316
    convection heat exchange, 298
    engineering controls for, 310–311
    evaluating thermal stress, 306–310
    evaporative heat loss (E), 298–299
    harmful effects of, 299–301
    heat, enhancing tolerance to, 312–313
    heat-alert program, 316
    heat and cold disorders, 299–301
    heat exchange, 295–299
    medical surveillance, 312, 316
    metabolic heat gain, 296–297
    overview, 294–295

physical fitness, 313–315
physiological monitoring, 308, 309–310
radiant heat, 303–304
radiant heat exchange, 297
relative humidity, 305
thermal balance, 295
thermal environment control methods, 310–316
thermal measurement, 301–305
training, 316
wet-bulb globe temperature index, 306–307
windchill index, 308
Thermistor, 302
Thermoanemometers, 303
Thermochemical detectors, 128
Thermochemical heat, 128
Thermocouple, 302
Thermoelectric thermometer, 302
Thermoluminescent dosimeters, 269
Third-priority inspections (OSHA), 38
Thoracic outlet syndrome, 324
Thoracic particulate mass, 142
Thorium, 276
Three-stage cassette, 151–152
Threshold dose, 79
in UV radiation, 279
Threshold Limit Values for Chemical Substances Committee, 19
Threshold Limit Values (TLVs), 19
noise exposure and ACGIH, 246–247
OELs and ACGIH, 98, 99–100, 102
permissible heat exposure, 306–307
skin notations and ACGIH, 227
UV radiation exposure, 360
Threshold screening, 208
Throat anatomy, 193, 194
Tidal volume (TV), 82
Time, in radiation protection, 274, 275, 281
Time-bound goals, 54
Time-weighted average (TWA), 97–100
Tingling, as symptom, 301, 322
Tissue damage, 66. *See also* Toxicology
Titanium dioxide, 82
TK assay, 90
TLVs. *See* Threshold Limit Values
TLV (total lung volume), 82
T lymphocyte cells, 86
T max rate (chemical absorption), 75
Toluene, 220
Tools (ergonomics), 334, 340
Total external dose, 75
Total lung volume (TLV), 82
Total particulate and mist concentration, assessing, 150–152
Total pressure (TP), 167–169
Total quality management (TQM), 55–57
Total sampled volume, 145–146
Toxicity, 6
dermal exposure and toxicity, 225, 226
lead paint and, 13–14
nanoparticles and, 10
Toxic materials
biosafety cabinets and, 363–366
decontamination and cleaning, 367–369
ducts and drawing toxic air into, 170
industrial hygienists and solutions for, 61
ventilation and, 160
Toxicokinetics, 72
Toxicology, 65–94. *See also* Aerosols
absorption, 75, 227
biomarkers of effects, 69–71
carcinogenesis and genotoxicity, 88–90, 93
distribution, 75–76
dose/response, 78–80
elimination, 77–78
immunotoxicity, 86–87, 93
kidney toxicity, 81–82, 92
liver toxicity, 80–81, 92
lung toxicity, 82–83, 92
mechanisms of action, 68–72
metabolism, 76–77
neurotoxicity, 83–84, 92
overview, 66
reproductive and developmental toxicity, 69, 70, 85–86, 93
resources on, 91
routes of exposure, 73–75
skin toxicity, 87–88, 93
toxic agent examples, organ system and effects on, 92–93
toxicokinetics, 72
understanding, importance of, 66–68
weight of evidence, 71–72
Toxic substances exposure (OSHA standards, Subpart Z), 37
TQM (Total quality management), 55–57
Tracer gases, 183, 184
Trachea (windpipe), 193, 194
Tracheobronchial region of respiratory system, 141, 142
Tractor manufacturing, 11
Trade publications, 19–20
Training and Employee Education (OSH Act, Section 21), 34
Transcellular permeation, 216, 217
Transient erythema, 271
Transportation and warehousing industries, 321
Transportation incidents, fatalities by, 40
Transportation industry, 17
noise in, 238
Transverse (horizontal) plane, 326
Traverse measurements, 184–185
Tremors, 84
Trench foot, 301
Triangle Shirtwaist factory fire (New York City; 1911), 28–29

Trigger finger, 325
Tritium (H-3), 266, 276
TSI Porta-Count, 209
Tuberculosis, 140, 351, 355, 357
Tubular element (kidney), 81
Tumors, 88–90
Turbinates, 194
Turpentine, air sampling for, 120
TV (tidal volume), 82
TWA. *See* Time-weighted average
Twisting of body, 333
Tympanic membrane (ear drum), 242, 243
Tympanic temperature, 309
Type 1 and type 2 microphones, 249

U

Ultrasonic/sonic vibration, as disinfectant, 367
Ultraviolet photometers, 130, 132
Ultraviolet radiation, 224
    protective skin creams for, 232
Ultraviolet (UV) radiation
    assessment and control of hazards, 280–282
    burns from, 277
    carcinogenesis and DNA damage, 89
    on electromagnetic spectrum, 260, 262
    in health care industry, 360
    laser hazards, 282
    measuring, 304
    sources and hazards, 278–279
Uncertainty factors (UFs), 103–105, 108
Universal precautions, 356
Unstable isotopes, 263
Uranium, 276
Uranium-238, 264
Urine and urinalysis, 69, 81–82
Urticaria, 222
UV dose calculation, 278
UV light-emitting diodes (ULEDs), 278

V

Vaccinations and workers, 354
Value Strategy Manual, 58
Vaneaxial fans, 175
Vaporization, heat of, 295
Vaporized peracetic acid, 367
Vapor pressure and density, 115–116. *See also* Gases and vapors
Vapors, as exposure pathway, 4. *See also* Gases and vapors
Vascular element (kidney), 81
VC (vital capacity), 82
Vectors, 4–5
Vegetative microorganisms, 367
Velocity pressure (VP), 167–169
Ventilation, 15, 35, 159–190. *See also* Gases and vapors
    aerosols and, 156–157
    air cleaning devices, 178–180
    in dermal exposure control, 229
    ducts, moving air in, 165–175
    fan laws, 175–178
    general ventilation systems, 180–182
    in health care environment, 356–357
    overview, 160–161
    supply system components, 161–164
    testing ventilation systems, 182–185
Ventilation systems (OSHA standards, Subpart G), 37
Vertical flow clean benches, 366
Vertical standards, 35
Vertigo, heat disorders and, 300
Very high frequency band (VHF), 285
Veterinary/animal husbandry hazards, 7, 350, 369
Vibration, as hazard, 6, 7, 325, 333, 336
Vibrations and noise reduction, 251, 252, 340
Vibration syndrome, 7, 325
Video imaging, 227, 335
Violations (OSHA), 32, 33, 39, 40–41
Violence and injuries by persons or animals, fatalities by, 40
Virulence, 353
Viruses, 350, 353, 355, 366, 367
Visible-light photometers, 130, 132, 133
Visible radiation (light), 260, 262, 277, 304
    hazards from nonlaser sources, 279–280
Vision impairment/loss, 7, 279–280, 282
Vital capacity (VC), 82
V max rate (enzymes and metabolic capacity), 76–77
Voice box, anatomy of, 193, 194
Volatility and chemicals, 214, 219
Volumetric flow rate, 163–165
Voluntary Protection Programs, 60–61
Voluntary Protection Programs (VPPs), 60–61
VPPs (Voluntary Protection Programs), 60–61

W

Wages, 29
Walk-behind powered rotary tillers-Definitions, safety requirements and test procedures (ISO 11449:1994), 59
Walking and working surfaces (OSHA standards, Subpart D), 36–37
Walsh-Healy Act (1936), 29
Warming shelters, 312
Washing and chemical exposure, 218, 220
Waste management industry, 18
Wastes, 11, 357–358, 369
Water, as exposure pathway, 351
Water-cooled garments, 315
Water flow and noise reduction, 252
Water intake, 311, 314
Water-reactive aerosols, 139
Waters, M., 98
Water-soluble gases, 73
Water vapor pressure in air, 296, 305

Watts per square meter (W/m2), 278, 284, 295
Watts (W) and sound, 241
Wavelengths, 239–240, 260–262, 277
WEELs (Workplace Environmental Exposure Levels), 98
Weight of evidence, 71–72
Welding arcs, 278, 280, 282
Welding fumes, 140
Wet-bulb globe temperature index, 306–307
Wet scrubbing filtration, 179, 180
Wet work, 224, 225
    effect on skin, 217
Whistleblowers, 32
White finger syndrome, 301, 325
Whole-body dose, 271
Whole-body dose or exposure, 273, 286
Whole-body heating, 285
Wholesale trades industry, 321
WHO (World Health Organization), 309, 362
Williams, Harrison A., Jr., 30
Williams-Steiger Occupational Safety and Health Act. *See* Occupational Safety and Health Act
Windchill index, 308
Wind speed/air speed, measuring, 296, 302–303
Wireless communications and RF radiation, 283, 285, 286
Witness testimony, 27
Workday/shift reduction, 311, 342
Worker job rotations, 342
Workers' Compensation, 29, 34
    hearing loss and, 238, 247
    musculoskeletal injuries, 321
Work heat, 296
Working provisional OELs, 110
Workload, 12–13
Workplace Environmental Exposure Levels (WEELs), 98
Workplace safety. *See also* Management systems; Occupational exposure limits; Occupational Safety and Health Administration; Toxicology
    acceptable levels or risk and, 9, 13
    goal-setting, 15, 48, 53, 54
    history and evolution of, 28–29
    labor organizations and, 29–30
    in medical and biological laboratories, 361–362
    NIOSH on, 34
    OSH Act and, 3, 24
    research on, 33–34
Workplace/workspace design, 320, 328–330, 337–341
Work practice controls, 230, 311–312
Work Practices Guide for Manual Lifting (NIOSH), 343
Work processes
    aerosol source control, 156–157
    changing, 13–14
    dermal exposure and, 228
    improving safety of, 61–62, 320
    knowledge about, 11
    personal protective equipment and, 16–17, 345
    reducing thermal stress, 311
    workplace and work process redesign, 337–338, 340
Workstation ergonomic procedures, 341
World Health Organization (WHO), 309, 362
Wounds and cuts, as exposure route, 352
Wrist vibration measurement, 336
Written procedures
    ergonomics program, 341
    on hazardous material handling, 230
    heat-alert program, 316

X

Xenobiotics, 80
Xenon lamps, 278
X-ray fluorescence instruments, 264
X-ray-generating machines, 267, 276
X-rays, 261. *See also* Radiation
    detecting and measuring radiation, 267–269

Y

Yttrium-90, 264

Z

Zinc oxide, 140
Zirconium-90, 264
Zoonotic diseases, 351